International Conference on Climate Change (ICCC 2017)

AA006482

Challenges and Opportunity on Environment Degradation Researches

IOP Conference Series: Earth and Environmental Science
Volume 129

Surakarta, Indonesia
24 – 26 October 2017

ISBN: 978-1-5108-6008-7
ISSN: 1755-1307

Printed from e-media with permission by:

Curran Associates, Inc.
57 Morehouse Lane
Red Hook, NY 12571

Some format issues inherent in the e-media version may also appear in this print version.

Copyright© (2017) by the Institute of Physics
All rights reserved. The material featured in this book is subject to
IOP copyright protection, unless otherwise indicated.

Printed by Curran Associates, Inc. (2018)

For permission requests, please contact the Institute of Physics
at the address below.

Institute of Physics
Dirac House, Temple Back
Bristol BS1 6BE UK

Phone: 44 1 17 929 7481
Fax: 44 1 17 920 0979

techtracking@iop.org

Additional copies of this publication are available from:

Curran Associates, Inc.
57 Morehouse Lane
Red Hook, NY 12571 USA
Phone: 845-758-0400
Fax: 845-758-2633
Email: curran@proceedings.com
Web: www.proceedings.com

Table of contents

Volume 129

International Conference on Climate Change: Challenges and Opportunity on Environment Degradation Researches
24–26 October 2017, Surakarta, Indonesia

Preface

Papers

Impact of depletion or enhance of capability of resources of air, water, soil, and vegetation

Strategy for environmental disaster reduction research

Climate model and uneven precipitation distribution

Pollution and contamination of land surface and atmosphere

Carbon footprint, greenhouse gas emission, recycle, and reuse energy research

Direct and indirect risks to wellbeing

Implication of climate adaptation and mitigation research

Infrastructures risks and planning on climate adaptation

Policy and legal aspect of climate change

The economic and and social elements of climate change

IOP Conf. Series: Earth and Environmental Science **129** (2018) 011001 doi:10.1088/1755-1315/129/1/011001

Preface

The International Climate Change Conference 2017 (ICCC 2017) is an event organized by Graduate School of Universitas Sebelas Maret (UNS), Indonesia to mediate the experts, researchers, practitioners, students, and societies to discuss the findings, problems, and solution about climate change. ICCC 2017 addressing the researches relate with climate change to the adaptation and mitigation strategy and the implementation to the societies. ICCC 2017 was carried out at Best Western Premier Hotel, Surakarta city, Indonesia from 24 to 26 October 2017.

ICCC 2017 develops new partnerships and associations with key decision makers across all sectors of climate, and accomodates the latest research findings, as well as the future impacts. The scope of subjects discussed in this conference are: Impact of depletion or enhance of capability of resources of air, water, soil, and vegetation; ecosystem and habitat destruction research; strategy for environmental disaster reduction research; thermal expansion research; climate model and uneven precipitation distribution; pollution and contamination of land surface and atmosphere; carbon footprint, greenhouse gas emission, recycle and reuse energy research; involuntary migration and forced displacement; direct and indirect risks to wellbeing; implication of climate adaptation and mitigation research; infrastructures risks and planning on climate adaptation; policy and legal aspect of climate change; and the economic and social elements of climate change.

The following are the pictures of the ICCC 2017.

Figure 1. Opening by Vice Rector of Sebelas Maret University, Indonesia

Content from this work may be used under the terms of the Creative Commons Attribution 3.0 licence. Any further distribution of this work must maintain attribution to the author(s) and the title of the work, journal citation and DOI.
Published under licence by IOP Publishing Ltd

Figure 2. Plenary Session 1 with invited speakers: Prof. Dr. Sutarno (Lecturer of Sebelas Maret University, Indonesia), Dr. Widada Sulistya (Indonesian Agency for Meteorology, Cimatology and Geophysics), Dr. Haris Gunawan (Indonesian Peat Restoration Agency) and Prof. Dr. MTH. Sri Budiastuti (moderator)

Figure 3. Plenary Session 2 with invited speakers: Dr. Takeo Onishi (Gifu University, Japan), Dr. Anthony Kent (Victoria University, Australia), Dr. Avishek Datta (Asian Institute of Technology, Thailand), and Dr. Komariah (moderator)

Figure 4. Plenary Session 3 with invited speakers: Prof. Dr. Patrick Van Damme (Ghent University, Belgium), Dr.James MacGregor (Eco Plannet, Canada), Dr.Keigo Noda (Gifu Univ., Japan), Dr. Anwar Fitrianto (Universiti Putra Malaysia, Malaysia), and Dr. Widyatmani Sih Dewi (moderator)

Figure 5. First Photo Session of participant, Steering Committee and invited speakers

IOP Conf. Series: Earth and Environmental Science **129** (2018) 011001 doi:10.1088/1755-1315/129/1/011001

Figure 6. Second Photo Session of participant, Steering Committee and invited speakers

Figure 7. Third Photo Session of participant, Steering Committee and invited speakers

Figure 8. Fourth Photo Session of participant, Steering Committee and invited speakers

Figure 9. Photo Session of participants at parallel session

Figure 10. Organizing Committee and invited speakers

Peer review statement

All papers published in this volume of *IOP Conference Series: Earth and Environmental Science* have been peer reviewed through processes administered by the proceedings Editors. Reviews were conducted by expert referees to the professional and scientific standards expected of a proceedings journal published by IOP Publishing.

Content from this work may be used under the terms of the Creative Commons Attribution 3.0 licence. Any further distribution of this work must maintain attribution to the author(s) and the title of the work, journal citation and DOI.
Published under licence by IOP Publishing Ltd

IOP Conf. Series: Earth and Environmental Science **129** (2018) 012001 doi:10.1088/1755-1315/129/1/012001

Water table depth fluctuations during ENSO phenomenon on different tropical peat swamp forest land covers in Katingan, Indonesia

A Rossita[1,3], A Witono[1], T Darusman[1], D P Lestari[1] and I Risdiyanto[2]

[1]Katingan-Mentaya Project, Ecosystem Restoration Concession Rimba Makmur Utama Co Ltd., Kantin Street No.8, Central Bogor, Bogor City, West Java, 16121 Indonesia

[2]Department of Geophysics and Meteorology, Faculty of Mathematics and Natural Science, Bogor Agricultural University (IPB), IPB Dramaga Campus, FMIPA Building, Wing 19, Level 4, Dramaga Street, Bogor, West Java, 16680 Indonesia

[3]Corresponding author: annurirossita@gmail.com

Abstract. As it is the main role to maintain hydrological function, peatland has been a limelight since drainage construction for agriculture evolved. Drainage construction will decrease water table depth (WTD) and result in CO_2 emission release to the atmosphere. Regardless of human intervention, WTD fluctuations can be affected by seasonal climate and climate variability, foremost El Niño Southern Oscillation (ENSO). This study aims to determine the correlation between rainfall in Katingan and ENSO index, analyze the pattern of WTD fluctuation of open area and forest area in 2015 (during very strong El Niño) and 2016 (during weak La Niña), calculate the WTD trendline slope during the dry season, and rainfall and WTD correlation. The result showed that open area has a sharper slope of decreasing or increasing WTD when entering the dry, compared to the forest area. Also, it is found that very strong El Niño in 2015 generated a pattern of more extreme decreasing WTD during the dry season than weak La Niña in 2016.

1. Introduction

Peatland is a unique ecosystem with high carbon stock that functions to maintain the hydrological condition in the surrounding areas. Disturbance to peatland ecosystem can alter its function from storing to emitting carbon. Based on the Biennial Update Report Republic of Indonesia, Land Use Change and Forestry (LUCF) sector contributes more than a half of national emission in the period 2000 – 2012, with the highest contribution, is from peat decomposition (45.5%) [1]. Emission from peat decomposition is allegedly sourced from drained peatland area with drainage construction for plantation activity purposes. Regardless of drainage and human intervention, fluctuation of water table depth (WTD) in peatland can be affected by seasonal and climate variability, foremost El Niño Southern Oscillation (ENSO). ENSO is a wide scale atmospheric-oceanic phenomenon cycled for 3 – 7 years in Equatorial Pacific [2]. ENSO occurrence will affect the onset and length of the dry season and rainy season, which will generate a different pattern of WTD in each El Niño and La Niña year [3]. However, though will not continuously occur, there is a possibility that ENSO amplitude will change in the future due to climate change. Therefore, it is necessary to analyze WTD fluctuation during ENSO, its trendline slope on the dry season, and its correlation with rainfall changes.

Content from this work may be used under the terms of the Creative Commons Attribution 3.0 licence. Any further distribution of this work must maintain attribution to the author(s) and the title of the work, journal citation and DOI.
Published under licence by IOP Publishing Ltd

2. Data and method

The historical rainfall data used in the present study are 30 years daily precipitation of Climate Hazard Group Infra-Red Precipitation with Station (CHIRPS) 1987 – 2017 with 5 km x 5 km spatial resolution, validated using available daily observation data of H. Asan Sampit weather station (located in the latitude of -2.55° and longitude of 112.93°). CHIRPS daily grid precipitation data was extracted using Ocean Data View (ODV) software and validated using Microsoft Excel and Minitab software. Monthly sea surface temperature (SST) anomaly of NINO 3.4 from 1987 – 2017 used in this study were obtained from NOAA.

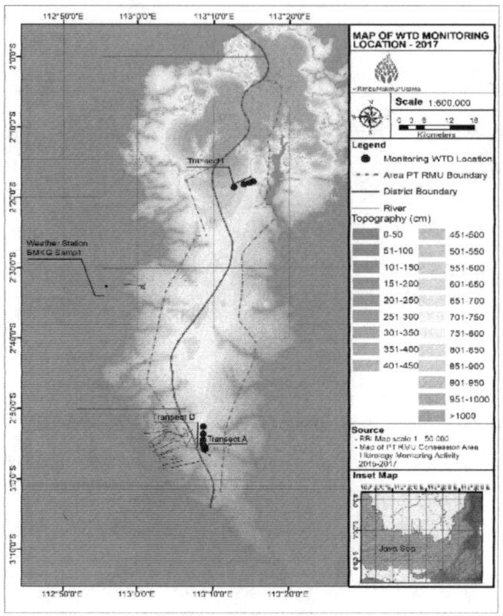

Figure 1. Map of Ecosystem Restoration Concession's project area and monitoring locations of WTD. Topography data obtained from SRTM with resolution 90m.

Determination of the onset of the dry season and rainy season in this study is based on the assumption of Indonesia's Bureau of Meteorology and Climatology of 10 days (known as *dasarian*) precipitation. The meteorological condition of equatorial Pacific Ocean visualized by Grid Analysis and Display System (GrADS) version 2.0.a9.oga.1, using data of 1000 m.bar zonal and meridional wind and SST were obtained from European Center for Medium-Range Weather Forecast (ECMWF) from January 2015 to January 2017.

Daily WTD were generated from hydrology monitoring activities in the study area from February 2015 to January 2017. Our study site in Katingan District, Central Kalimantan, Indonesia, covers 149.800 ha project area of PT Rimba Makmur Utama that is flanked by two main watersheds (figure 1). WTD in the project area is monitored in 3 transects, transect A that represents the open area, and transect D and I that represent forest land cover. The sequence of monitoring point of transect A from the nearest to farthest from canal are A1, A2, and A3. The sequence of monitoring point of transect D from the nearest to farthest from the open area are D1, D2, D3, and D4, whilst for transect I, from the nearest to farthest from canal are I1, I2, I3, and I4. Missing values of WTD were corrected using weighted interpolation method, and only the most complete data during the 2 years monitoring period were used for seasonal and correlation analysis (I1 and I3 were excluded from this analysis).

3. Result and Discussion

ENSO and rainfall variability in the study area is proved that SST anomaly of NINO 3.4 zone (170°W-120°W and 5°N-5°S) has a stronger correlation with monthly anomaly precipitation in Indonesia if compared to another zone [4]. High correlations of rainfall index (1987 – 2017) and SST NINO 3.4 anomaly were showed in July – November, where the highest correlation was -0.744, in September

IOP Conf. Series: Earth and Environmental Science **129** (2018) 012001 doi:10.1088/1755-1315/129/1/012001

(figure 2a). Period of July – November showed the most significant correlation with a p-value less than 0.1. Based on the historical data of SST NINO 3.4 anomaly and monthly rainfall in the study area, the positive value of SST NINO 3.4 anomaly always followed by the decrease of rainfall, while the opposite during negative value of SST NINO 3.4 (figure 2b).

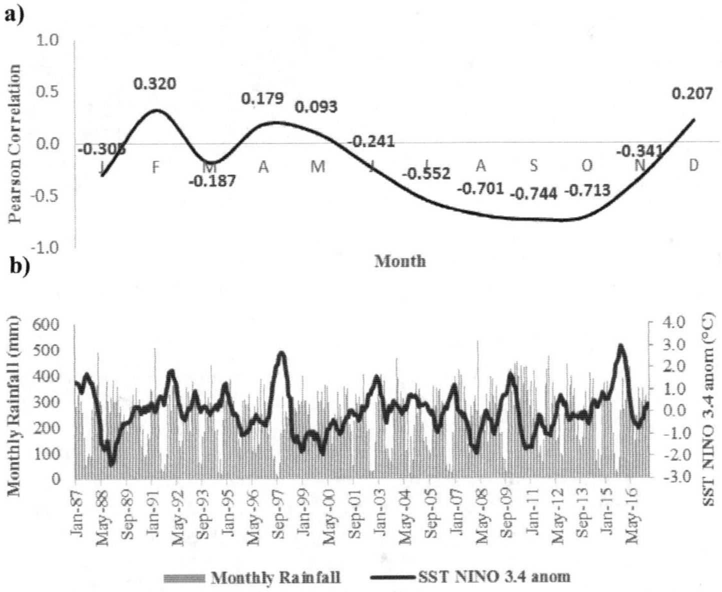

Figure 2. (a) Correlation of monthly rainfall index in the study area and monthly SST NINO 3.4 anomaly **(b)** Monthly rainfall in the study area and monthly SST NINO 3.4 anomaly Jan 1987 – April 2017.

In the year of 2015, very strong El Niño occurred with the highest SST anomaly in zone NINO 3.4 reached 2.95°C on November 2015, while in the year of 2016, weak La Niña occurred with the lowest SST anomaly in zone NINO 3.4 reached -0.73°C on October 2016. El Niño in 2015 started in February-March-April (FMA) 2015 and ended in April-May-June (AMJ) 2016, whilst La Niña in 2016 started directly a few months later and ended in November-December-January (NDJ) 2016 (figure 3). El Niño to La Niña transitions usually occurs much faster than La Niña to El Niño where almost all El Niño to La Niña transitions occur within one year [5].

Besides the strong influence of rainfall variability in Indonesia, ENSO events also influence inter-annual rainfall variability in Indonesia. End of the dry season tends to occur later than normal during El Niño and earlier than normal during La Niña years, and the onset of the wet season is delayed during El Niño and advanced during La Niña years [6]. Based on the result of precipitation analysis year 2015 and 2016, the onset of the dry season in 2015 occurred earlier in the 3rd 10 days of June, compared to 2016 that occurred later in the 1st 10 days of August. On the contrary, the onset of the rainy season in 2015 moved backward in the 3rd 10 days October, while in 2016 moved forward in the 2nd 10 days of September. Moreover, it was found that dry spell in 2015 occurred longer than in 2016, with the value 21 and 8, respectively. A wet spell in 2015 on the other hands, occurred shorter than in 2016, with value 24 and 40, respectively.

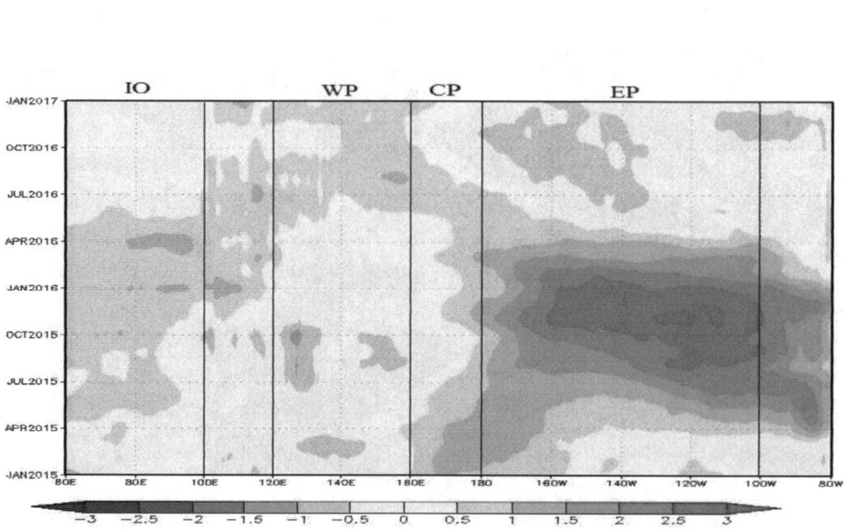

Figure 3. Development of SST anomaly (°C) in equatorial Pacific (5°N – 5°S) on January 2015 – January 2017 along 60°E - 80°W, straight line showed the boundary of Indian Ocean (IO), West Pacific (WP), Central Pacific (CP), and East Pacific (EP). (Graph generated from ECMWF data using GrADS).

Water table depth fluctuation as shown in figure 4a, water table depth in A3 was logged only during the rainy season and almost never logged in A1 and A2. As shown in figure 4b, there was a clear difference of the maximum, minimum, and average value of WTD in 2015 and 2016. Minimum water table depth in transect A always on the below ground and reached -2 m on October 2015 (average -1.5 m), while in September 2016 only reached -1.2 m (average -0.6 m). Meanwhile, based on the graph of maximum WTD, it is shown that on 2015 water table were logged for 2 months, while on 2016 water table were logged for 6 months. But on average condition, water table depth in transect A was never logged neither on dry season nor rainy season.

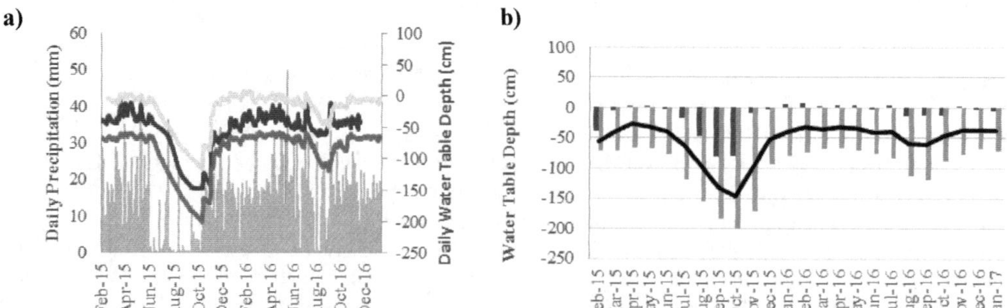

Figure 4. Graph of WTD in Open Area **(a)** Daily Fluctuation **(b)** Monthly min, max, and average.

Variations of water table depth fluctuations of each monitoring point in transect A are more affected by the elevation rather than its position towards the canal. Based on the slope data collected from the field, the low slope of A3 to A2 result in slow water flow from A3 to A2. Meanwhile, the high slope of A2 to A1 result in high water flow from A2 to A1 that makes water table depth in A2 become low. The close distance from A1 towards the canal could interfere water table in A1 by holding up outflow from open area to the canal. As a result, water flow from A2 and A3 accumulated in A1 and leads to a high water table depth in A1.

As shown in figure 5a, water table depth in transect D was logged during the rainy season and partially on dry season, except in D1. Based on figure 5b, there was a difference of the minimum, maximum, and average value of WTD transect D in 2015 and 2016, especially during dry season. Moreover, it is shown that in 2015 water table were always logged except in September and October where El Niño developed became stronger as sea surface temperature of NINO 3.4 increase approximately 2.46°C.

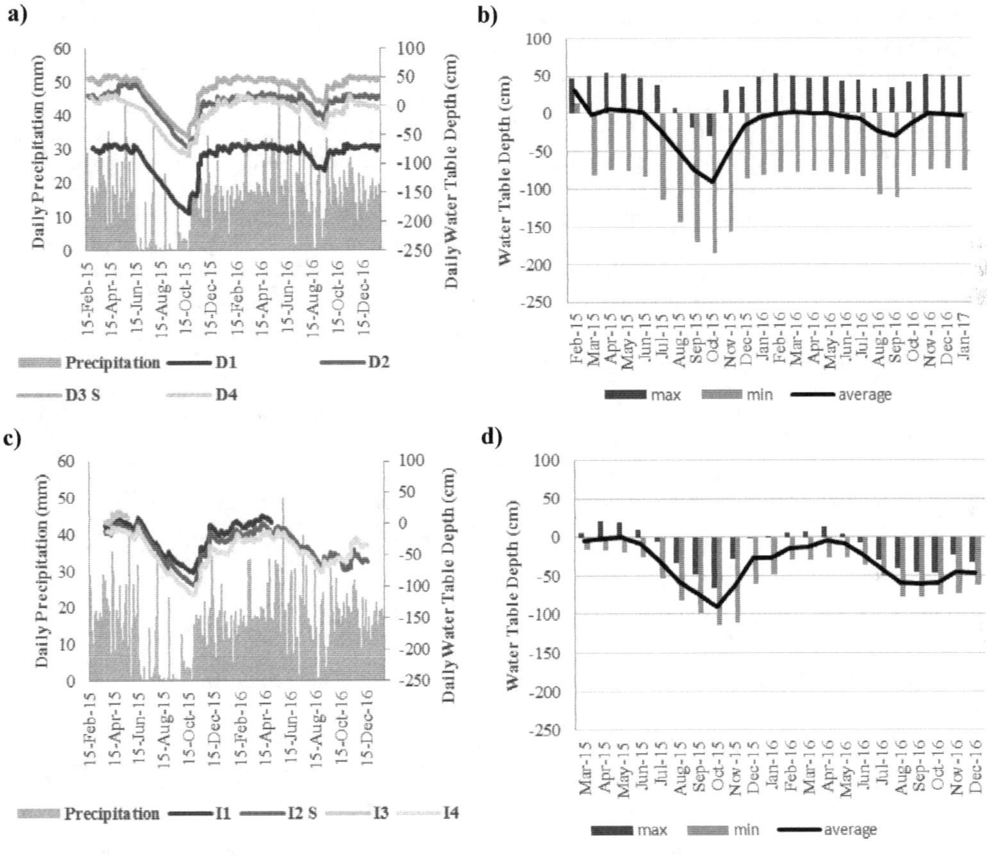

Figure 5. Graph of WTD in Forest Area **(a)** Daily Fluctuation of Transect D, **(b)** Monthly min, max, and the average of Transect D, **(c)** Daily Fluctuation of Transect I, and **(d)** Monthly min, max, and the average of Transect I.

WTD on transect D in 2016 were logged longer than 2015 due to the longer occurrence of the dry season in 2015 than 2016. As ENSO phenomenon affect the length of the dry season and the rainy season, it will also affect the length of water table depth was logged in transect D. According to another study, the lowest value of WTD also depends on the length of dry spell [7]. WTDs in point D1 were considerably lower than those in D2, D3 and D4, likely caused by its closer proximity to open area. Therefore, it is likely that point D1 tend to represent WTD fluctuations in forest edge rather than forest area, showing a wide gap between monitoring points inside forest areas (D2, D3, D4) that are relatively similar to each other (figure 5a).

Based on figure 5c, WTD in transect I were logged for a while during the rainy season and decreased on the below ground during the dry season. According to figure 5d, there was less difference on minimum, maximum, and average value of WTD Transect I in 2015 and 2016, showing that ENSO occurrence was less apparent on WTD fluctuation in transect I. In the year of 2016, WTD decrease sooner than 2015 and based on the WTD data on 2016, it is also shown that transect I has earlier downturn point before the onset of the dry season. However, the lowest WTD in transect I was not as low as transect D, showing that WTD fluctuation in transect I tend to be more stable than transect D (not too dry, not too wet), which is explained by the location of transect I on the upstream position, near to Klaru river (figure 1).

Based on the data obtained, lowest WTD always occurs at the end of dry season, and if compared to two different ENSO occurrences, lowest WTD during the dry season in 2015 was lower than during dry season in 2016. Hence, the difference of lowest minimum WTD in two different ENSO phenomena will result in additional emission decomposition [8]. As WTD in each transect is also affected by elevation and its location towards the canal or open area, it is too soon to conclude that type of land cover affected the value of water table depth in general, but certainly land cover could affect the difference between highest and lowest value of WTD. However, based on the other study, when comparing site factors, bare peatland consistently emits higher CO_2 than under forested land [9].

Seasonal analysis of water table depth's decrease during the dry season (figure 6a, b) is observed from the onset of the dry season until the end of dry season. Based on the result in figure 6, transect A tends to have steeper trend than transect D and I, while transect I tend to has slightly trend than transect D (Table 1). Regardless of its water table depth, the types of land cover are more affected in the trend line slope of water table depth decrease, where it can be seen a clearly different value of slope in the different type of land cover (Table 1). Trendline slope can describe the characteristic of each land cover, especially its capability to maintain and play its hydrological function. The slighter the slope, the better water storage condition.

Table 1. Trendline slope of WTD decrease or increase in all monitoring point.

Monitoring point	Trendline Slope								
	A1	A2	A3	D1	D2	D3	D4	I2	I4
Dry Season 2015	-1.07	-1.11	-0.89	-0.92	-0.89	-0.82	-0.75	-0.75	-0.79
Dry Season 2016	-0.15	-0.95	-0.80	-0.72	-0.70	-0.73	-0.80	-0.59	-0.48

Based on the different year of seasonal analysis, the decrease water table depth on the dry season 2015 was steeper than dry season 2016. The trend of WTD during the dry season in 2016 quite similar in transect A and D (Table 1). It is estimated that the smallest trend in transect A shown in monitoring point A1 is caused by an interfering from the canal. The lowest trend during dry season 2016 results in transect I.

Rainfall and water table depth correlation Cross-correlation is used to analyze the time-lagged relationship between rainfall and WTD and assess the sensitivity and responsiveness of water table fluctuation to the rainfall change, described by the highest value of cross-correlation (CC) [10]. The highest value of CC occurred on the varied lag time of all monitoring point (Figure 7). Highest CC in A1, A2, and A3 occurred on the lag time 35, 30, and 23 respectively. Later lag time in A1 and A2 than A3 expected caused by its close distance to the canal. Highest CC of all monitoring point of transect D occurred on the lag 24, except D2 that occur on the lag 30, meanwhile transect I have the longest lag time, where both I2 and I4 have lag time 32 and 35. Based on the other study, the lag of water table changes due to addition or reduction of rainfall occurs a month [11].

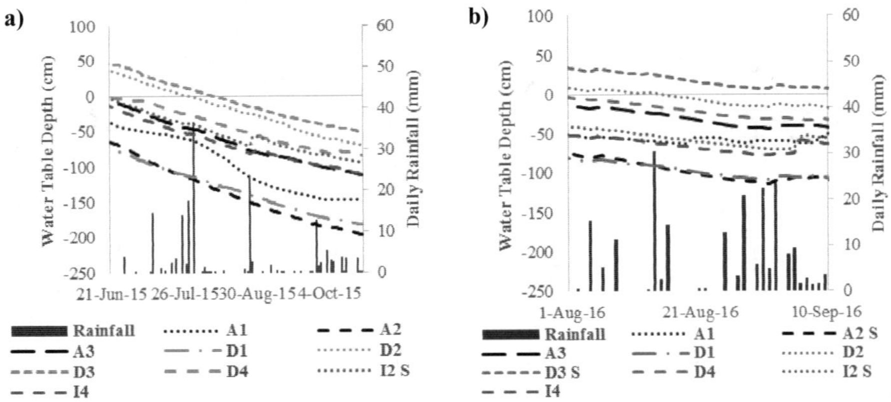

Figure 6. Water Table Depth **(a)** Decrease during dry season 2015 **(b)** Decrease during dry season 2016.

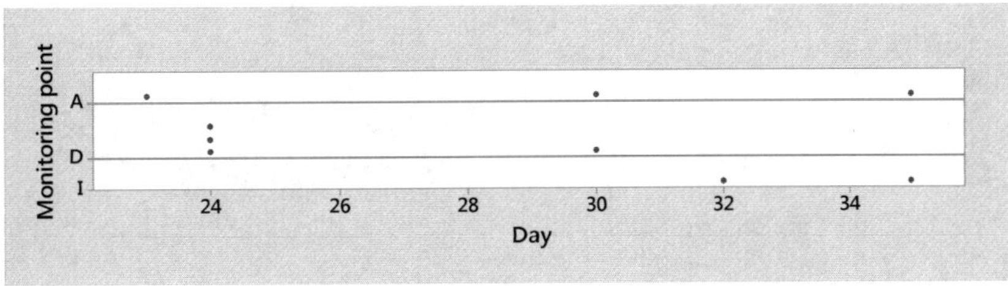

Figure 7. Dot plot of lag time in monitoring point A (open area) and D and I (forest area)

Based on the regression equation (figure 8), the value of slope showed how changes in addition or reduction of 10 days rainfall affect the changes of average 10 days water table depth in the lag time (t+30). Even though transect A and D have quite similar time lag (ignoring the interfere from a canal in A1 and A2), the slope of WTD changes due to additional and reduction of rainfall is different. Based on its land cover, transect A has a higher value of slope than transect D and I. Therefore, transect A will experience the highest change of WTD due to any change (addition or reduction) of rainfall compared to others transect. Meanwhile, the lowest value of slope measured in transect I that indicated the water table in transect I tend to be more stable than another transect.

Based on the other study Qalbi [12], the result of projected future sea surface temperature anomaly by four model Global Climate Model Couple Model Intercomparison Project phase 5 (GCM CMIP5) that has a different condition of SST anomaly in each of the model, generally showed that ENSO phenomena in period 2006 – 2035 is estimated to be more intensive and strong with SST anomaly that can reach up to approximately 4°C. Based on the result of the scenario of occurrence single El Niño event, it is predicted that monthly rainfall will decrease 50 – 100 mm/month in Central Kalimantan if compared to rainfall variability in period 1981 – 2010 [12].

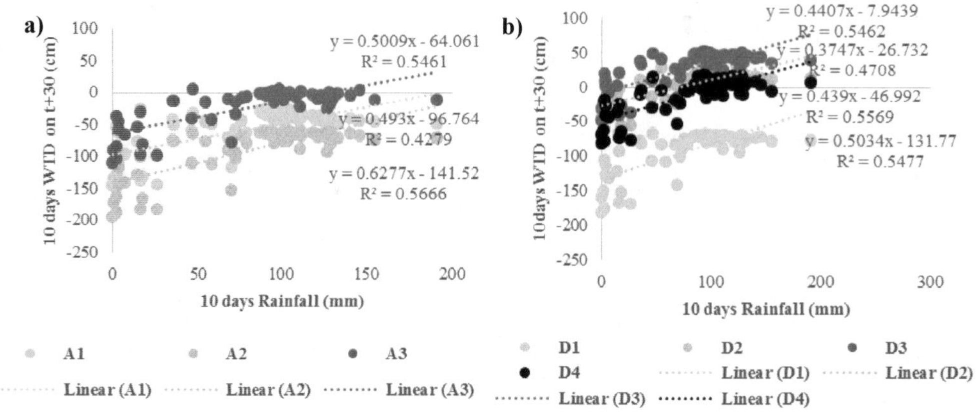

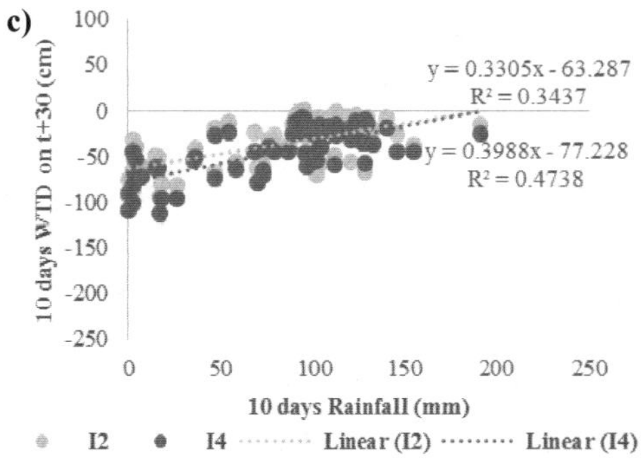

Figure 8. 10 Days Rainfall and 10 days Water Table Depth Correlation in **(a)** Transect A (open area) at (t+30) and **(b)** Transect D (forest area) at (t+30) **(c)** Transect I (forest area) at (t+30).

4. Summary and concluding remarks

The highest and most significant correlation of SST NINO 3.4 anomaly and rainfall index in the study area occurred during the period of July – November. ENSO occurrence evidently affected the onset and length of the dry season and rainy season, which will generate a different pattern of WTD in each El Niño and La Niña year. ENSO occurrence will also affect the trendline of water table fluctuation during the dry season and wet season, where its trendline during the dry season in El Niño year was more abrupt compared to La Nina year. However, regardless of the land cover, there are other factors that affect the variation of WTD in each transect such as elevation and distance towards the canal. A land cover factor tends to influence more on the difference of maximum and minimum WTD, trend line slope of WTD during the dry season, and the slope of regression that describes changes of WTD to the rainfall change. The impact of land cover on lag time remains conjectural due to the interference factor from the nearest canal, particularly in transect A. ENSO occurrence clearly affects water table depth fluctuation and contributes to an additional emission of peat decomposition. Thus, with the possibility that ENSO amplitude will change in the future due to climate change, there should be an attempt to increase WTD by rewetting to mitigate the GHG emission on degraded peatland especially when ENSO occurs.

Acknowledgement
The authors wish to thank Katingan Mentaya Project, Rimba Makmur Utama Co Ltd. for the support on this research, the hydrology team of Rimba Makmur Utama Co Ltd. for the field data support, Anita Pratiwi, an expert of statistician, for the support on statistical analysis, also lecturers and colleague of Applied Meteorology IPB for kindly help with climate data analysis.

References

[1] Ministry of Environment and Forestry 2015 *Indonesia First Biennial Update Report* (Jakarta: Directorate General of Climate Change, Ministry of Environment and Forestry, Republic of Indonesia) p 239

[2] Mcphaden M J 2003 El niño and la niña : causes and global consequences *Encyclopedia of Global Environmental Change* ed M C Maccracken, J S Perry and T Munn (United States of America: Wiley) p 353–70

[3] Mezbahuddin M, Grant R F and Hirano T 2014 Modelling effects of seasonal variation in water table depth on net ecosystem CO_2 exchange of a tropical peatland *Biogeosciences* **11** 577–99

[4] Boer R, Notodiputro K A and Las I 2007 Prediction of daily rainfall characteristics from monthly climate indices *Agromet. Indones.* **21** 12–20

[5] Larkin N K and Harrison D E 2002 ENSO warm (el niño) and cold (la niña) event life cycles : ocean surface anomaly patterns, their symmetries, asymmetries, and implications *J. Clim.* **15** 1118–40

[6] Aldrian E and Dwi S R 2003 Identification of three dominant rainfall regions within Indonesia and their relationship to sea surface temperature *Int. J. Climatol.* **23** 1435–52

[7] Ritzema H and Wösten H 2002 "Hydrology of Borneo's peat swamps" *http://www.splu.nl* [online] Available: http://www.splu.nl/strapeat/download/2%20hydrology.pdf [accessed: 21-08-2017]

[8] Agus F, Handayani E, van Noordwijk M, Idris K and Sabiham S 2010 "Root respiration interferes with peat CO_2 emission measurement" *iuss.org* [online] Available: http://iuss.org/19th%20WCSS/Symposium/pdf/0739.pdf [accessed: 20-07-2016]

[9] Astiani D, Burhanuddin, Taherdjadeh M and Curran L M 2016 Effects of water table level on soil CO_2 respiration in West Kalimantan forested and bare peatland : An experimental stage. *Nusant. Biosci.* **8(2)** 201–6

[10] Cryer J D and Chan K-S 2008 Time series regression models *Time Series Analysis (second edition)* ed G Casella, S Fienberg, and I Olkin (New York: Springer) p 249–74

[11] Runtunuwu E, Kartiwa B, Kharmilasari, Sudarman K, Nugroho W T and Firmansyah A 2011 Water elevation dynamics on land and channels on peatland *Geol. Min. Res.* **21** 63–74

[12] Qalbi H B, Faqih A and Hidayat R 2017 Future rainfall variability in Indonesia under different ENSO and IOD composites based on decadal predictions of CMIP5 datasets *IOP Conf. Ser. Earth Environ. Sci.* **54** (Jakarta: IOP Publishing)

Linking climate change to water provision: greywater treatment by constructed wetlands

S Qomariyah[1], AH Ramelan[2], P Setyono[3] and Sobriyah[1]

[1]Department of Civil Engineering, Faculty of Engineering, Sebelas Maret University, Jl. Ir. Sutami 36A Kentingan, Surakarta, Central Java 57126 Indonesia
[2]Department of Physics, Faculty of Mathematics and Natural Sciences, Sebelas Maret University, Surakarta, Central Java 57126 Indonesia
[3]Graduate School of Environmental Science, Sebelas Maret University, Surakarta, Central Java 57126 Indonesia

Abstract. Climate change has been felt to take place in Indonesia, causing the temperature to increase, additional drought with more moisture evaporates from rivers, lakes, and other bodies of water, and intense rainfall in a shorter rainy season. One of the major concerns is the risk of severe drought leading to water shortages. It will affect water supply and agriculture yields. As a country extremely vulnerable to the climate change, Indonesia must adapt to the serious environmental issues. This paper aims to offer an effort of water provision by recycling and reusing of greywater applying constructed wetland systems. The treated greywater is useful as water provision for non-consumptive uses. A recent experiment was conducted on a household yard using a single horizontal subsurface flow type of constructed wetland. The experiments demonstrated that the constructed wetland systems reduced effectively the pollutants of TSS, BOD, COD, and detergent to the level that are compliant with regulatory standards. The constructed wetland has been established for almost two years however the system still works properly.

1. Introduction

Due to the rise of the greenhouse gas emissions in Indonesia, the impacts of climate change will be heavily felt by this country [1]. The impacts have been characterized by an increase of temperature, change of precipitation, a rise in sea level, and a threat to food security. One of the potential impacts of climate change would be the decrease in water availability [2]. As outlined clearly in the report of World Bank [3], the effect of climate change will come through the water cycle and are already visible.

Despite the abundance of water resources, in the dry season, water shortages are common in some parts of Indonesia due to varying climatic and geographic conditions [4]. Therefore, it is reasonable if this country receives a lot of attention as a vulnerable country in the face of climate change [1]. The most pressing area of concern at climate change is that of water security. The climate change hazards for water supply shortages had been studied based on the six indicators of vulnerability to water supply i.e., population density, land cover, water demand, confined aquifer and aquifer potential, water infrastructures, and water quality. The study found that the risk level of water supply shortage in Java-Bali is the highest among all regions in Indonesia. It was due to its dense population, built up areas domination, highest water demand for population and industrial growth [2].

Content from this work may be used under the terms of the Creative Commons Attribution 3.0 licence. Any further distribution of this work must maintain attribution to the author(s) and the title of the work, journal citation and DOI.
Published under licence by IOP Publishing Ltd

The vulnerability of water supply shortage for irrigation will reduce the production of agriculture which in turn could pose a threat to food security in Indonesia [1]. The poor communities who rely on agriculture will greatly be affected by the impacts of climate change associated with the drought or water shortage [3]. The food and drinking water for daily needs will be expensive and inaccessible to the poor.

Environmental stresses like severe drought are most likely to continue. Therefore, adaptation to climate change through reducing the vulnerability of water supply shortage needs to be prepared. One of the possible efforts would be recycling and reusing of wastewater, and the cost-effective method of wastewater treatment is constructed wetland systems [5]. Water recycling and applicable technologies are key aspects to adapt to the impacts of climate change [6]. In addition, concern over water shortage issue is not only limited to the amount available to meet demand but also the quality of water. Hence, this paper aims to promote the constructed wetlands by providing a water supply for non-potable purposes as an effort to adapt to climate change and present the result of the constructed wetlands study.

1.1. Constructed wetlands

Constructed wetland systems are engineered systems which are designed and constructed to utilize the natural processes in treating wastewater [5]. These systems comprised of aquatic vegetation, substrates, water, micro-organisms and utilized complex processes involving physical, chemical, and biological mechanisms to improve water quality or remove such pollutants [7]. They have advantages such as low cost of construction and maintenance, low energy requirement compared to conventional technical wastewater treatment systems, and provide green aesthetic views [6].

Classifications of wetlands based on wetland hydrology are surface flow and sub-surface flow. Figures 1 and Figure 2 present the scheme of the types of the constructed wetlands.

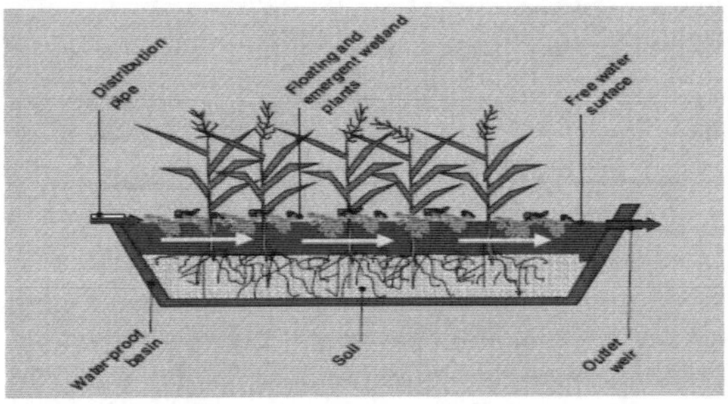

Figure 1. Surface flow [8]

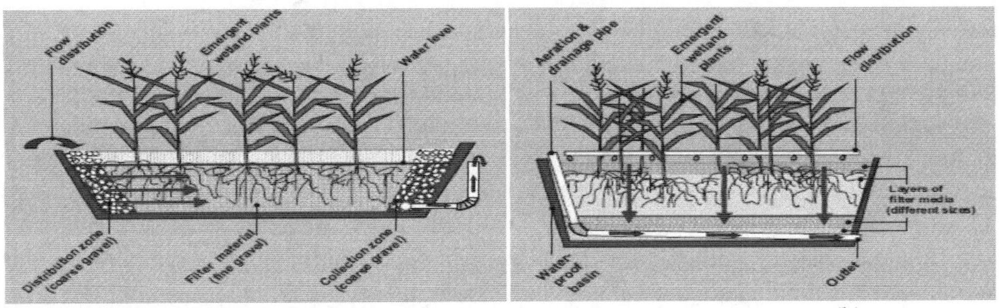

(a) (b)

Figure 2. (a) Horizontal sub-surface flow; (b) Vertical sub-surface flow [8]

The free surface flow systems are similar to natural wetlands in which shallow flow of wastewater over saturated substrates. Based on the water flow direction, the subsurface flow wetlands are classified into horizontal and vertical sub-surface flow systems. The horizontal systems allow wastewater to flow horizontally from inlet to outlet. The vertical systems facilitate wastewater to move down vertically to the bed. To achieve maximum removal efficiency of pollutants, horizontal and vertical-flow wetlands are combined as a single stage known as hybrid systems [9]. Constructed wetlands for wastewater treatment are potentially a good solution for treating domestic and industrial wastewater due to the simple operation and low implementation cost, and they are flexible systems which can be used for single households or for communities [6].

1.2. Aquatic plants
Aquatic plants commonly used in constructed wetlands are emergent plants, floating-leaved plants, submerged plants, and free-floating plants. Among the mentioned plants, emergent plants are the main aquatic plants used for wastewater treatments in constructed wetlands [10]. The functions of aquatic plants are to release oxygen from their root, to provide microorganisms with a source of carbon, to allow solids to settle out of the water column, to assimilate nutrients, to provide a substrate for microorganisms who are the most important processors of wastewater contaminants. Hence, plants play an important role in wastewater treatment wetland [11,12]. Large aquatic plants growing in wetlands, called macrophytes, have an additional site-specific value by making wastewater treatment systems aesthetically pleasing [6,10].

1.3. Substrates
Substrates also play an important role in constructed wetland systems. They provide a suitable growing medium for plants, allow the movement of wastewater flow, support biochemical and chemical transformation, and provide sites of removed pollutants [9]. Types of substrates commonly used for constructed wetlands wastewater treatment are natural materials such as sand, gravel, clay, limestone, zeolite, etc. Recently, an industrial by-product of substrates like fly ash, coal cinder, alum sludge, oil palm shell, etc. and artificial products like activated carbon, lightweight aggregates, compost, calcium silicate hydrate, and ceramic were also used [7].

1.4. Micro-organisms
There are various micro-organism communities in a wetland, both in aerobic and un-aerobic conditions, comprising many types of bacteria, fungi, algae, and protozoa, including pathogenic viruses. They live in the natural conditions of water, substrate, and the roots of aqua plants. They consume organic and nutrient loads in the wastewater, so that the pollutant existing in the wastewater could be reduced, destroyed, or even removed [13,14]. Among the micro-organisms, *Escheria coli* is commonly used as an indicator of faecal contamination [15]. Removal of pathogen indicated by coliform bacteria (fecal coliform and total coliform) is to be the priority in wastewater treatment in developing countries [16]. In addition, removal of coliform bacteria more than 99% would be a target of constructed wetland design criteria [9].

1.5. Application of constructed wetlands
Today, the application of constructed wetlands in various wastewater purification systems is popular all over the world, especially in developed countries [5,6,7]. In Denmark, the application of constructed wetlands for two decades of experience shows that the systems are generally efficient in removing TSS and BOD concentration to be less than 20 mg/L [17]. The use of surface water flow and subsurface flow constructed wetlands for various wastewater originated in a range of industries, aqua farms, agricultural, animal, and petrochemical wastes, highway runoff, etc were reviewed in [5]. Greywater treatment systems for households and neighborhoods in low and middle-income countries were discussed in detail by [13]. The application of constructed wetland system in developing countries is however still very limited [16] including In Indonesia.

2. Greywater Treatment

Greywater is the wastewater generated from domestic activities in bathrooms, laundry, and all washing in the kitchen in households and other buildings [15]. The wastewater generated from the toilet is considered blackwater which may not be reused due to the high concentration of harm full viruses and bacteria. The composition of greywater greatly depends on the quality and type of water supply and household activities [13,15]. Bathroom greywater contains soaps, shampoos, toothpaste, and other body care products, traces of urine and feces. Greywater from laundry contains chemicals from detergents. Kitchen greywater contains food residues, oil & fat, and dishwashing detergent. Greywater nonetheless contains micro-organisms as a result of its contact with human body. However, the concentrations of micro-organisms are substantially lower than that of blackwater and are therefore much safer to be recycled and reused. In addition, although greywater is generally less polluted than other wastewater, it may still contain pathogenic microorganisms, suspended solids, and other household substances or chemicals [13]. Even though greywater is less contaminated than blackwater, the large volume of greywater produced will contribute to the total pollution of domestic wastewater discharged into existing drainage systems. In Indonesia, domestic wastewater contributes to the amount of 60% pollution in the receiving natural water systems [18].

In general, households, apartments, and business areas in cities in Indonesia already separate greywater from blackwater. Due to one of the most serious impacts of climate change is drought or water shortages, recycling and reusing greywater for providing non-consumptive water source would be a significant effort to anticipate the drought. The advantage of recycling greywater is that it is a large source with a low organic content [6]. In addition, greywater is produced regularly, available throughout the year, and can be readily reused for domestic application. Therefore, the treatment of greywater would be an innovative water provision concerning water supply shortages impact for climate change.

3. Material and methods

The study was performed within a household yard located in the city of Surakarta, Central Java, Indonesia. The site is characterized by a humid tropical climate region. A single horizontal subsurface flow constructed wetland was used to investigate pollutant removal efficiency of greywater from a single house. The constructed wetland was made with a dimension of 1.7 m x 0.7 m x 0.7 m (L x W x D) and filled with river sand and gravel as its substrate. At the inlet and outlet zone of the constructed wetland, gravels with a diameter of 1-3 cm were filled to 50 cm deep and 20 cm long. In the plantation zone, river sands with its permeability of 5.48 x 10^{-5} cm/sec were set up for 130 cm long. The locally available macrophyte plant of Cyperus was used and planted in the plantation zone at 25 cm interval.

Before entering the constructed wetland, the greywater was mechanically pre-treated in a sedimentation tank. The plants need to be acclimatized by discharging the greywater in the constructed wetland with gradual concentration from 25 to 100% within four weeks. Greywater was distributed across the width of the system. The influent and effluent samples were collected at the inlet and outlet respectively through a Ø25 mm pipe. Water samples were collected in 1.5 liters of plastic bottles, stored at 4°C, and transported to the laboratory. Household greywater is produced daily and directly accumulated in the sedimentation tank, therefore this study applied hydraulic retention time of one day. The parameters to be monitored were total suspended solids (TSS), the power of hydrogen (pH), 5 days biochemical oxygen demand (BOD_5), chemical oxygen demand (COD), detergent, and oil & grease. The first and the second study were carried out in November 2015 and August 2017 respectively.

4. Result and discussion

The efficient removal of the greywater treatment was checked by testing the influent and effluent water quality. The result of the effluent testing would be confirmed with the Regulation of Ministry of Environment, the Republic of Indonesia, number 5 the year 2014, about the standard of wastewater quality, and the Regulation of Government, Republic of Indonesia, number 82 the year 2001, about management of water quality and controlling of water pollution.

The second experiment aims to get to know how the performance of the constructed wetland after two-year establishment. Results of the first and second experiments were presented in table 1.

The study in 2015 resulted in a high removal of suspended solids (TSS). Suspended solid is removed primarily through the mechanism of sedimentation and interception [6]. The solid removal efficiency ranged from 95.5% to 99.6% with an average reduction of 98.1%. TSS at the outlet varied between 2-10 mg/L. In 2017, the TSS removal still achieved a high rate of 95.4%. The level concentration of 39 mg/l at the outlet is due to the very high level of the TSS concentration at the inlet. However, based on the Environment Ministry standard, the TSS level is still much safer (< 100 mg/l) to the environment.

Table 1. The removal percentage of greywater treatment

Parameter		Year 2015									Year 2017		
		3/11	4 /11		15/11	16/11		26/11	27/11		22/8	23/8	
		Inlet	Outlet	%	Inlet	Outlet	%	Inlet	Outlet	%	Inlet	Outlet	%
TSS	mg/l	461.0	2.0	99.6	238.0	2.0	99.2	222.0	10.0	95.5	840.0	39	95.4
pH	-	5.2	6.9		6.1	6.5		5.3	6.8		7.1	5.6	
BOD	mg/l	496.0	9.5	98.1	850.2	17.2	97.9	528.7	2.2	99.6	1372.0	142.0	89.6
COD	mg/l	1254.0	26.6	97.8	1862.5	37.2	98.0	1252.0	6.1	99.5	4067.5	414.2	89.8
Detrgt	mg/l				76.4	0.1	99.9				9825.6	95.5	99.0
O&G	mg/l	2.2	3.6					4.8	2.6		-		

The pH of water at the inlet and outlet remains in the allowable level (5 – 9) of the regulatory standards. Both studies, in 2015 and 2017, demonstrated that the removal of BOD and COD are very high. The treatment efficiency of BOD and COD in 2015 varied between 97.9-99.6% and 97.8-99.5% respectively. In 2015, the outlet BOD concentrations were 9.5 mg/l, 17.2 mg/L, and 2.2 mg/l. All values are below the threshold of 100 mg/L as set by the Environment Ministry standard. The values of 9.5 and 2.2 mg/l, which are less than 12 mg/l, fulfill the requirement of irrigation water standard of the Government Regulation. However, the one of 17.2 mg/l is a little bit higher than that of 12 mg/l need. The outlet COD concentrations ranged between 6.1-37.2 mg/L. It fulfills the quality standard of class IV (< 100 mg/l) for irrigation; moreover, it is compliant with class III (< 50 mg/l) for irrigation, livestock, fish farm, and other outdoor uses as set by the Government Regulation. In 2017, the constructed wetland, which has been established for almost two years, removed efficiently both parameters of BOD and COD, achieving more than 89% of removal. The concentration levels of both parameters at the outlet, which are higher than 100 mg/l, are certainly due to the very high level of both BOD and COD concentrations of the household greywater to be treated at that time. However, the magnitude of 89% removal after two years establishment indicates the constructed wetlands still work properly.

Typically, household greywater contains high concentrations of detergent. The constructed wetlands removed the detergent successfully, resulting in a very high-efficiency percentage of 99.9% and 99% in the study of 2015 and 2017 respectively. Unfortunately, the removal of oil & grease in the study could not be well defined because their removal is not consistent. The Government Regulation did not set such conditions of detergent and oil & grease parameters for class IV (irrigation use) water quality, however, further studies should examine the removal of both parameters by which the quality standard of water for class III and II can be determined.

Water for gardening or irrigation purposes (class IV or even class III) does not need such water qualities as high as the drinking water – class I. Therefore, the treated greywater resulted from the constructed wetland system could provide a water source for agriculture and other non-consumptive uses. In addition, the treated greywater will remain available as long as people use water for their activities during the period of climate change.

5. Conclusion

The separation of greywater and blackwater which is commonly applied in most households and business areas would be a big step toward water provision regarding water shortages. Using treated greywater is one of the methods which can be helpful in dealing with environmental climate change. By recycling greywater and reusing for irrigation/agriculture and other non-consumptive needs, saving a large amount of drinking water can be obtained.

The study of greywater treatment using constructed wetland system demonstrated that the system can effectively reduce contaminants of TSS, BOD, COD, and detergent to the levels that are compliant with regulatory standards for agriculture. After two years of its establishment, the constructed wetland system is still capable to remove the greywater pollutants in a significant amount of removal, hence, using the power of nature and not manmade chemical the constructed wetlands still work properly. The reuse of treated greywater considered as a new water resource will reduce water consumption and decrease water cost from water-piped supply services. Thus, constructed wetland systems are a viable option for removing contaminants from wastewater, simultaneously providing water resource for non-potable uses, and getting a green aesthetic environment.

The constructed wetland systems can be applied on the individual household yard or on a larger scale open space for communities. However, public awareness campaign should be carried out related to water shortage issues due to climate change impact. In addition, public campaigns on the application of constructed wetlands technology need to be deployed to cope with the water shortages.

Acknowledgement

This work was financially supported by a research grant from the Sebelas Maret University, to whom the authors are very grateful.

References

[1] Mariah M 2010 *G. Majority E-J.* **1** 31-45
[2] Djoko S A S, Oman A and Budhi S "Impacts of climate change on the sustainability of water supply in Indonesia" *http://www.researchgate.net* 2010 [online] available: http://www.researchgate.net/publication/283056995 [Accessed: 10-04-2016]
[3] World Bank "Indonesia: Climate Change" (Washington, DC: The World Bank) *http://www.//go.worldbank.org* 2010 [online] Available: http://go.wordlbank.org/HQQFW5MV70 [Accessed: 20-11-2015]
[4] Mohamad AF 2014 *Water* **6** 2000-2020
[5] Jan V 2010 *Water* **2** 530-549
[6] Heike H, Christopher P, Martina W and Elizabeth von M 2011 *Technology review of constructed wetlands* (Germany: Gesellschaft fur Internationale Zusammenarbeit/GIZ) pp 9-17
[7] Haiming W, Jian Z, Huu H N, Wenshan G, Zhen H, Shuang L, Jin L F and Hai L 2015 *Biores. Tech.* **175** 594-601
[8] Martin G 2008 *Constructed Wetland: A Promising Wastewater Treatment System for Small Localities* (Grafica Biblos, Peru) pp 7-8
[9] Robert H K and Scott D W 2009 *Treatment Wetlands* second ed. (Boca Raton: Taylor and Francis Group CRC Press) p 65
[10] Jan V 2011 *Hydrobiologia* **674** 133-156
[11] Hans B 1994 *Water Sci. Tech.* **29** 71-8
[12] Hans B 1997 *Water Sci. Tech.* **35** 11-7
[13] Antoine M and Stefan D 2006 *Greywater Management in Low and Middle-Income Countries* Sandec Report No14/6 (Switzerland: Eawag) pp 1-7
[14] Peter F R 2006 *J. Env. Bio.* **2** 78-89
[15] Eva E, Karina A, Mogen H and Anna L 2002 *Urban Water* **4** 85-104
[16] Amelia K 2001 *Ecol. Eng.* **16** 545-60

IOP Conf. Series: Earth and Environmental Science **129** (2018) 012002 doi:10.1088/1755-1315/129/1/012002

[17] Hans B, Schierup H, and Arias C A 2007 *Water Sci. & Tech.* **56.** 63-8
[18] Diana H, Sulistyoweni W, Setyo S M and Robertus W T 2013 *European Sci. J.* **9** 17 230-1

Effects of goat manure liquid fertilizer combined with AB-MIX on foliage vegetables growth in hydroponic

Y Sunaryo[1], D Purnomo[1], M T Darini[2] and V R Cahyani[1,3]

[1] Graduate School, Sebelas Maret University, Jl. Ir. Sutami 36A Kentingan, Surakarta, Central Java 57126 Indonesia
[2] Faculty of Agriculture, University of Sarjanawiyata Tamansiswa, Jl. Batikan 6 Tempel, Wirogunan, Yogyakarta 55167 Indonesia
[3] Corresponding author: vitaratri@staff.uns.ac.id

Abstract. Hydroponic as one of the protected cultivation practices is very important to be developed in Indonesia due to not only the reduction of arable agricultural lands in lines with increasing of residential demand and other public facilities but also due to the negative influences of climate change as well global warming to plant growth. The effects of liquid fertilizer made from goat manure (LFGM) in combination with AB-Mix on three kinds of foliage vegetable growth was examined in hydroponics. The research was conducted by 3 x 4 factorial experiment and arranged in Completely Randomized Design with 3 replications. The first factor was foliage vegetable consisting of 3 levels: Mustard Green, Lettuce, and Red Spinach. The second factor was the mixture composition of nutrient solution consisting of 4 levels: LFGM + AB-Mix (v/v: 1:1), LFGM + AB-Mix (v/v: 1:3), LFGM + AB-Mix (v/v: 3:1), and A/B mix as control. Results indicated that the application of LFGM + AB-Mix (v/v: 1:3) resulted in similar plant growth as control (AB-Mix application), and also resulted in the highest chlorophyll content of Mustard green.

1. Introduction

The reduction of arable agricultural lands in lines with increasing of residential demand and other public facilities, and also the negative influences of climate change as well global warming to plant growth can be serious obstruct for food production in the future all over the world. It has been projected that 50% of the world's fertile land may be unfeasible in near future [1]. In several countries, population growth is increasing rapidly and therefore there is an increase in food consumption, but the areas for agricultural crops do not increase at the same proportion [2].

In recent years, hydroponic crop cultivation significantly increased and allows a more efficient use of water and fertilizers, as well as control the climate and pest factors [3]. The hydroponic system can induce highest vegetative growth, yield, and quality of plants [4]. The benefits of hydroponically agricultural practices are numerous [5,6] and it can be applied for soil conservation and preservation of water resources [2,7]. The practice of hydroponic in a controlled environment, resulting in higher yield and water efficiency, also support continuous production throughout the year [5]. Hydroponically plant cultivation has been found to be a good alternative and is commercially used in many western countries [8].

The production of organic liquid fertilizer is a very important strategy to support the development of hydroponic cultivation. The liquid organic fertilizer can substitute inorganic chemical substances which is more expensive and difficult to obtain in several countries such as Indonesia. Fertilizers as a

Content from this work may be used under the terms of the Creative Commons Attribution 3.0 licence. Any further distribution of this work must maintain attribution to the author(s) and the title of the work, journal citation and DOI.
Published under licence by IOP Publishing Ltd

source of nutrients are very important for plant growth, and the quality of liquid fertilizer is determined by pH and the content of nutrient elements. The quality of organic liquid fertilizer is also determined by the content of another chemical substance such as growth regulator and organic acid [9,10]. The use of electrical conductivity (EC)-based nutrient control in hydroponic can afford excessive or inadequate nutrients for plant growth [11]. An equilibrium nutrient supply is vital for better plant growth in the hydroponic system.

The quality of liquid fertilizer is determined by not only the pH but also the nutrient elements content. The use of effectiveness microorganisms (EM) during incubation of raw materials is suggested to obtain the high quality of liquid organic fertilizer. In order to give energy and nutrients to accelerate the development of EM during incubation, it can be added molasses, NPK, and urea in the production of liquid organic fertilizer [12]. EM is the probiotic microorganisms used to ferment organic materials and applied to produce better plant growth [13]. This research was objected to determine the appropriate combination of liquid fertilizer made from goat manure and AB-mix to can accelerate plant growth and yield. AB-Mix is well known as a commercial nutrient solution for hydroponic with macro-elements (N, P, K, S, Ca, and Mg) and micro-elements (Fe, B, Mn, Zn, Cu, and Mo) nutrients composition.

2. Materials and Methods

2.1. Research Site and Period

Fermentation of liquid fertilizers made from goat manure (LFGM) was carried out from the first week until the fourth week of February 2017. The application of liquid fertilizer was conducted from March until August 2017 in Wonosari, Yogjakarta, Indonesia.

2.2. Fermentation of the materials

Fermentation of the materials consisted of the following steps: (1) 0.5 kg Sugar + 1 kg ZA were diluted by 20 liters water in the 25-liter volume plastic bucket, then added with 20 ml of EM and 2 kg of air-dried goat manure; (2) The mixed materials were stirred manually around 5 minutes and then the plastic bucket was capped properly; (3) The plastic was placed in the room temperature condition; (4) The mixed fermented materials were stirred manually around 2 minutes, every day in the morning and afternoon; (5) After three weeks of fermentation, the pH, EC, and TDS of the solution was observed using portable pH/EC/TDS/Temperature meter Hanna Instruments H19811-5.

The experiment of the application of liquid fertilizer on foliage vegetables (mustard green, lettuce, and spinach) was carried out by 3 x 4 experimental factorial arranged in Completely Randomized Design (CRD) using three replications. The first factor was foliage vegetables consisted of 3 levels: P1: Mustard green, P2: Lettuce, and P3: Red spinach. The second factor was the combination of nutrient solution consisted of 4 levels: A: LFGM + AB-Mix (v/v: 1:1), B: LFGM + AB-Mix (v/v: 1:3), C: LFGM + AB-Mix (v/v: 3:1), and D: A/B mix as control.

Seeds of mustard green (*Brassica rapa* L), lettuce (*Lactuca sativa L),* and red spinach (*Alternanthera amoena* Voss) were cultivated in the compost media during for two weeks until seedlings, then selected to get the homogenous seedlings. The seedlings then planted in the hydroponic connected pipes installation constructed to appropriately shallow flow technique (SFT). The plants were irrigated and fertilized by the combination of nutrient solution in an appropriate treatment. The EC of the nutrient solution was consistently set in the range of 1600-1650 μScm^{-1}, whereas the pH of the nutrient solution was in the range of 5.5-6.5.

Four weeks after planting, leaf number, shoot fresh and dry weight, root dry weight, chlorophyll content, and root/shoot ratio from 5 plant samples at each treatment were observed. Chlorophyll content was measured on the intact leaves of plants with no destructive assays taken from lower, middle, and upper leaves of the plants using digital chlorophyll meter called *CCM 200 plus Chlorophyll Content Meter*.

IOP Conf. Series: Earth and Environmental Science **129** (2018) 012003 doi:10.1088/1755-1315/129/1/012003

3. Results and Discussion

Table 1 indicated there was no significant effects of liquid fertilizers on leaf numbers, shoot fresh weight, and root dry weight of Lettuce, also on chlorophyll contents of lettuce and red spinach. The application of LFGM + AB-Mix (v/v: 1:3) resulted in the highest shoot fresh weight, dry weight, and chlorophyll content of Mustard green. In general, the application of goat manure liquid fertilizer combined with AB-Mix in ratio 1:1-3 (v/v) resulted in the similar plant growth compared to control (AB-Mix application) on foliage vegetables. These results indicated that the quality of goat manure liquid fertilizer is good. An experiment conducted by Sunaryo *et al.* [14] showed that the nutrient content of the goat manure was in the range of standard recommended by the Regulation Minister of Agriculture Republic Indonesia [15].

The application of LFGM + AB-Mix (v/v: 3:1) resulted in the lowest shoot fresh weight, shoot dry weight, and root dry weight of Mustard green as well as shoot dry weight of Lettuce. The application of LFGM + AB-Mix (v/v: 3:1) on Red spinach resulted in the lowest shoot fresh weight and shoot dry weight.

There was an interaction between the application of liquid fertilizers and kinds of foliage vegetables on R/S (root/shoot) ratio (Table 2). Red spinach had the lowest R/S ratio, whereas the highest R/S ratio was gained on lettuce with both applications of LFGM + AB-Mix (v/v: 3:1) and application of A/B mix. These results indicated that the application of LFGM + AB-Mix (v/v: 1:1, 1:3, and 3:1) and AB-Mix on Red spinach resulted in better assimilate partitioning to the shoot compare than that of Mustard green and Lettuce.

Table 1. Effects of liquid fertilizers on Mustard green, Lettuce, and Red spinach growth

Plants	Treatments	Leaf numbers	Shoot fresh weight (g)	Shoot dry weight (g)	Root dry weight (g)	Chlorophyll content (%)
	A	16.44 ab	132.47 b	11.59 b	1.60 b	17.12 b
P1	B	18.44 a	155.23 a	14.63 a	2.08 a	20.00 a
	C	15.22 b	99.07 c	9.27 c	1.27 c	16.88 b
	D	17.67 a	125.20 b	11.13 b	1.56 b	16.69 b
ρ >F		0.0284	0.0031	0.0005	0.0006	0.0578
	A	13.44 a	79.03 a	3.30 ab	0.47 a	4.50 a
P2	B	12.77 a	81.66 a	3.80 a	0.58 a	6.11 a
	C	13.11 a	66.72 a	2.51 b	0.45 a	4.90 a
	D	13.32 a	75.33 a	3.17 ab	0.58 a	4.80 a
ρ >F		0.9809	0.8799	0.0626	0.3101	0.2794
	A	18.11 bc	108.61 a	16.44 a	2.01 a	15.39 a
P3	B	20.66 a	126.28 a	19.19 a	2.10 a	16.84 a
	C	16.78 c	83.31 b	12.69 b	1.56 b	15.27 a
	D	19.33 ab	106.51 a	16.23 a	1.86 ab	15.25 a
ρ >F		0.0011	0.0067	0.0064	0.0529	0.2935

Mean followed by the same letter in the same column is not significantly different at α=5% level by DMRT
P1: Mustard green, P2: Lettuce, P3: Red spinach
A: LFGM + AB-Mix (v/v: 1:1)
B: LFGM + AB-Mix (v/v: 1:3)
C: LFGM + AB-Mix (v/v: 3:1)
D: A/B mix

Table 2. The interaction between the application of liquid fertilizers and kinds of foliage vegetables on R/S ratio

Treatment combination	P1 (Mustard green)				P2 (Lettuce)				P3 (Red spinach)				Interaction
	A	B	C	D	A	B	C	D	A	B	C	D	ρ >F
R/S ratio	0.138 bc	0.142 b	0.137 bcd	0.139 b	0.139 b	0.152 b	0.181 a	0.182 a	0.122 de	0.109 e	0.123 cde	0.114 e	0.0003

Mean followed by the same letter is not significantly different at α=5% level by DMRT A: LFGM + AB-Mix (v/v: 1:1)
B: LFGM + AB-Mix (v/v: 1:3)
C: LFGM + AB-Mix (v/v: 3:1)
D: A/B mix

4. Conclusion

The combination of goat manure liquid fertilizer with AB-Mix in ratio 1:1-3 (v/v) resulted in similar plant growth as control (AB-Mix application) at all vegetables,and the highest chlorophyll content at Mustard green, while the combination with AB-Mix (v/v: 1:1, 1:3, and 3:1), and AB-Mix resulted in better assimilate partitioning to the shoot of red spinach.

References

[1] Okemwa E 2015 Effectiveness of aquaponic and hydroponic gardening to traditional gardening *International Journal of Scientific Research and Innovative Technology* **2(12)** pp 21-52

[2] Oliano de Carvalo R, Weymar Jr L C N, Zanovello C B, Silva da Luz M L G, Gadotti G I, Silveira da Luz and Gomez M C 2015 Hydroponic lettuce production and minimally processed lettuce *Agric Eng Int CIGR Journal* special issue 2015 18th World Congress of CIGR pp 290-293

[3] Trejo-Tellez L I and Gomez-Merino F C 2012 *Nutrient Solutions for Hydroponic System, Hydroponics – A Standard Methodology for Plant Biological Researches* 'ed Dr. Toshiki Asao (Montecillo, Texcoco, State of Mexico, Mexico: In Tech) p 224

[4] Wahome P K, Oseni T O, Masarirambi M T and Shongwe V D 2011 Effects of different hydroponics system and growing media on the vegetative growth, yield and cut flower quality of gypsophila (*Gypsophila paniculata* L.) *World Journal of Agricultural Sciences* **7(6)** pp 692-8

[5] Barbosa G L, Gadelha F D A, Kublik N, Proctor A, Reichelm, Weissinger F, Wohlleb and Haldem RU 2015 Comparison of land, water, and energy requirements of lettuce grown using hydroponic vs. conventional agricultural methods *Int J Environ Res Public Health* **12 (6)** pp 6879-91

[6] Sardare M D and Admane S V 2013 A review on plant without soil – hydroponics *International Journal of Research in Engineering and Technology* **2(3)** pp 299-304

[7] Patil P, Kakade S, Kantale S and Shinde D 2016 Automation in hydroponic system using PLC *International Journal of Scientific and Technical Advancements* **2(2)** pp 69-71

[8] Mugundhan M R, Soundaria M, Maheswari V, Santhakumari P, and Gopal V 2011 Hydroponics a novel alternative for geoponic cultivation of medicinal plants and food crops *International Journal of Pharma and Bio Sciences* **2(2)** pp 286-96

[9] Arancon N Q, Edwards C A, Lee S and Byrne R 2006 Effects of humic acids from vermicomposts on plant growth *European Journal of Soil Biology* **42** S65-S69

[10] Campitelli P, Velasco M and Ceppi S 2012 Characterization of humic acids derived from rabbit manure treated by composting-vermicomposting process *Journal of Soil Science and Plant Nutrition* **12(4)** pp 875-91

[11] Lee J Y, Rahman A, Azam H, Kim HS and Kwon M J 2017 Characterizing nutrient uptake kinetics for efficient crop production during *Solanum lycopersicum* var. cerasiforme Alef.

growth in a closed indoor hydroponic system *PLoS ONE* **12(5)**
[12] Desyane H K and Wiyana A F 2012 Proposed quality improvement of liquid fertilizers "Herbafarm" to meet national standards in Indonesia *The Indonesian Journal of Business Administration* **1(6)** pp 343-52
[13] Higa T and Parr J F 1994 Beneficial and effective microorganism for a sustainable agriculture and environment *International Nature Farming Research Center Atami* Japan pp 16
[14] Sunaryo Y, Purnomo D, Darini M Th and Cahyani V R 2017 *Nutrients Content and Quality of Liquid Fertilizer Made from Goat Manure* The 1st International Conference on Science, Mathematics, Environment and Education (ICoSMEE) Faculty of Teacher Training and Education Sebelas Maret University
[15] The Regulation Minister of Agriculture Republic Indonesia No.70/Permentan/SR.140/10/2011 "Organic Fertilizer, Biofertilizer, and Soil Ameliorant" p 88 (in Indonesia) *http://perundangan.pertanian.go.id* 2011 [Online] Available: http://perundangan.pertanian.go.id/admin/file/Permentan-70-11.pdf [Accessed: 10-08-2017]

IOP Conf. Series: Earth and Environmental Science **129** (2018) 012004 doi:10.1088/1755-1315/129/1/012004

High-zinc rice as a breakthrough for high nutritional rice breeding program

U Barokah[1,5], U Susanto[2], M Swamy[3], D W Djoar[4] and Parjanto[4]

[1]Master student Agronomy Program Graduate School of Sebelas Maret University, Jl. Ir. Sutami 36 A, Kentingan, Surakarta, Central Java 57126 Indonesia
[2]Indonesian Center of Rice Research, Sukamandi, Jl. Raya, Patok Besi, Subang No.9, Rancajaya, Patokbeusi, Subang,West Java 41256 Indonesia
[3]International Rice Research Institute, Pili Drive, UPLB, Los Baños, 4031 Laguna, Philippines
[4]Department of Agrotechnology, Faculty of Agriculture, Sebelas Maret University, Surakarta, Central Java 57126 Indonesia
[5]Corresponding author: barokahumi@yahoo.com

Abstract. WHO reported climate change already takes 150,000 casualties annually, due to the emergence of various diseases and malnutrition caused by food shortages and disasters. Rice is the staple food for almost all of Indonesian citizens, therefore Zn *biofortification* on rice is expected to be effective, efficient, massive, and sustainable to overcome the Zn nutritional deficiency. This study aims to identify rice with high Zn content and yield for further effort in releasing this variety. Ten lines along with two varieties as a comparison (*Ciherang and Inpari 5 Merawu*) were tested in Plumbon Village, Mojolaban Subdistrict, Sukoharjo Regency during February-May 2017. The experiment was designed in a Randomized Completely Block Design with four replications on a 4 m x 5 m area, with 25 cm x 25 cm plant spacing using seedling transplanting techniques of 21 days old seeds. The results showed that the plant genotypes treated had differences in yield characteristics, heading date, harvest age, panicle number, filled and un-filled grain per panicle, seed set, 1000 grains weight, Zn and Iron (Fe) content in rice grain. *B13884-MR-29-1-1* line (30.94 ppm Zn, 15.84 ppm Fe, 4.11 ton/ha yield) and *IR 97477-115-1-CRB-0-SKI-1-SKI-0-2* (29.61 ppm Zn, 13.49 ppm Zn, 4.4 ton/ha yield) are prospective variety to be released. *Ciherang* had Zn content of 23.04 ppm, 11.93 ppm Fe, and yield of 4.07 t/ha.

1. Introduction

Climate changes characterized by rising air temperatures and changes in the magnitude and distribution of rainfall have had a wide impact on many aspects of human life [1]. Increased air temperature directly affect the production of cereals including rice, the staple food of the Indonesian population. Indonesia Country Study on Climate Change 1998 [2] reported the vulnerability of agricultural production systems by climate change, as in 1991 and 1994, climate anomaly caused Indonesia to import rice (600,000 tons in 1991 and more than one million tons in 1994). Rice and other cereals are very sensitive to temperature change even in small degree rise. Rice reproductive part called spikelet will become sterile if the air temperature increase, which will affect its productivity [3]. FAO states that climate change, as well as changes in disease patterns and pests, will affect how food production systems will be done in the future. It will also have a direct impact on food security, and poverty levels, especially in countries with dependence on the agricultural sector. WHO reports that

Content from this work may be used under the terms of the Creative Commons Attribution 3.0 licence. Any further distribution of this work must maintain attribution to the author(s) and the title of the work, journal citation and DOI.
Published under licence by IOP Publishing Ltd

climate change caused 150,000 casualties annually due to the emergence of various diseases and malnutrition due to food shortages and disasters [4].

Indonesia needs additional food production along with increasing population [5]. Based on Statistics Bureau data, the national rice demand reaches the number of 28 million tons per year or 139 kg per capita per year or 380 grams per day while the total population of Indonesia in 2015 was 254.9 million with the poor population was 28.51 million people [6]. Poor people living in this condition will experience health problems because they cannot afford to buy rice with high nutrition such as those containing iron and zinc (lack of micronutrient) due to its expensive price.

Zinc is a component with more than 300 enzymes to repair cell damage, retain fertility, synthesize proteins, and boost immunity among many important functions in human health [7]. Symptoms of zinc deficiency in both small and large level can lead to stunted growth, eczema, hair loss, delayed sexual maturity and mental development disorders. Micronutrition study in Indonesia found that the prevalence of zinc deficiency children is 36.1%, with the highest percentage occur in West Nusa Tenggara province of 46.6% and the lowest in West Sumatera province of 11.7% while the percentage for South Sulawesi province was 22.7%.

An efforts to overcome the lack of micronutrients supply is urgent to bring the provision of supplementation (fertilization) and *biofortification* sustainably. Provision of supplementation through zinc fertilization of rice can increase zinc levels in rice [8]. However, this method is less effective due to nutritional loss from runoff, leaching, and the evaporation process. Therefore, a new strategy to overcome malnutrition of micronutrient is by the mean of *biofortification. Biofortification* provides a cost-effective and sustainable solution in tackling the lack of nutrients supply [9]. This method is one of the plant breeding strategies to increase the zinc content in rice while improving the nutrition capacity with relatively low cost. Breeding materials are conventionally formed (hybridization and selection) or non-conventional (another culture and gene transformation). High zinc levels rice produced by these, inexpensive cost, production can be consumed directly by the middle to lower community as a source of energy and sources of nutrients [10]. These results provide rice with high micronutrient content. Rice with high micronutrient content can improve the micronutrient supply for consumers which will overcome the micronutrient deficiency.

In this regard, IRRI (International Rice Research Institute) and Indonesian Center of Rice Research (ICRR) has assembled rice lines that have high zinc content. Selection and yield test stages have been done to obtain high-zinc rice lines. Multi-location tests are being conducted in 16 locations in various parts of Indonesia to obtain the adaptive high-zinc variety to meet the national needs. The purpose of this study was to identify high-zinc rice genotypes with high yield which will be proposed for the release of new rice variety.

2. Methods

The experiment of high-zinc rice lines was carried out in 2017 during the rainy season (February to May 2017) in Plumbon Village, Mojolaban Subdistrict, Sukoharjo Regency, Indonesia with an altitude of ± 120 meters above sea level. A total of 10 lines of high zinc content from IRRI and ICRR along with two comparative varieties (*Inpari 5 Merawu* and *Ciherang*) were tested. The experiment was designed in a Randomized Completely Block Design with four replications in 4 m x 5 m plots with planting space of 25 cm x 25 cm. Transplanting was applied to 21 days old seedlings. Observations including the agronomic characteristics such as plant height, number of productive vegetative and generative phases, heading date, harvest age, panicle number, filled and un-filled grain per panicle, weight of grain per seed, *seed set*, 1000 grain weight, yield, Zinc content and Fe content in brown rice grains with XRF tool. The cultivation technique was carried out in accordance with the guidelines of integrated crop management. The observed data were analyzed using *crop start* statistic software and the mean difference between lines were analyzed using the LSD method at the threshold level of error at 5%.

3. Result and discussion

Variance analysis showed differences between genotypes in agronomic character and genotype results of line observed (Table 1). The results showed that the tested genotypes had differences on yield, heading date, harvest age, panicle number, filled and un-filled grain per panicle, the weight of filled grain per clumps, *seed set*, 1000 grain weight, Zn and Iron (Fe) content in brown rice grains. This indicates that the differences occur between genotypes, not simply because of environmental influences. The appearance of a variety or line will be influenced by two factors: the genetic properties it carries and the environment where it is cultivated; if the environment is uniform, the different appearance of the plant character will be influenced only by the genetic properties.

Table 1. Variance analysis of results, zinc and Fe content and agronomic properties of 10 high zinc rice lines along with 2 check varieties, Sukoharjo, MH 2017.

No.	Properties	F Value	Significance
1	Yield	8.15	0.000**
2	Zink Content	12.94	0.000**
3	Fe Content	3.75	0.002**
4	Vegetative Plant Height	0.89	0.556
5	Generative Plant Height	0.63	0.788
6	Number of Vegetative Tillers	1.42	0.21
7	Number of Generative Tillers	1.18	0.336
8	Heading date	27.01	0.000**
9	Harvest date	70.36	0.000**
10	Panicle number/plant	3.89	0.001**
11	Filled grain/panicle	4.25	0.001**
12	Unfilled grain/panicle	6.14	0.000**
13	Weight filled grain/plant	1.59	0.015*
14	Seed Set	6.21	0.000**
15	1000 grain weight	13.25	0.000**

** = significantly different at α=0.01; * = significantly different at α = 0.05

Observation on the yield variables of high zinc rice line in Sukoharjo shows significant differences (Table 2) as shown *in IR 95133:1-B-16-14-10-GBS* (5.12 ton/ha) and *IR 97477-115-1-CRB-0-SKI-3-SKI-0-2* (4.84 ton/ha) which gives a higher yield than the best of *Ciherang* check variety yield (4.07 ton/ha). *IR 97477-115-1-CRB-0-SKI-3-SKI-0-2* shows highest results compared to all checks. This is because the strain has a lot of panicles (17 panicles) and more resistant to pest attack (majorly aphis and stem borer). The more the number of panicles formed, the more results will be obtained. However, *IR 99284-15-3-2* (2.59 ton/ha) and *IR 99284-15-3-2* (2.73 ton/ha) performed a lower yield than the best of *Ciherang* check variety yield (4.07 ton/ha). This is because *IR 99284-15-3-2* at the time of the initial vegetative phase and the final vegetative were hit by aphis attack while the other lines are relatively resistant to aphis attack. Efforts to overcome by means of pesticide application have been done, but the strain still shows less growth compared to another line. *B13884E-MR-30-2* line shows a lower yield than the best of *Ciherang* check variety because it has fewer numbers of tillers.

Table 2. Average yield, Zinc and Fe content of 10 high zinc rice line with 2 check varieties, Sukoharjo, MH 2017

No.	Genotype lines	Yield	Heading date	Harvest date	Panicle number/plant	Filled grain/panicle	Filled grain weight/plant	Unfilled grain/panicle	Seed Set	1000 grain weight	Zn content	Fe content
1	IR 95133:1-B-16-14-10-GBS	**5.12**	81	105	14	89	28.06	31	73.59	22.97	25.03	13.11
2	IR 99270-34-2-1	4.19	*93*	*119*	17	54	21.47	52	52.78	23.63	24.13	13.23
3	IR 97477-115-1-CRB-0-SKI-1-SKI-0-2	4.40	84	107	18	92	33.79	24	79.66	20.92	29.61	13.49
4	IR 99284-15-3-2	*2.59*	*90*	117	17	75	29.72	36	65.36	23.92	27.45	13.43
5	IR 97477-115-1-CRB-0-SKI-3-SKI-0-2	**4.84**	*86*	107	17	82	31.18	33	71.19	22.88	27.59	12.70
6	IR 97646-108-1-CRB-0-SKI-1-SKI-2-2	4.59	84	108	16	95	32.64	22	80.71	22.44	26.79	12.48
7	IR 99680-3-CRB-0-SKI-1-SKI-2-5	3.98	84	106	20	79	36.03	27	74.31	22.83	27.13	12.84
8	B13884E-MR-22-3-1	3.62	81	104	16	98	37.85	24	80.72	24.67	25.73	13.40
9	B13884-MR-29-1-1	4.11	**73**	**97**	15	113	44.22	24	82.05	25.97	30.94	15.84
10	B13884E-MR-30-2	*2.73*	87	108	13	113	30.69	29	79.21	21.39	28.25	13.23
11	Inpari 5 Merawu	4.01	81	105	13	80	28.64	35	68.74	27.46	23.61	12.88
12	Ciherang	4.07	84	106	12	91	28.67	32	73.83	25.85	23.04	11.93
	LSD 5%	0.76	3	2	4	23	13.07	10	9.70	1.54	1.93	1.40
	CV (%)	13.20	2.30	1.30	15.80	17.50	27.90	22.10	9.00	4.50	5.00	7.40

Description:
Yellow print = significantly higher than the best check
Green print = significantly lower than the best check
Red Print = best check

B13884-MR-29-1-1 line resulted heading and harvest date 73 and 97 Days After Planting (DAP) respectively, earlier than *Inpari 5 Merawu* of 81 DAP and 105 DAP. *IR 99270-34-2-1* line (93 DAP and 119 DAP), *IR 99284-15-3-2* (90 DAP and 117 DAP), and *B13884E-MR-30-2* (87 DAP and 108 DAP) showed longer heading and harvesting date time than *Inpari 5 Merawu*, 81 DAP and 105 DAP, respectively. Similarly, *the IR 97477-115-1 CRB-0-SKI-3-SKI-0-2* (86 DAP) line having a heading date longer than *Inpari 5 Merawu* (90 DAP) and *IR 97646-108-1-CRB-0-SKI-1-SKI-2-2* (108 DAP) which has a longer harvest date than *Inpari 5 Merawu* of 105 DAP. The lines that have a faster harvest date (maturity) will be harvested faster compared to the line that has a longer physiological lifetime (length). Differences in flowering age are caused by genetic differences in varieties and line, characterized by different vegetative phase and different plant responses to the growing site [12].

IR 99270-34-2-1 (17 panicles), *IR 97477-115-1-CRB-0-SKI-1-SKI-0-2* (18 panicles), *IR 99284-15-3-2* (17 panicles), *IR 97477-115-1-CRB-0-SKI-3-SKI-0-2* (17 panicles) and *IR 99680-3-CRB-0-SKI-1-SKI-2-5* (20 panicles) have more number of panicles than the best check of *Inpari 5 Merawu* (13 panicles). However, *IR 99270-34-2-1* (52 grains) have more number of filled grain per panicle than the best check of *Ciherang* (32 grain) but the number of filled grains per panicle is less than the best check of *Ciherang* (91 grains). *IR 97646-108-1-CRB-0-SKI-1-SKI-2-2* (22 grains) line has less number of unfilled grains per panicle than *Ciherang* (32 grains), unlike the *B13884-MR-29-1-1* line which has a higher weight of filled grain per plant as much as 44.22 grams from the best check *Ciherang* (28.67 grams). However, for seed set properties, *IR 99270-34-2-1* (52.78%) had the lower fertility of the best check of *Ciherang* (73.83%). The percentage of unfilled grain is affected by environmental conditions before, during or after flowering, especially by climate (rainfall and number of rainy days) [13]. In addition, the difference in the number of failed grain per panicle produced from each genotype is due to genetic factors of each genotype. The number of filled grain per panicle is influenced by genetic factors [14]. Besides, the environmental factors play a role in the high number of filled grain per panicle, because the sunny weather conditions can increase the rate of photosynthesis, the energy of light used to overhaul the water and the gas of carbon dioxide is converted into food, the resulting photosynthate will be stored in stem and leaf tissue, then it will be translocated to grain in the maturation stage. All of the tested lines have 1000 grains weight smaller than the best check of *Inpari 5 Merawu* (27.46 grams). This is because the form of grain lines tested in the form of long and somewhat lean is different from the form of grain *Inpari 5 Merawu* which has a long and slightly fat grain.

All line tested had zinc content in rice higher than the best check of *Inpari 5 Merawu* (23.61 ppm). Only *IR 95133: 1-B-16-14-10-GBS* (25.03 ppm) and *IR 99270-34-2-1* (24.13 ppm) line showed no significant difference in their zinc content compared to *Inpari 5 Merawu*. *B13884-MR-29-1-1* line contains the highest zinc content (30.94 ppm) and iron content (15.84 ppm) compared to the other lines and the best checks. Rice contains beneficial active components for physiological functions of the body in order to maintain healthiness [15]. Rice with high zinc content required to meet the needs of micro-minerals in the body [16]. Zinc is included in the group of micronutrients that the body absolutely needs. Zinc intake provides great benefits to the human body. The physiological functions that depend on zinc are cell growth and division, antioxidants, sexual development, cellular and humoral immunity, dark adaptation, tasting, and appetite. Zinc is especially needed for growth acceleration process. This is not only due to the effects of cell replication and nucleic acid metabolism but also as a mediator of growth hormone activity. Thus, zinc plays an important role in the synthesis and degradation of carbohydrates, fats, proteins, nucleic acids and embryo formation. Currently, Nutritional Adequacy Ratio (NAR) for zinc is 9-12 mg/day for women and 12-17 mg/day for men (depending on age group). If the intake of zinc in the body was not sufficient, it will be more susceptible to disease attacks. Even according to statistical data, the increasing death rate, in particular, were caused majorly by the lack of zinc intake in children and adults [17].

B13884-MR-29-1-1 is prospective to be proposed as a new rice variety with high nutritional value because this line has high zinc content, iron, and high yields beyond the study checks. Rice with high zinc content is very useful in improving the nutrition of the community. In addition, consuming high-zinc rice can prevent low appetite, dermatitis, growth disorder, slow sexual maturity, infertility, and immunodeficiency. Severe Zn deficiency is characterized by decreased immune response function and increased incidence of infection [18]. In addition, this strain also has the advantage of having a heading date and shorter harvest date than other genotypes and has the highest filled grain weight per plant compared to other genotypes. Early maturing rice has high economic value because of its rapid growth, able to compete with weeds and requires relatively little water during growth thereby reducing management costs [19 in 20]. Early-age rice can also increase the intensity of planting and land use.

IR 97477-115-1-CRB-0-SKI-1-SKI-0-2 is also perspective to be proposed to be new rice variety with high Zn and yield characteristics. As is shown that the line has a zinc content of 29.61 ppm, Fe 13.49 ppm and the result of 4.4 ton/ha, higher than the reference check.

4. Conclusion

Climate changes characterized by rising air temperatures and changes in the magnitude and distribution of rainfall causing decreased productivity of agricultural products, especially rice. In addition, climate change causes the emergence of pests and diseases that directly affect food security and the level of poverty. This is worsened by malnutrition problem due to food shortages consequently to cause death. Therefore, it is necessary to create high-zinc rice varieties to overcome these problems. The results showed that the plant genotypes determined the yield characteristics, heading date, harvest age, panicle number, filled and unfilled grain per panicle, seed set, weight per 1000 grains, Zinc (Zn) and Iron (Fe) content in rice grain. *B13884-MR-29-1-*1 (30.94 ppm Zn, 15.84 ppm Fe, 4.11 ton/ha yield) and *IR 97477-115-1-CRB-0-SKI-1-SKI-0-*2 (29.61 ppm Zn, 13.49 ppm Fe, 4.4 ton/ha yield) are prospective to be proposed as new high yielding varieties with high Zn and Fe content.

Acknowledgements
This research was conducted on the cooperation of Indonesian Center of Rice Research Team (ICRR) with International Rice Research Institute (IRRI).

References

[1] Izrael T A " Report from working group II to IPCC; potential impacts of climate change". *http://www.ciesin.org* 2005 [Online] Available: http://www.ciesin.org/docsl001-0111001-Oll.html [Accessed: 08-09-2017]
[2] "31th Session of the committee on world food security; special event on impact of climate change, pests, and diseases on food security and poverty reduction" *http://www.fao.org/* 2005 [Online] Available: ftp://ftp.fao.org/docrep/fao/meeting/009/j5411e.pdf [Accessed: 08-09-2017]
[3] Harmoni A 2005 *Proceeding Seminar Nasional PESAT 2005* (Jakarta: Universitas Gunadarma) pp E62-E68
[4] Rosenblith L "WHO study of the impact of climate change on human health". *http://environmentabout.com* 2003 [Online] Available: http://environmentabout.com!cslglobalwarming/alhealthclimatep.html [Accessed: 30-7-2017]
[5] Mahmud Y and Sulistyo S P 2013 Agronomic diversity of some new superior varieties of rice plants (*Oryza sativa* L.) in the integrated crop management model *Journal scientific solutions* **1(1)** pp 1-10
[6] "Production, harvested area and productivity of rice and palawija in Indonesia" *https://www.bps.go.id/* 2016 [Online] Available: https://www.bps.go.id/Subjek/view/id/53#subjekViewTab3 [Accessed: 23-12-2016]
[7] Mares-Perlman J A, Subar A, Block G, Greger J L and Luby M H 1995 Zink intake and sources in the US adult population: 1976- 1980 *J Am Coll Nutr* **14** pp 349-57
[8] Welch R M 1986 Effects of nutrient deficiencies on seed production and quality *Adv. Plant Nutr.* **2** pp 205–47

[9] Bouis H 2004 *Proceedings of The 1st International Conference on 'Rice for The Future'* (Bangkok, Thailand: Kasetsart University) pp 43-64
[10] Indrasari S D *et al.* 2004 *Indonesian Final Report Year III. Breeding for Iron Dense Rice: a Low Cost, Sustainable Approach to Reducing Anemia in Asia International Food Policy* Research Institute (IFPRI) and Indonesian Center for Food Crops Research and Development (ICFORD) (nutrition aspect) (unpublished)
[11] Surowinoto S, Samad and Harjuno 1984 *Furnished Rice Planting* (In) (Jakarta: Jakarta Foundation) p 199
[12] Edi S 2013 The Performance of upland paddy varieties and lines in Batang Asai watershed Sarolangun Jambi (In) *BPTP Jambi* **2(3)** pp 113-21
[13] Susanto U, Barokah U and Ali J 2016 Yield potential green super rice (GSR) line (In). *Prosiding Seminar Nasional Agroteknologi (PAGI)* (Surakarta: Universitas Sebelas Maret)
[14] Guswara A 2007 *Increased Rice Crop Production Through Development of Hybrid Rice: In the Collection RDTP/ROPP (*Sukamandi, Indonesia: Indonesian Center of Rice Research)
[15] Wijayanti S 2004 Potential and prospect of indigenous Indonesian functional food (In) *Presented at Seminar Nasional: Pangan Fungsional Indigenous Indonesia. Potensi, Regulasi, Keamanan, Efikasi dan Peluang Pasar* (Bandung: Litbang Pertanian)
[16] Susanto U, Barokah U, Hidayatullah A, Satoto and Swammy M 2017 Yield and Zn Content of biofortified rice genotypes in an Indonesian rice agro-ecosystem *Nusantara Bioscience* **9(3)** pp 288-94
[17] Hidayat A 1999 Zink (zinc) essential for health *Journal of Medicine Trisakti* **18(1)** 19-27
[18] Widhyari S D 2012 The Role and Impact of Zinc Deficiency (Zn) Against The Immune Response System *WARTAZOA* **22(3)** pp 141-8
[19] Khush G S 1995 Modern varieties-their real contribution to food supply and equity *Geojournal* **35(3)** 275-284
[20] Suardi D 2005 Potential of red rice to improve food quality *Journal of Agricultural Research,* **24(3)** pp 93-100

Leachate pollution management to overcome global climate change impact in Piyungan Landfill, Indonesia

Harjito[1,3], Suntoro[1], T Gunawan[2], M Maskuri[1]

[1]Department of Environmental Science, Sebelas Maret University, Ir. Sutami street No. 36A Kentingan, Surakarta, 57126, Indonesia
[2]Department of Geography, Gadjah Mada University, Sekip Utara Yogyakarta 55281, Indonesia
[3]Corresponding author: jitoo_2007@yahoo.co.id

Abstract. Environmental problems associated with the landfill system are generated by domestic waste landfills, especially those with open dumping systems. In these systems, waste degrades and produces some gases, namely methane gas (CH_4) and carbon dioxide (CO_2), which can cause global climate change. This research aimed at identifying the areas that experience groundwater pollution and the spread pattern of leachate movement to the vicinity as well as to develop a leachate management model. The Electricity Resistivity Tomography (ERT) survey is deployed to assess the distribution of electrical resistivity in the polluted areas. In this study, the groundwater contamination is at a very low in the aquifer zone, i.e., 3-9 Ωm. It is caused by the downward migration of leachate to water table that raises the ion concentration of groundwater. These ions will increase the electrical conductivity (EC), i.e., up to 1,284 μmhos/cm, and decrease the electrical resistivity. The leachate spreads westward and northward at a depth of 6-17 m (aquifer) with a thickness of pollution between 4 and11 m.The recommended landfill management model involves the installation of rainwater drainage, use of cover and baseliner made of waterproof materials, and massive waste treatment.

1. Introduction

Global warming is no longer a problem of the future but a problem that the world's population is facing today. Various studies prove that global warming and climate change affect the annual precipitation in different countries. Viglizzo et al. [1] revealed that these two phenomena increase the annual precipitation in Argentina. In line with these results is the study performed by Plummer et al. [2], which found a corresponding increase in the annual precipitation in Australia and New Zealand. On the contrary, Mason [3] reported that global warming decreases the annual precipitation in Africa. The studies carried out by Zhai et al. [4] in China and Masoudi and Afrough [5] in Iran also proved the negative correlation between climate change and precipitation.

The search in the International Disaster Database results in a considerable number of natural disasters in Java Island that occurred in the rainy season, as discovered by Naylor et al. [6]. This study concludes that in the next 40 years, global warming will cause an early end to the rainy season in Central Java. Therefore there will be a shorter-duration of the rainy season in that region. On the other hand, while declining in the dry season, the amount of precipitation in rainy season will likely tend to increase.

The quantity of the resultant leachate is significantly influenced by water input from outside; most of which comes from rainwater. At the same time, the presence of water in landfills also relies heavily

Content from this work may be used under the terms of the Creative Commons Attribution 3.0 licence. Any further distribution of this work must maintain attribution to the author(s) and the title of the work, journal citation and DOI.
Published under licence by IOP Publishing Ltd

IOP Conf. Series: Earth and Environmental Science **129** (2018) 012005 doi:10.1088/1755-1315/129/1/012005

on the operational aspects, such as the application of soil-based cover, the slope of the surface, the climatic condition, and the ability of soil and waste to retain water vapor (i.e., 'field capacity'). When storage capacity is exceeded, gravity causes water to flow downward and out of landfills and thereby, forms leachate [7].

Having a closer look at the characteristics of the surrounding materials leachate can become a problematic phenomenon because it can flow both in lateral and vertical ways. The discharge and the quality of leachate formed in waste piles highly fluctuate depending on precipitation and the characters of the waste itself. The relationship between the amount of rainfall and leachate discharge needs to be identified before designing the capacity of leachate treatment plan. Furthermore, the pollutant load (of leachate) used in the design has to be carefully defined.

2. Methods

This research used the Electrical Resistivity Tomography (ERT), particularly the Schlumberger electrode array, to obtain a primary data from direct measurement in the field. ERT assesses the materials of the underground layers based on the nature of their resistivity [8]. The electrical resistivity (ρ) was calculated from electrical current (I) and potential difference (V). Both of which were measured directly in the field [9] by injecting an electrical current beneath the ground through pairs of current electrodes (C1, C2) and potential electrodes (P1, P2) [10]. The location of the geoelectrical sounding is presented in Figure 1.

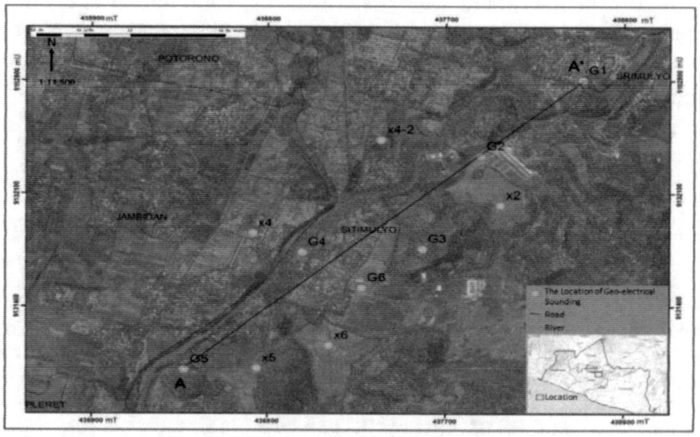

Figure 1. The location of the geoelectrical sounding

2.1. Data Acquisition

The electrical current was injected through electrodes arranged in Schlumberger configuration (Figure 2). As depicted in Figure 2, it flows through a circuit in which the current electrode pairs (C1, C2) were positioned in a purposed distance than the potential electrode pairs (P1, P2). The wider distance between the current electrode pairs (AB or L) was wider to allow for measuring the electrical resistivity of deeper materials. Meanwhile, the distance between the potential electrodes (MN or a) was widened only when the potential difference becomes difficult to measure or, in other words, when the sensitivity of the measuring instrument decreases [11].

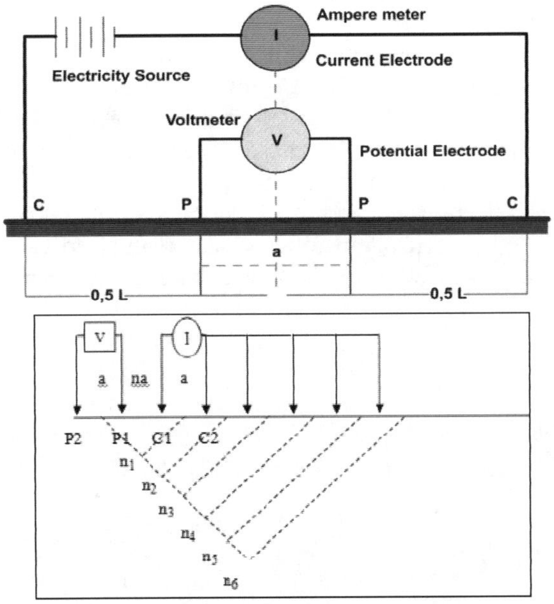

Figure 2. An electrical circuit formed by the arrangement of electrodes in Schlumberger configuration

2.2. Data Processing

The data acquisition above resulted in a set of data on electrical current (I) and potential difference (V) [12]. This research however required pseudo-electrical resistivity instead of these two data. Hence, these two data was formulated into the following equations to obtain the values of the pseudo-electrical resistivity (ρa):

$$R = \frac{V}{I}, \quad (1)$$

$$\rho a = \frac{V}{I}.k, \quad (2)$$

$$\rho a = R.k. \quad (3)$$

The pseudo-electrical resistivity was converted to the actual electrical resistivity using an inverse modeling technique provided by software Res2DINV.

2.3. Data Analysis Technique

The leachate pollution is indicated when the measured electrical resistivity is lower than the expected range of related materials. For instance, groundwater (freshwater) aquifer with an electrical resistivity of 10-100 Ωm will present a reading of lower than 10 Ωm if it is contaminated. Aside from depicting the distribution of resistivity value, the pseudo-section also provides the information in depth. The interpretation of the analysis results referred to the Regulation of the Ministry of Health No.492/MenKes/Per/IV/2010 on the Requisite of Drinking Water Quality.

3. Results and Discussion

The area of study involves the settlements situated in the north of the landfill. It represents the typical influence of landfill on the surrounding settlements. Although strategic, the location only left a small open space for measurement. Therefore, the maximum line for geoelectrical sounding was 121 m. Figure 1 shows the location of the geoelectrical sounding lines around Piyungan Landfill. In general, the water table of the study area was quite shallow, i.e., 3-15 m from the surface. The farther the area from the river system on the west is, the narrower the water table will be. Following those findings, further analysis is required to discover the trend of groundwater flow pattern and the influence of river on the groundwater system around Piyungan Landfill.

The results of the geoelectrical sounding in the study area revealed that the indicator of pollution is presumably caused by the leachate of Piyungan Landfill seeping into the intermittent river system. This particular river system allows the infiltration of leachate-contaminated water into the soil. Infiltration then triggers percolation slowly, i.e., a process of fluid or liquid entering the groundwater system vertically. The contamination was indicated by the distribution of extremely low resistivity values, i.e., 3-9 Ωm, in the aquifer zone. A significant decline in electrical resistivity is caused by the addition of leachate into the groundwater system that increases the ionic contents and hence, the electrical conductivity (EC) of the groundwater. Groundwater pollution in the study area was found at a depth of 6-17 m with a thickness of pollution between 4 and 11 m. The distribution and interpretation of resistivity values in this zone are presented in Figure 3.

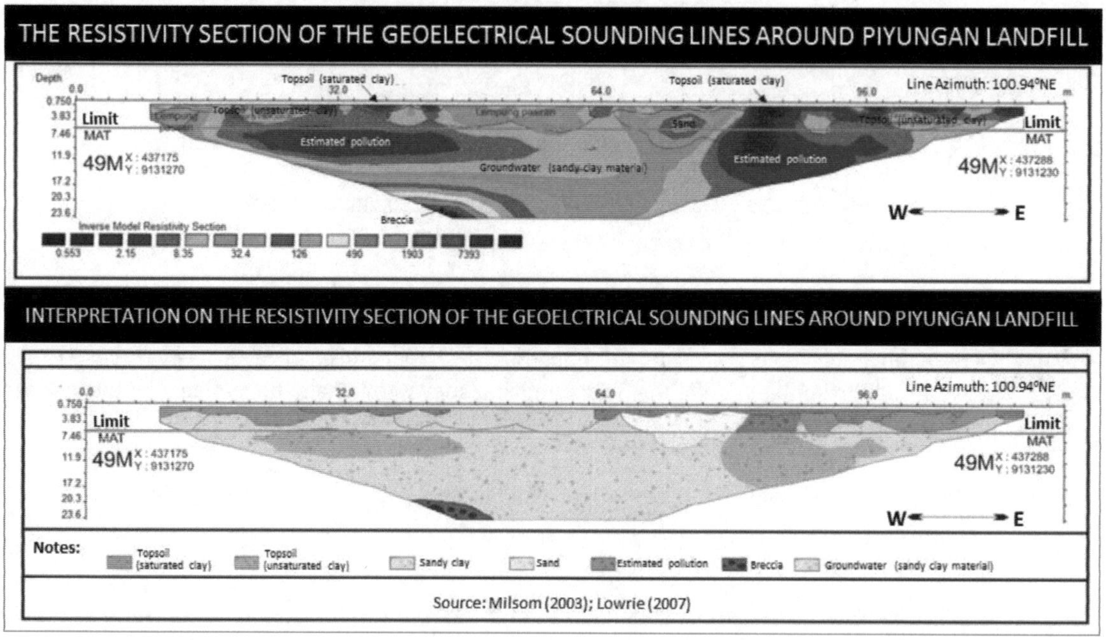

Figure 3. The resistivity section and the interpretation results of geoelectrical sounding lines around Piyungan Landfill

The results of geoelectrical sounding discovered the pollution indications in the groundwater around Piyungan Landfill. These indications were then validated by groundwater quality analysis. The groundwater was sampled at a specified distance to the pollutant source, namely 100 m, 200m, 500m, 700 m, and 1,000 m. Pollution is observable from the presence of physical, chemical, and biological changes in the groundwater. Physically, groundwater pollution presents a change in electrical conductivity. Meanwhile, the chemical and biological changes are identifiable from the excessive presence of dissolved chemicals (i.e., nitrate) and microorganisms (i.e., pathogenic bacteria), respectively.

3.1. The description of groundwater properties in the study area is as follows:

3.1.1. Physical Properties. Based on the laboratory analysis results, the electrical conductivity in Piyungan Landfill varied between 982 and 1,284 μmhos/cm. It can be concluded that the groundwater in the study area contained many soluble and ionizable salts that increase the electrical conductivity (EC). Electrical conductivity (EC) depends on the presence of inorganic ions, valence, temperature, and total and relative concentration. The measurement results indicated a high concentration of minerals in the groundwater. Referring to the Decree of the Indonesian Ministry of Health No.907/MENKES/SK/VII/2002, the maximum level of electrical conductivity allowed in drinking water is 1,000 μmhos/cm. Accordingly, the groundwater in some parts of the landfill does not meet the quality standard for drinking water.

3.1.2. Chemical Properties. Nitrate (NO_3) is the main form of nitrogen in natural waters and a major nutrient for the growth of aquatic plants and algae. In addition to its stable character nitrates are highly soluble in water. This compound is produced by the perfect oxidation process of nitrogen compounds in the water. Nitrification is the oxidation of ammonia to nitrite (nitritation) by Nitrosomonas bacteria, followed by the oxidation of nitrite to nitrate (nitration) by Nitrobacter. This aerobic process is a major step in the nitrogen cycle. These two genera of nitrifying bacteria are chemotrophs, meaning that they obtain energy from chemical processes. The reaction of nitritation is presented in chemical equation (4), while the stoichiometry of nitration is presented in equation (5) [13].

$$2\ NH_3 + 3\ O_2^- \rightarrow 2\ NO_2^- + 2\ H^- + 2\ H_2O \quad (4)$$

$$2\ NO_2^- + O_2^- \rightarrow 2\ NO_3^- \quad (5)$$

Based on the results of the laboratory analysis, the concentration of nitrate in Piyungan Landfill was 21.7-176 mg/L. This concentration is higher than 5 mg/L, indicating the presence of anthropogenic pollutants from human activities and animal feces. Consuming groundwater with high nitrate content reduces the oxygen-binding capacity of blood, particularly in infants younger than five months. This condition causes blue coloration of the skin (cyanosis), known as methemoglobinemia or blue baby syndrome [14]). Based on the Decree of the Indonesian Ministry of Health No.907/MENKES/SK/VII/2002, the maximum concentration of nitrate in drinking water is 50 mg/L. Therefore, the groundwater in some parts of Piyungan Landfill does not meet the quality standard for drinking water.

3.1.3. Biological Properties. Inspection of microbiological characteristics is necessary to assess the concentration of microorganisms in groundwater. Various types of pathogenic bacteria can be found in clean water supply system even though in low concentration. The contaminants from warm-blooded animals and human feces are highly detectable from the existence of these bacteria, which in this case function as indicator organisms. If these indicator organisms are found in water samples, there is a high possibility of water contamination by feces and pathogenic. On the contrary, the absence of indicator organisms signifies the minimum presence of feces and pathogenic bacteria. Water quality test using indicator organisms is the easiest way to determine water contamination by pathogenic bacteria routinely. Coliform, belonging to the family Enterobacteriaceae and the genus Escherichia, is a rod-shaped, gram-negative, highly motile, and non-spore forming bacterium. It is also facultatively aerobic, meaning that it can respirate oxygen in aerobic condition and switch to fermentation in an anaerobic environment.

The presence of *E. coli* in the water is harmless, but it indicates the presence of other pathogenic bacteria. Some strains of *E. coli* can cause stomachache, such as diarrhea, when ingested. The mixed acid fermentation in E. coli occurs in the following stages:

- The conversion of pyruvate to the end products, such as acetyl-coenzyme A (acetyl-CoA) and formate,
- The reduction of acetyl-CoA to ethanol,
- The inability to convert pyruvate to acetoin and 2.3-butanediol, and
- The conversion of formate to carbon dioxide and hydrogen only in an anaerobic condition and with the help of pyruvate formate-lyase (an enzyme) as a catalyst.

Enterobacteraerogenes, a non-fecal coliform, is found in soil and other environments outside of the intestinal tract of mammals and warm-blooded animals. The fermentation of this bacterium, which distinguishes it from *E. coli*, involves the production of acetoin and 2.3-butanediol from pyruvate and the production of Succinate. Although categorized as a fecal coliform, *E. coli* is not always pathogenic. One of the dangerous strains of *E. coli* is *E. coli* O157:H7 that produces a harmful Shiga toxin. When the contaminated food is not cooked properly, this strain leads to severe digestive disorder, such as diarrhea and dysentery, as well as death if not treated immediately. The results of the laboratory analysis showed that the fecal coliform in Piyungan Landfill was around 28-1,100 MPN/100 mL. According to the Decree of the Indonesian Ministry of Health No. 907/MENKES/SK/VII/2002, the maximum concentration of fecal coliform in drinking water is 0 MPN/100 mL. As a conclusion, all groundwater in the landfill exceeds the quality standard for drinking water.

3.2. The proposed management model of waste disposal site (landfill) associated with the vulnerability of groundwater system to leachate movement is as follows:

3.2.1. Base Lining(Liner). Infilling needs a baselining system to reduce the mobility of leachate in the groundwater. An effective liner prevents pollutants from migrating into the environment, especially groundwater. However, there is empirically no 100% efficient liner system. The discharge of leachate is inevitable, indicating the need for leachate collection channel aside from liner system. Therefore, the bottom of the landfill has to be sealed with layers of liner materials to prevent leachate from seeping out of the landfill and equipped with leachate collection channel.The liner is composed of natural materials (e.g., clay and bentonite) or synthetic materials. Depending on the desired functions, it may consist of one material (single) or a combination of natural and synthetic materials, commonly referred to as geocomposite. The layer formations and the types of liner materials vary according to the characteristics of the solid waste piled onto the landfill. The recommended base lining is geosynthetic or known as flexible membrane liner (FML). The commonly used geosynthetic for baselining is composed of Geotextile (functioning as the filter), Geonet (drainage channel), and Geomembrane and geocomposite(buffer layer). Geomembrane, as an impermeable coating, is a geosynthetic made of impermeable polymer. The best polymer is high-density polyethylene (HDPE) that is resistant to chemical reactions occurring in B3 waste.

3.2.2. Leachate Collection Channel. Leachate collection system is recommended to have holed pipes placed inside the rock-coated channels. It also requires coated channels that contain hollow river stones. In general, the necessary facilities for pipe-equipped leachate collection channels are as follows:

3.2.2.1. Terraced Slope. The base of the landfill is formed into terraces with a certain slope (i.e., 1-5%), which prevents leachate from accumulating at the bottom of the landfill but flowing into the collection channels (i.e., 0.5-1%) instead. Every collection channel is equipped with holed pipes so that leachate can flows into the collection or re-circulation unit. The designs of the maximum slope and the length of the collection channels are based on the capacity of the collection channel facility. To assess this capacity, the research uses Manning's equation.

3.2.2.2. Piped Bottom. The bottom of the landfill is divided into several rectangles with clay-based divides. The width of the divide depends on the width of the cells. Leachate collection pipes are placed parallel to the length of cells and directly on geomembrane.

3.2.3. Final Cap. The final cap of landfill consists of several sections. The upper part is mainly soil that functions as protection and plant-growing medium (topsoil). When the soil in the area does not correspond to these requisites, it needs an improvement, i.e., by mixing or replacing it with soils from other areas. The supposed thickness of the topsoil is 60 cm. The layer beneath it functions as a drainage system, which drains as much precipitation as possible so that rainwater does not seep into the layer beneath it. The materials used for composing this layer are porous, e.g., sand, gravels, and synthetic materials like geonet. The thickness of this layer is around 30 cm. The second part is a leachate-retaining layer, which is commonly composed of geocomposite (geomembrane and compacted clay). The thickness of geomembrane is recommended to be more than 2.5 mm; while the thickness of clay is more than 50 cm. Beneath this layer is the gas ventilation system. It is a requisite for city waste treatment because most of the city waste is organic matters that decompose biologically. In aerobic condition, the resultant gas is mainly carbon dioxide and methane. Therefore, this biogas can be utilized as an alternative source of energy. The layer of gas ventilation system consists of porous media like sand/gravel or pipe system. The lowest part of the final cap is subgrade layer that functions to increase the stability of the landfill surface. Furthermore, it forms a necessary slope for accelerating lateral drainage and reducing the hydraulic level. Its thickness is 30 cm. Aside from this final cap, reducing the amount of overland flow that enters landfill involves slope arrangement, equipped with surface drainage and sowing.

3.2.4. Leachate Treatment. The leachate treatment system is divided into two levels, namely secondary and tertiary processing, as depicted in Figure 4. As for the secondary leachate processing, it is represented by the stabilization pool (facultative and anaerobic) and aeration ponds.

Figure 4. The perspective of leachate processing plan

4. Conclusion

The analysis results of the geoelectrical sounding data proved that there are pollution indications in the groundwater around Piyungan Landfill. The pollutants were detected from the very low resistivity (i.e., 3-9Ωm) in aquifer zone. The interpretation of geoelectrical sounding data was validated using laboratory analysis of groundwater samples. The results of the laboratory analysis showed that the groundwater around Piyungan Landfill was contaminated. The electrical conductivity, as well as the concentration of nitrate and fecal coliform, have already exceeded the quality standard for drinking water according to the Decree of the Indonesian Ministry of Health No. 907/MENKES/SK/VII/2002.

Acknowledgment

The author would like to thank Dr. M. Widyastuti, M.T, the head of Hydrology and Environmental Climatology Dept., for his encouragement and constant support. Thanks are also due to Dr. Rika Harini, M.P, Head of the Environmental Geography Department, and to the author's colleagues Erik Febriarta, S.Si., M.Sc and Ajeng Larasati, S.Si for reading and improving the paper. Thanks also are due to Okta Rama, S.Si and Sambodo, S.Si for their help in the field.

References

[1] Viglizzo E F, Roberto Z E, Filippin M C and Pordomingo A J 1995 Climate variability and agroecological change in the Central Pampas of Argentina *Agriculture, Ecosystems and Environment* **55** pp 7-16.

[2] Plummer N, Salinger M J, Nicholls N, Suppiah R, Hennessy K J, Leighton R M, Lough J M

1999 Changes in climate extremes over the Australian Region and New Zealand During the Twentieth Century *Climatic Change* **42** pp 183-202.
[3] Mason S J 1996 *Climatic Change over the Lowveld of South Africa* **32(1)** pp 35–54
[4] Zhai P M, Sun A J, Ren F M, Liu X L, Gao B and Zhang Q 1999 Changes of climate extremes in China *Climatic Change* **42** (1) pp 203–18
[5] Masoudi M and Afrough E 2011 Analyzing trends of precipitation for humid, normal and drought classes using standardized precipitation index (SPI), A Case of study: Fars Province, Iran. *International Journal of AgriScience* **1(2)** 58-69.
[6] Naylor R L, Battisti D S, Vimont D J, Falcon W P and Burke M B 2007 Assessing risks of climate variability and climate change for Indonesian rice agriculture *Proceeding of the National Academic of Science* **114** pp 7752-57
[7] Damanhuri E 2008 *Landfilling waste* (Bandung: Bandung Institute of Technology Press) pp 8-9.
[8] Milsom J 2003 *Field Geophysics: The Geological field guide* series (New York: John Wiley & Sons Ltd) pp 102-3
[9] Allred B J, Daniels J J and Ehsani M R 2000 *Handbook of agricultural geophysics* (Boca Raton: CRC Press) pp 123-5
[10] Loke M H "Electrical imaging survey for environmental and engineering studies" *http://pages.mtu.edu* 2000 [Online] Available: http://pages.mtu.edu/~ctyoung/LOKENOTE.PDF [Accessed: 21-9-2017]
[11] Lowrie W 2007 *Fundamentals of geophysics* (New York: Cambridge University Press) pp 99-100
[12] Telford W M, Geldart L P and Sheriff R E 2004 *Applied geophysics* (London: Cambridge University Press) pp 171-2
[13] Effendi H 2003 *Water quality study for resource management and water environment* (Yogyakarta: Kanisius Press) pp 66-68

A vulnerability assessment for water availability related to the impacts of climate change in Banyuasin Valley, South Sumatra, Indonesia

Y Hamdani

Department of Civil Engineering, Universitas Tamansiswa Palembang, Jl. Tamansiswa No.261, Palembang, South Sumatera, Indonesia.

Abstract. Banyuasin Valley region is located in lowland areas that is potentially subjected to hazard of flooding and submergence. The indication of reduction hazard in water availability is in the period of 2010 with decreasing value of Total Run Off at - 500 mm/year and in the period 2030 is the lowest decline of the region Banyuasin with a value of -100 mm/year. Tanjung Api-api port, built in this area, is feared to be vulnerable to the availability of clean water due to the impact of climate change. The vulnerability components consist of exposure, sensitivity, and adaptive capacity. The formula means vulnerability to a certain hazard is strengthened by its exposure and its sensitivity and decreased by its adaptive capacity. The results of this study showed that water availability in the baseline condition is in low vulnerability (47.91%) whereas, at the projection condition, vulnerability is in the category of moderate vulnerability (81.28%).

1. Introduction

Global climate change is an issue that is increasingly influencing the development of policy at national and regional levels. It is not an exaggeration, as some studies suggest that the history of climate change impacts could be fatal to the survival of a nation. Some of the historical studies of the extinction of the Maya civilization in Central America around the year 950 AD concluded that the prolonged drought has affected people who are vulnerable to climate change [1]. The phenomenon of climate change has now become a necessity that needs to be anticipated and mitigated by all the parties in view of 83% from total emissions [2]. One of the causes of the phenomenon of climate change, Indonesia is generated from the change in land use and forestry [3]. Referring to the law No. 32 of 2009, the protection plan and environmental management requires a strategic environmental assessment on article 16 of the need for assessment of vulnerability and adaptive capacity to climate change. Based on the report of the IPCC, climate change impacts on coastal areas/river have the highest risk such as tropical storms, floods and rising sea levels [4]. In general, the changes of rainfall patterns and freshwater runoff are potential to cause some significant changes in the physical environment, among others, changes in the hydrological cycle, the effect on the availability of water in coastal areas and changes in water circulation [5].

Measurement of water availability effect on the coast is not easy to conduct and very necessary because of the availability of clean water in the coastal areas that has become a social and environmental issue disrupted by the tidal conditions. As the impacts of climate change, it needs to protect the public from the potential dangers posed by rising surface seawater, flooding, changes in rainfall, and other negative impacts [6]. Unless adaptive measures are not taken immediately, an analysis predicts that Indonesia could experience freshwater scarcity, declining crop yields, as well as

Content from this work may be used under the terms of the Creative Commons Attribution 3.0 licence. Any further distribution of this work must maintain attribution to the author(s) and the title of the work, journal citation and DOI.
Published under licence by IOP Publishing Ltd

the loss or destruction of habitat and ecosystem in coastal areas. As one approaches that the measurement can be done is with a vulnerability assessment [7]. Research on the vulnerability assessment, in general, has been widely performed, especially the vulnerability of social communities, such as the Social Vulnerability Index developed by [8].

South Sumatra Province is one of the areas in Indonesia who suffered exposure to climate change triggered by global warming. At 91.774.99 km^2, approximately 30% of the province area, which is low lying areas, are exposed to climate change. The exposure also occurs through the medium of the atmosphere and ocean (Bangka Strait), which is based on the study of the scientific basis of climate and sea level rise. The scientific basis of oceanographic studies concludes the rise in sea level to approximately 13.5 ± 6,15 cm at year 2030 relative to current conditions (scenario of greenhouse gas emissions are moderate - SRESa1b) [9]. The results report also mentions that the general conditions of water availability in the forecast period are relatively unchanged and actually increased when compared to the baseline period of 1970-2000. However, it could be the indication of hazard reduction in water availability in some areas, such as in the period of 2010 Musi Rawas and Lubuk Linggau with decreasing value of Total Run Off at - 500 mm / year and in the period 2030 which is the lowest decline of the region Musi Rawas and Musi Banyu Asin with a value of -100 mm / year. A region port of Tanjung Api-api as an area of strategic potential for the port area of the ocean and various supporting facilities, is expected to develop various sectors of development in the Province of South Sumatra, and in turn it could attract and generate growth in various sectors of development and will influence and impact significantly to Banyuasin. The results of these studies in micro can be illustrated by a map that can be predicted by region and area that has a low start up the vulnerability of the highest rate of climate change.

At the macro level study, Banyuasin Valley region located in lowland areas is potentially subjected to the danger of flooding and submergence [10]. This macro scale hazard assessment needs to be studied further in micro scale, to provide more detailed information concerning vulnerabilities in Banyuasin Valley region to climate change.

2. Research Method

This research is a field survey analysis. The parameters or indicators of vulnerability are employed to analyze the vulnerability of Banyuasin Valley and its surrounding areas. The parameter of vulnerability is exposure (water needs, land use) and, sensitivity (population density/population, water resources), as well as the adaptive capacity (welfare of the population). The degree of vulnerability is determined based on an index of vulnerability resulting from the parameters/indicators that are owned by the elements that have the potential risk to the impacts of climate change [11].

Surveying the measurement/marking has already done to see changes in land use in Banyuasin Valley region based on the maps of land cover/land use existing and the condition of land cover Banyuasin Valley region. GPS *(Global Positioning System)* is used to perform measurements, and determine the location of spatial and elevation measurements. Meanwhile, the questionnaire is administered in residential areas/housing in Banyuasin Valley Region. The survey activities include discussions and interviews with local people guided to fill the questionnaire. The data collected from the interview are arranged to identify the conditions that exist in the field. The collection of secondary data that aims to obtain the data of the number of the population, the population growth rate, topography, administration maps, land use maps and land use planning is collected from the Detail Spatial Plan Area Support Tanjung Api–api and the supporting data related to this research.

2.1. Research Location

The research location is located in Banyuasin Valley with limits on the north bordered by Bangka Strait, Sebalik canal at the South, the West by the Banyuasin River and East by the Telang River. The location was selected because the region will build the ocean port of Tanjung Api-api and will be used as a special economic zone in South Sumatra Province.

IOP Conf. Series: Earth and Environmental Science **129** (2018) 012006 doi:10.1088/1755-1315/129/1/012006

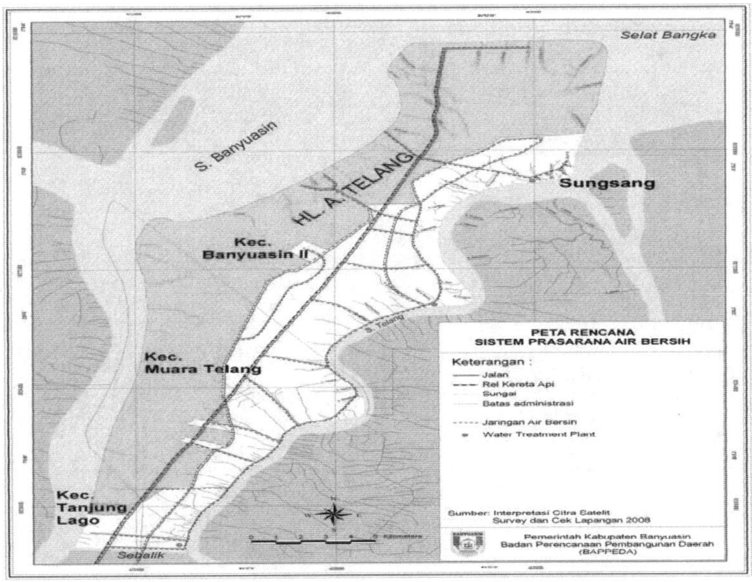

Figure 1. Map location research in Banyuasin Valley Region

2.2. The Framework of Vulnerability Assesment.

Vulnerability in this study is obtained from the preparation of the component parameters/indicators of vulnerability composed of topographic maps, population density, land use, water requirements, and welfare of the population. It will then be over load to get a map of vulnerability in the existing conditions (baseline) and the projection condition in year 2030.

The components of vulnerability consist of exposure (E), sensitivity (S), and adaptive capacity (AC). The function initially is a multiplication between E with S factors and divided by AC factor as in the following formula:

$$V = (E \times S) / AC \qquad (1)$$

Where V = vulnerability
E = exposure
S = sensitivity
AC = adaptive capacity

The formula of vulnerability to specific hazards is strengthened by exposure and sensitivity and decreased its adaptive capacity. In this assessment, the function of V is a pseudo multiplication, because, in practice, the vulnerability is obtained from the addition of the exposure component to the sensitivity component and reducing adaptive capacity component. Vulnerability index has a value of 0-,1, with the highest index value is indicated as the highest vulnerability. The Index itself is a combination of proxies scores indicators of vulnerability. Proxie scoring on each parameter is different, that takes into account how much influence these parameters are. The greater the influence of parameters, the greater the value is and otherwise.

The level of vulnerability is classified into five classes of vulnerabilities including very low vulnerability, low vulnerability, moderate vulnerability, high vulnerability, and very high vulnerability. The parameters of vulnerability analysis will determine the accuracy of the resulting level of vulnerability of the region and the adaptation strategies to be formulated later. It can be seen from the level of vulnerability in each level such as macro, meso and micro level.

The framework of vulnerability assessment can be seen below.

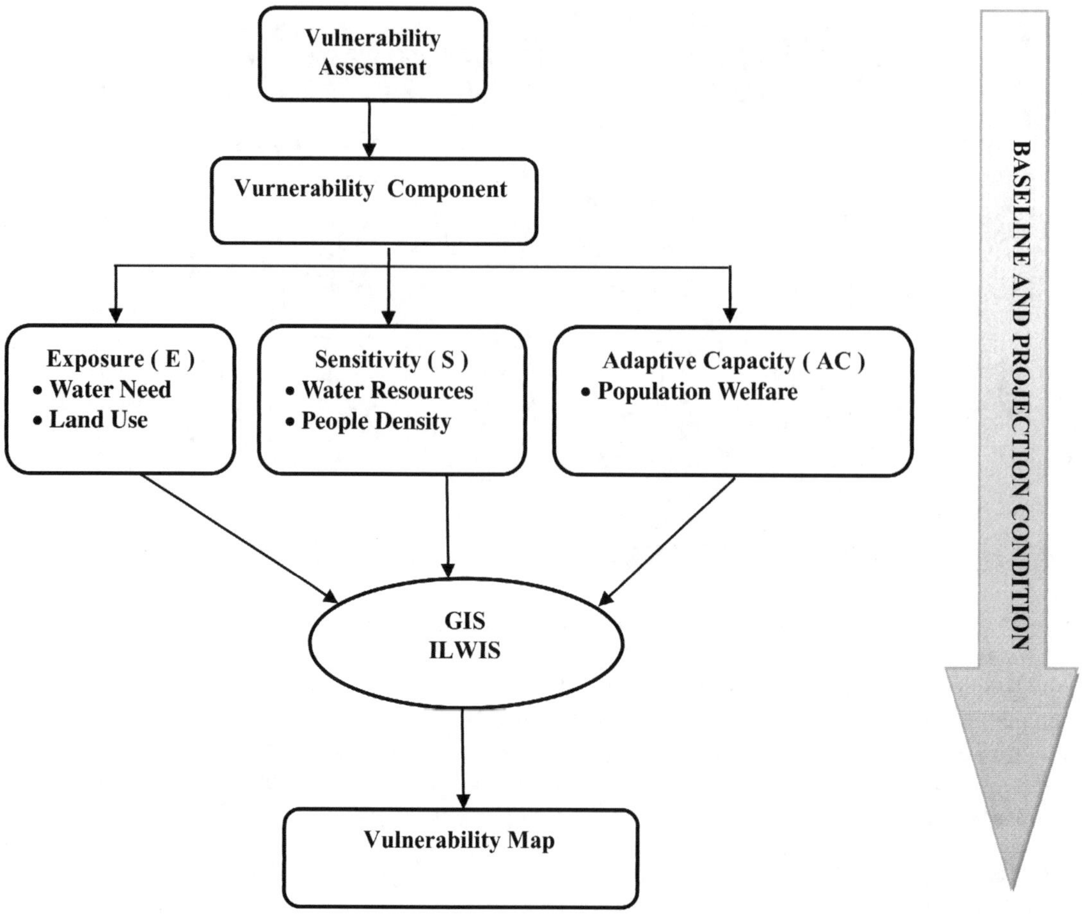

Figure 2. Framework of vulnerability assessment

Vulnerability in this study is obtained from the preparation of the component indicators of vulnerability composed of topographic maps, population density, land use, water requirements, and welfare of the population that will be an overlay to get a map of vulnerability in the baseline condition and the projection condition. Administratively based on the Detail Spatial Plan of Banyuasin Valley, this region consists of three districts including the District of Banyuasin II, Muara Telang, and Tanjung Lago, with an unevenly distributed population. The topography of Banyuasin Valley Region is a lowland swamp elevation. Based on the current condition of land cover, most of the area is covered by rice fields and swamps and forests. A vulnerability assessment on these activities is done with qualitative and quantitative analysis based on the weighting of indicators that influence the susceptibility of water availability at the area of research, in which the level of vulnerability weights is determined by the pairwise comparison method is used to get a vulnerability map. The modeling results were processed using GIS based program that will be used in data processing to obtain a vulnerability map.

Weighting is based on the land use element of the ability of the land use to deal with the hazard of climate change. In this case, it is the hazard of flooding and water storage capacity that influences the reduction in water availability. Satellite imagery from the map results obtained from the Regional Development Planning Agency Banyuasin, known that the existing land use on location Banyuasin Valley is grouped on the secondary mangrove forests, plantations, settlements, dry land agriculture mixed shrubs, fields, shrub swamps and water bodies. In the projection conditions, parameters of land use change into 8 kinds of spatial planning, namely the industry, the logistics center/warehouse, urban, residential and campus, central park, recreational water, swamp forest, and river border. Granting land use parameter weights based on the vulnerability reduction in water availability are grouped in residential areas and vital infrastructure, agricultural areas, aquatic ecosystems, and non-agricultural forest areas then are ranked and given weight.

The level of water needs are classified into three categories including low, medium, and high in every district, The calculation of water demand in the region is the sum of the domestic and non domestic water needs. Its weighting parameter is based on the population density of each district. Classification weighting is based on population density, low, medium and high density. Population density index is distributed into the region in accordance with the land use map of Banyuasin Valley Region. In the analysis of vulnerability to the condition of the projection using the data, the average percentage of population growth per year is calculated based on the capacity of a regional population of Banyuasin Valley. Water resources are one of the indicators of vulnerability that are classified based on the source of water that is used such as rainwater, wells, water river/lake, bottled water, local water installations, and make the determination to obtain the rank of the parameter weights. The population welfare as a vulnerability in terms of the economics of the population is obtained from the survey and interviews in the field to the baseline condition. It is then classified based on the percentage of the prosperous and underprivileged population. For the projection conditions, the population welfare is assumed by the industrial area. Meanwhile, the level of welfare is expected to increase that it can contribute to reducing vulnerability.

2.3. Data and Material

The data included in this study is based on the spatial data for each of the necessary data. The results of studies are based on spatial or in the form of a map. Nevertheless, the qualitative data are also available when it is still involved in the study. For it, a variety of secondary data collection has been done to obtain the complete data.

The data and tools used in the analysis are shown in Table 1 and Table 2

Table 1. Types and sources of data required

No	Data properties	Data type	Sources
1	Water quality	Primary	PDAM Tirta Musi Lab.
2	Number of inhabitants	Primary	Field
3	Condition of the house	Primary	Field
4	Water consumption	Primary	Field
5	Income per capita	Primary	Field
6	Existing land use map	Primary	Field
7	Land use plan map	Secondary	BAPPEDA
8	Rain data	Secondary	BMG
9	Map of Digital Elevation Model (DEM) SRTM	Secondary	http://srtm.csi.cgiar.org
10	Map Region Banyuasin Valley	Secondary	BAPPEDA Banyuasin

Table 2. List of tools used in the study

No	Name of tools	Number of uses
1	Stationery 2 pieces	Writing to recording data
2	Computer (RAM 2 GB) 1 unit	Performing general modeling
3	Printer 1 unit	Displaying form of report
4	ILWIS 3.6 Program, ArcGIS, Global Mapper 16	Performing modeling and data processing
5	GPS Tracker	Tracking research location
6	Laptop and Printer 1 pieces	Assisting to preparing reports

3. Result and Discussion

The vulnerability assessment in Banyuasin Valley and its surrounding areas are classified as a vulnerability assessment of local level where global climate change is translated into local events in response to geography and local environmental, economic and social factors. It results from the interaction between the many socio-ecological factors and processes such as income level, and individual habits. Vulnerability and adaptive capacity are also manifested locally. The vulnerability of water shortages consists of three indicators, where each indicator represents a different component of vulnerabilities. People's water needs and land use are considered as exposure component indicators, whereas water resources and people density belong to sensitivity component indicators. Meanwhile, the welfare of the population is considered as adaptive capacity component indicators. Physically, the land use of natural areas that changes into an area of man-made, which is often found in urban areas, will reduce the capacity of the surface water infiltration into the soil. Changes in land use will also be accompanied by an increase in water demand, and the risk of flooding as well as the degradation of areas as a result of erosion, resulting in increased sedimentation of rivers.

The comparison components of vulnerability can be seen in Table 3.

Table 3. Comparison of vulnerability component

Component	Indicator	Ratio
Exposure (E)	People's water needs Land use	0,50
Sensitivity (S)	Water resources People density	0,32
Adaptive Capacity (AC)	Population welfare	0,18

Based on the indicators above, the vulnerability map drawn up will be overlaid so that the resulting vulnerability level map of Banyuasin Valley region were analyzed at baseline and projections conditions. The results of the vulnerability map can be seen in Figure 3.

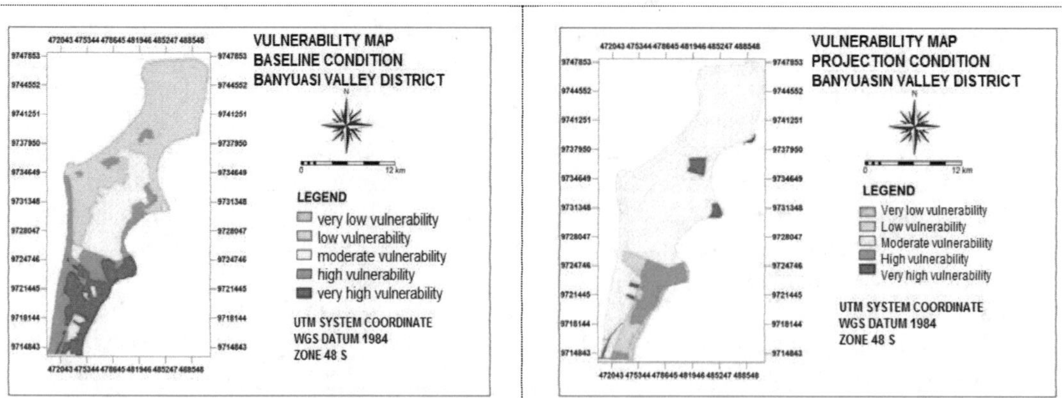

(a) (b)

Figure 3. Vulnerability map at (a) Baseline Condition and (b) Projection Condition

Table 4. Vulnerability area per district at Banyuasin Valley

District	Vulnerability area (Ha) Baseline condition (left) and Projection condition (right)									
	Very low		Low		Moderate		High		Very high	
Banyuasin II	33.5	0	13156	0	5258.5	19372.5	1215.5	0	0	559.25
Muara Telang	0	0	10	2002.5	681.5	0	955.75	2447.5	2802.5	0
Tanjung Lago	0	0	0	0	3.75	3184.5	2198	0	1167.25	184.75

According to Table 4. at baseline condition, Banyuasin Valley and its surrounding areas consist of a very low vulnerability level of 33.5 ha, low 13 166 ha, moderate 5943.75 ha, high 4369.25 ha, very high 3969.75 ha. The high and very high vulnerability that has the most extensive area lies in the sub-district of Tanjung Lago and Muara Telang. Meanwhile, viewed from the projection condition, Banyuasin Valley and its surrounding areas consist of a low level of vulnerability of 2002.5 ha, moderate 22 557 ha, high 2447.5 ha, very high 744 ha. The high and very high vulnerability that has the most extensive area lies in Muara Telang, and Banyuasin II. In general, susceptibility to water availability in the baseline condition is in the category of low vulnerability, which is mostly located on Banyuasin II, There are still a lot of industries and settlements built so that water demand yet too much and people in the region still mostly rely on rainwater reservoirs for clean water and river water for the daily usage. On projection condition, the greatest degree of vulnerability is moderate vulnerability, which is administratively located in Banyuasin II and Tanjung Lago district. According to spatial planning port of Tanjung Api – Api, in this area settlements and the various industries will be built so that it will require much water that needs to be anticipated from now on, as well as in Muara Telang. Thus, these regions are considered as a high vulnerability.

Both conditions above indicate that the level of vulnerability in the region of Banyuasin Valley and its surrounding likely tends to decline in the extent of the area of the baseline condition to projections, especially at the level of vulnerability that is very low, low, high and very high, while an increase in the extent of which is significant at the level of vulnerability. This indicates that the potential for an increased risk of water availability is in the medium category.

4. Conclusion

From the analysis of the study, it can be summarized that the results of total vulnerability at the site showed that the highest level of vulnerability at baseline condition is low vulnerability whereas the projected conditions level of vulnerability occurring is a moderate vulnerability.

Physically, the land use of natural areas changes into an area of man made, which is often found in urban areas, will reduce the capacity of the surface water infiltration into the soil. Changes in land use will also be accompanied by an increase in water demand and the risk of flooding, as well as the degradation of areas due to erosion, resulting in increased sedimentation of rivers.

The main concept is built without putting pressure on the surrounding environment and green open space requires at least 30%. Therefore, the existing land use and prediction of future changes should be considered carefully. The increase of the development in urban areas is expected to result in the increasing welfare of urban living. Short term adaptation strategy is to increase the water storage capacity by creating a retention pond, or reservoir, and expansion of rainwater, increasing the efficiency of water use by recycling and rainwater harvesting in areas that do not have groundwater. As for the long-term adaptation strategy are the development of artificial water catchment area in the protected Telang forest, the development of surface water resources by supplying water from the local water company, the desalination of seawater if necessary, the control of the population, and the supervision of the spatial planning.

Acknowledgments

The author would like to thank the Province Government of South Sumatra and District Government of Banyuasin for giving the opportunity to involve in this study and use a part of the results to be written for this paper.

References

[1] Peterson L and Haug G 2005 Climate and the collapse of Maya civilization *American Scientist* **93(4)** pp 322-9

[2] Downing T E and Patwardhan A. "Vulnerability assessment for climate adaptation" *http://www.start.org*2003 [Online] Available: http://www.start.org/Program/advanced_institute3_web/p3_documents_folder/Downing_Pat

wardhan_TP3.pdf [Accesed 17-10-2016]

[3] Yunan H, Budhi S, Dwi S and Azhar K A 2013 *Proc Intl Workshop on Sustainable Management of Lowland* **1** p165

[4] IPCC"Summary for policymakers climate change 2001 impacts, adaptation and vulnerability" *https://www.ipcc.ch* 2001 [Online] Available: https://www.ipcc.ch/ipccreports/tar/wg2/pdf/wg2TARspm.pdf [Accesed 19-10-2015]

[5] Alistair H and Paul W 2011Climate change impacts and adaptation in cities: a review of the literature *Climatic Change* **104(1)** 13–49

[6] Messner F and Meyer V 2005 "Flood damage, vulnerability and risk perception – challenges for flood damage research"*https://www.ufz.de* [Online] Available: https://www.ufz.de/export/data/2/26222_Disk_Papiere_2005_13.pdf [Accesed 20-04-16]

[7] Yunan H, Budhi S, Dwi S and Azhar K 2014 *Journal of Sustainable Development* **7(6)** p8

[8] Rygel L, O'sullivan D and Yarnal B 2006 A Method for constructing a social vulnerability index:an application to hurricane storm surges in a developed country *Mitigation and Adaptation Strategies for Global Change* **11(3)** pp 741–64

[9] Emma R 2012 *Risk and adaptation climate change study tarakan city, south sumatera, and malang raya*, Summary for Policy Makers 'ed Haneda S M, Tri W, *et al* (Jakarta:KLH) p19

[10] Budhi S, Norma P and Ambyar S 2010 *Proc Seminar Nasional Teknik Sipil VI(10)* **II** p61

[11] Brooks N "Vulnerability, risk and adaptation, a conceptual framework" *https://www.researchgate.net/* 2003 [Online] Available: https://www.researchgate.net/publication/200032746_Vulnerability_Risk_and_Adaptation_A_Conceptual_Framework [Accessed 12-06-2014]

IOP Conf. Series: Earth and Environmental Science **129** (2018) 012007 doi:10.1088/1755-1315/129/1/012007

Mycorrhizal diversity of stevia (*Stevia rebaudiana* Bertoni) rhizosphere in Tawangmangu, Indonesia

D Y Astuti[1,2], Parjanto[1] and V R Cahyani[1,3]

[1] Graduate School, Sebelas Maret University, Jl Ir Sutami 36A Kentingan, Surakarta Central Java 57126 Indonesia
[2] Directorate General of Plantations, Ministry of Agriculture, Republic of Indonesia Jl. Harsono RM 3 (building C) Ragunan,South Jakarta 12550 Indonesia
[3] Corresponding author: vitaratri@staff.uns.ac.id

Abstract. Mycorrhizal fungi is a group of soil fungi with mutualistic symbiosis between fungi and plant roots. The diversity on mycorrhiza contributes the maintenance of plant biodiversity, ecosystem function, and plant productivity. Climate change may affects the distribution and diversity of mycorrhizal fungi, and thus the study on mycorrhizal diversity is important to develop the information about mycorrhizal function and utilization. The present study investigated mycorrhizal diversity in the rhizosphere of stevia at four locations in different altitudes and soil types. The samples taken from Tlogodlingo (Andisols 1), Kalisoro (Andisols 2), Nglurah (Alfisols 1) and Ledoksari (Alfisols 2) in Tawangmangu, Karanganyar, Central Java, Indonesia. The result showed that *Glomus* sp. and A*caulospora* sp. were the common genus found at all locations, whereas *Gigaspora* sp. was the only species found in the acidic Alfisol soil. Statistical analysis indicated that altitude, soil pH, and P availability significantly positively correlated with mycorrhizal spore density. The increase of altitude, soil pH and P availability, also increase the mycorrhizal spore density. Mycorrhizal infectivity negatively correlated with C/N ratio.

1. Introduction

Mycorrhizal fungi is a group of soil fungi performs mutualistic symbioses between fungi and plant roots. These symbioses improve biomass and the host plants growth rate, pathogen resistance, mineral nutrient uptake, rate of photosynthesis and secondary metabolites content when compared with symbioses [1,2]. Overall, a symbiosis of mycorrhiza may improve productivity and quality of host plant particularly in the sustainable agricultural system [3].

Tawangmangu is an area located in Karanganyar District, Central Java Province, Indonesia. The altitude is from 800 to 2,000 m asl [4], mean annual rainfall 3,188 mm with mean annual rainy days is 132 days (2003-2011) [4,5,6], and thus it suitable for stevia (*Stevia rebaudiana* Bertoni) culture [7,8]. Stevia is one of a potential alternative sweeteners plant [9] in Central Java, and dominant in Tawangmangu, Indonesia.

Stevia produces secondary metabolites called *steviol glycoside.* Stevioside and rebaudioside-A are the main components of steviol glycosides [10]. According to Mandal et al. (2013)[1], mycorrhizal symbioses potentially augment the production of *steviol glycoside* main components. However, many factors can affect the success of mycorrhizal symbioses including the host plant [11], the environment, and correlation levels with other soil organisms [12]. Climate change has major impacts on the environment, especially related with biodiversity and ecosystem function changes. The association of mycorrhizal fungi and host plant is very important to cope with climate

Content from this work may be used under the terms of the Creative Commons Attribution 3.0 licence. Any further distribution of this work must maintain attribution to the author(s) and the title of the work, journal citation and DOI.
Published under licence by IOP Publishing Ltd

change effects, such as water shortage delay, revegetation of degraded ecosystem [2,13,14] and also reduce symptoms of Mn^{2+} and Al^{3+} toxicity in acid soils [15].

The diversity on mycorrhiza may contribute to plant biodiversity maintenance, ecosystem function and plant productivity [16]. According to Bellgard and Williams (2011) [17], climate change may affect the distribution and diversity of mycorrhizal fungi, and thus the study on mycorrhizal diversity is important to develop the information about mycorrhizal function and utilization. In this study, we used stevia root and soil rhizosphere to determine the diversity of mycorrhiza in two soil types at four different altitudes in Tawangmangu, Indonesia.

2. Materials and Methods

2.1. Location of sampling area

This research used rhizosphere soils and roots of Stevia (*Stevia rebaudiana* Bertoni), which were collected from four different locations and two soil types in Tawangmangu (Table 1).

Table 1. Details of the sampling area.

Location	Soil type	Altitude (m asl)	Latitude	Longitude
Tlogodlingo	Andisols 1	1,772	07°40′5.28″S	111°10′43.11″E
Kalisoro	Andisols 2	1,121	07°39′49.48″S	111°8′6.77″E
Nglurah	Alfisols 1	1,049	07°40′31.59″S	111°7′55.22″E
Ledoksari	Alfisols 2	956	07°40′33.16″S	111°6′58.43″E

2.2. Sampling of rhizosphere soil and stevia root

Both rhizosphere soil and plant root samples were taken randomly at 20-25 cm depth. Approx 1-2 kg rhizosphere soil samples were collected by carefully separating the root with adhering soils. Root samples were collected by dipping the roots into the water to remove soil particles and then inserted into 50% alcohol solution.

2.3. Chemical analysis of rhizosphere soil

Soil samples were air dried and divided into two portions, to be sieved with 0.5 mm and 2 mm filter, respectively. 0.5 mm filtered soil sample was employed to measure C/N ratio, obtained from the ratio of soil organic carbon and soil total nitrogen. Soil organic carbon (C-organic) was determined by spectrophotometry and soil total nitrogen (N total) was determined by the Kjeldahl method. 2 mm filtered soil sample was prepared to measure soil pH using pH meter and phosphorus availability (P-availability) using Olsen method [18].

2.4. Mycorrhizal infectivity

The roots in alcohol solution were carefully washed with water, then cleaned roots were cut into pieces about 1 cm long, and then into a glass beaker containing KOH 10% at 70 °C for 10 minutes, rinsed with distilled water. Then neutralize the roots in 1 N HCl and then rinsed with distilled water. The cleared roots were stained with trypan blue, prepared in lacto glycerol [19]. Infection roots were viewed under the microscope, the percentage of root colonization was assessed and scored by observing the presence of arbuscular and vesicles in the observed roots.

2.5. Isolation and identification of mycorrhiza

Isolation of mycorrhiza spore was intended to measure spore density per 100 gram of soil and to assist in the identification process. Isolation of spores using wet-sieving and decanting methods, modified from Gerdermann and Nicolson [20], followed by adding 60% sucrose solution. Air-dried soil for about 100 gram was dissolved in 1 liter water and then the mixture was poured into a sieve with 250 μm, 90 μm and 45 μm filter. From this method, only the fine soil particles along with mycorrhizal spores were collected on the 45 μm sieve. Then the fine soil was placed in a centrifuge tube and added with a sucrose solution, then centrifuged at 2000 ppm for 7 minutes. Sucrose-containing spores were poured on cleaned 45 μm sieve and immediately washed to remove the sucrose. The spores were placed on cleaned Petri dish containing distilled water and observed under a microscope. Identification of spore was done by placed the spore at the prepared glass and identified under the microscope. Identification of spore-based on the character of spore morphology including spore size, spore walls, hyphae, and spore color especially discoloration that occurs after dropping it with Melzer reagent. The spore was compared with *Species Description of* INVAM (*International Culture Collection of Vesicular-Arbuscular Mycorrhizal Fungi*) [21].

2.6. Statistical analysis
Standard deviations were calculated for mean values and the Pearson correlation coefficient was used to determine the relationship of altitude, soil pH, P availability, C/N ratio, mycorrhizal spore density and mycorrhizal infectivity.

3. Result and Discussion

Abiotic variables such as soil pH, C/N ratio, and phosphorus may determine the diversity of natural mycorrhizal [22]. From figure 1 it can be figured out that both of Andisols 1 and Andisols 2 were included to neutral soils at 6.6±0.02 and 6.5±0.11, respectively, and Alfisols 1 and Alfisols 2 were included to acid soils with each pH was 5.5±0.11 and 5.7±0.07. It can be seen too that Andisols 1 and Andisols 2 performs very low P availability (2.1±0.10 ppm and 2.5±0.01 ppm), while acid soils (Alfisols 1 and Alfisols 2) shows lower P (1.2±0.07 and 0.7±0.01 ppm).

Correlation analysis between soil pH and P availability levels resulted positive correlation as seen in table 2 with a correlation coefficient (r) 0.84. It means that the increase of pH also increases P availability. Altitude and soil pH positively correlated (r= 0.775), but altitude and P availability is not significantly correlated with r= 0.488.

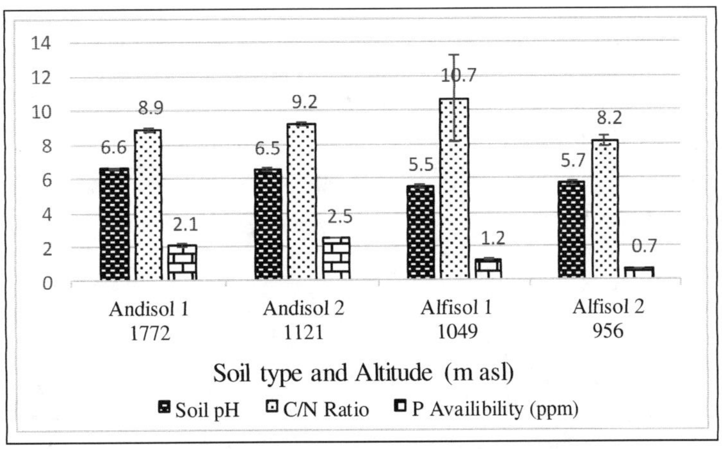

Figure 1. soil pH (H_2O), C/N ratio and P Availability of stevia rhizosphere

Figure 1 shows the Alfisol 1 soil has the highest C/N ratio (10.7±2.49), and the lowest is Alfisol 2 (8.2±0.31). All soil are categorized as soil with low C/N ratio. According to Wu et al. (2001) [23], low soil C/N ratio could accelerate the process of microbial decomposition of organic matter and nitrogen, which is non conducive for carbon absorption.

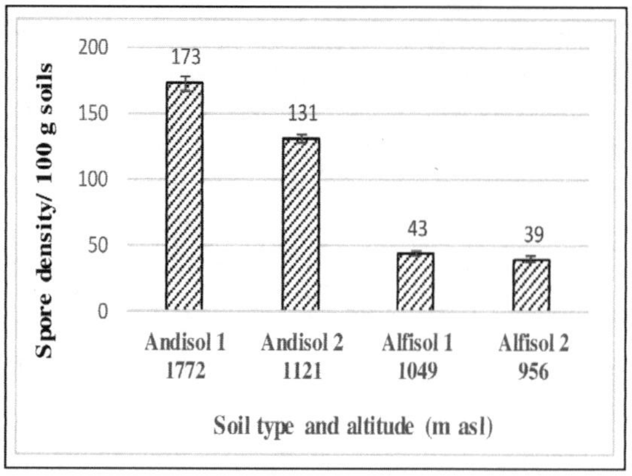

Figure 2. Mycorrhizal spore density/100 g soils at different soil type and altitude (m asl)

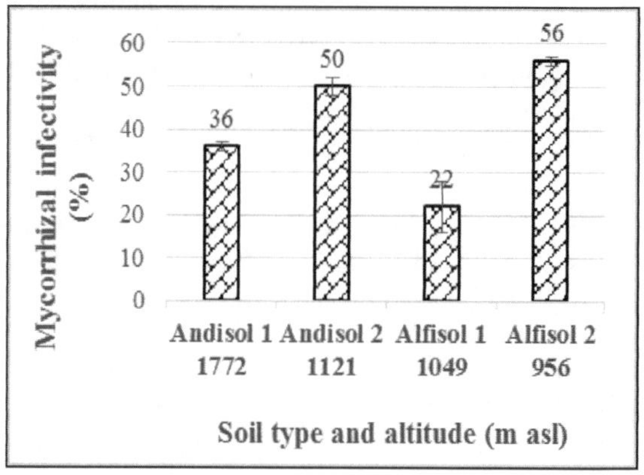

Figure 3. mycorrhizal infectivity (%) at different soil type and altitude (m asl)

Table 2. Correlation between altitude, soil pH, P availability, C/N ratio, spore density and mycorrhizal infectivity calculated using Pearson Correlation

	Pearson correlation (r)					
	Altitude (m asl)	Soil pH	P availability (ppm)	C/N Ratio	Spore density	Mycorrhizal infectivity (%)
Altitude (m asl)	1	0.775**	0.488	-0.194	0.848**	-0.07
soil pH	0.775**	1	0.84**	-0.14	0.967**	0.198
P avaibility (ppm)	0.488	0.84**	1	0.069	0.864**	-0.067
C/N Ratio	-0.194	-0.14	0.069	1	-0.121	-0.711**
Spore density	0.848**	0.967**	0.864**	-0.121	1	0.027
Mycorrhizal infectivity (%)	-0.07	0.198	-0.067	-0.711**	0.027	1

**Correlation is significant at level 0.01

Mycorrhiza spore's density observation resulted in various numbers as shown in Figure 2. Neutral soils (Andosols 1 and Andisols 2) had higher numbers of spore density at 173±5.57 and 131±2.65, respectively, while acid soils were 43±2.00 spores (Alfisols 1) and 39±2.00 spores (Alfisols 2). The correlation analysis found that spore density had a positive correlation with soil pH at 0.967 and P availability at 0.864 (Table 2). This indicated that increases of pH and P availability related with increases of mycorrhizal spore density. The other study also reported that spore density significantly correlated with soil pH [24]. Figure 2 also shows higher spore density was observed at higher altitude, similarly with the correlation analysis (table 2), that altitude positively correlated with mycorrhiza spore density ($r = 0.848$).

Figure 2 and 4 shows the higher spore density and altitude irrespective with mycorrhiza diversity. The neutral soils with higher spore density had less diverse genus than the acid soils with lower spore density. The neutral soils had two types of mycorrhizal genus namely *Glomus* sp and *Acaulospora* sp., while the acid soils which have lower spore density and soil pH had three types of mycorrhizal genus, i.e. *Glomus* sp, *Acaulospora* sp. and *Gigaspora* sp. Likewise, other research of Gou et al (2012) [25] found that the lower soil pH had higher mycorrhizal diversity compared to higher soil pH on perennial pastures with lime treatments. Figure 4 also shows that mycorrhizal less genus diversity found at higher altitude. Some studies also reported similar results, that mycorrhizal diversity is lower at higher altitude [24,26].

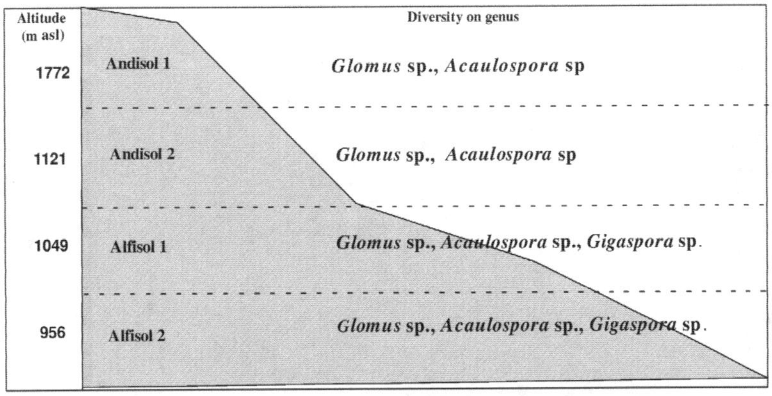

Figure 4. Mycorrhizal diversity by genus on different soil types and altitudes (m asl)

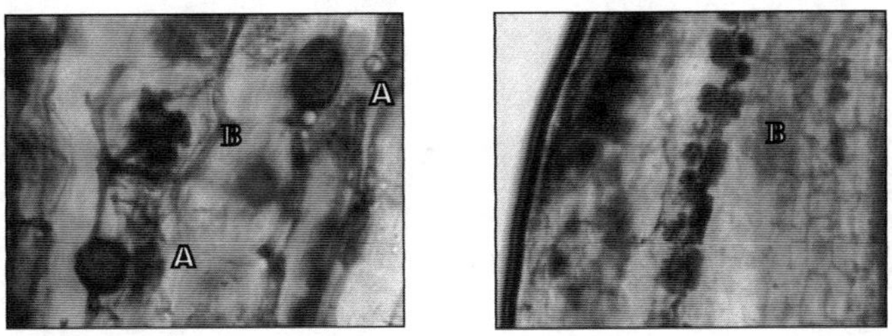

Figure 5. Mycorrhizal colonization of stevia's root. (A) Vesicle colonization form. (B) Arbuscular colonization form

Observations on root found that mycorrhizal colonization as vesicles and arbuscular structures as displayed in Figure 5. The mycorrhizal infectivity (figure 3) showed that the neutral and acid soils had various numbers. The highest infectivity belongs to Alfisols 2 at 56±1.00 %, followed with Andisols 2 at 50±2.00 %, Andisols 1 at 36±1.00 % and the lowest were Alfisols 1 at 22±6.00 %. However, there was reduction trend of mycorrhizal infectivity at a higher altitude at the same soil type. Pearson correlation resulted a negative correlation between mycorrhizal infectivity and soil C/N ratio at -0.711 (table 2). It means that the increasing of mycorrhizal infectivity related to decreasing of soil C/N ratio.

From identification of mycorrhizal spores on four different altitudes and two types of soil, including Andisols 1, Andisols 2, Alfisols 1 and Alfisols 2, obtained three different type of genus consisted of *Glomus* sp., *Acaulospora* sp. and *Gigaspora* sp.

According to table 2, *Glomus* sp. were existed in all soils types. Andisols 1 and Andisols 2 had two kinds of *Glomus* sp. each whereas the acid Alfisols 1 and Alfisols 2 only had one. The spores had their own different characteristics, such as colors consisted of hyaline, light yellow, brown to dark red-brown; the majority had globose shape but the red one had globose, subglobose to ellipsoid shapes; had smooth and rough surfaces. Each spore had special characters including an ornament, subtending hyphae and reaction with Melzer reagent. However, the main approaches of spores included in *Glomus* sp. were the thin spore walls and hyphae in spores directly stick on the spore walls with the same color [21].

IOP Conf. Series: Earth and Environmental Science **129** (2018) 012007 doi:10.1088/1755-1315/129/1/012007

Table 3. Morphology of *Glomus* sp. from different soils types (Andisols 1, Andisols 2, Alfisols 1, and Alfisols 2)

	Shape	Colour	Spores surface texture	Special characteristics	Reaction with Melzer
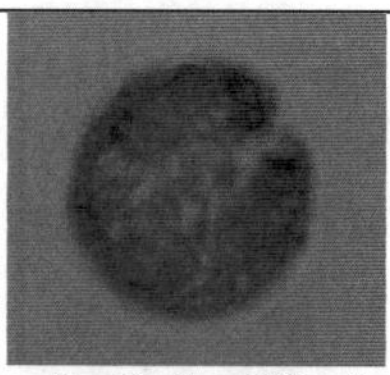 *Andisols 1, Alfisols 2*	Globose	Hyalin	Rough	Have an ornament	Change from hyaline to pale yellow
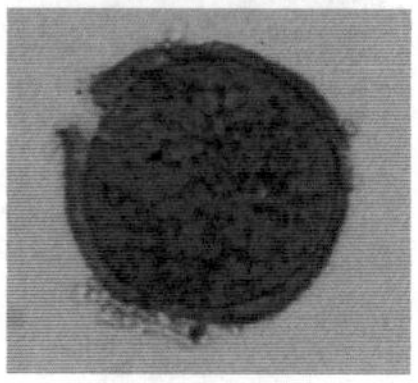 *Andisols 1, Andisols 2*	Globose	Brown	Rough	-	-
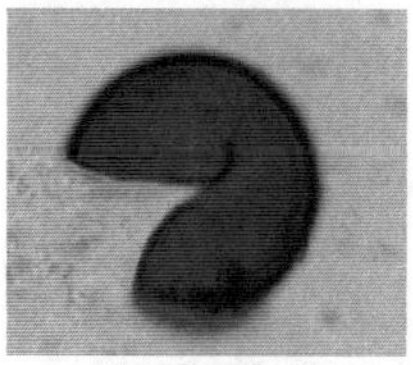 *Andisols 2*	Globose, subglobose, ellipsoid	Dark red-brown	Smooth	-	-
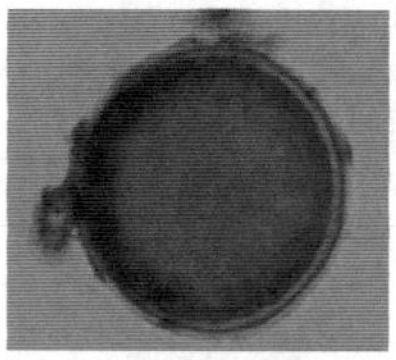 *Alfisols 1*	Globose	Hyalin	Smooth	Subtending hyphae	-

Table 4. Mycorrhizal spore morphology of *Acaulospore* sp.

	Shape	Colour	Spores surface texture	Special characteristics	Reaction with Melzer
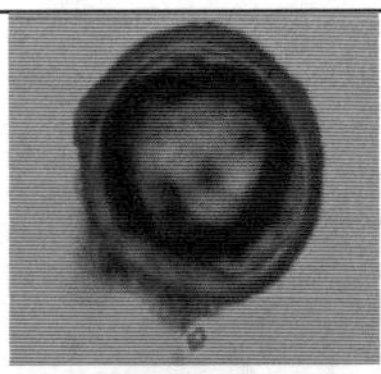 *Andisols 1*	Globose	Golden brown	Smooth	cicatrix	-
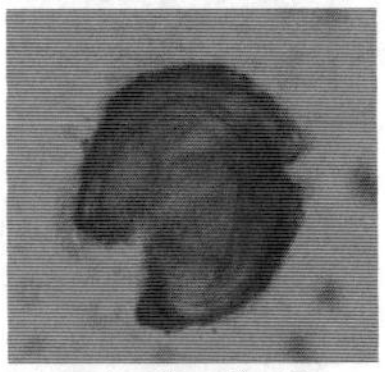 *Andisols 2*	Subglobose, irregular	Subhyalin	Smooth	-	Slight change from sub hyaline to pale yellow
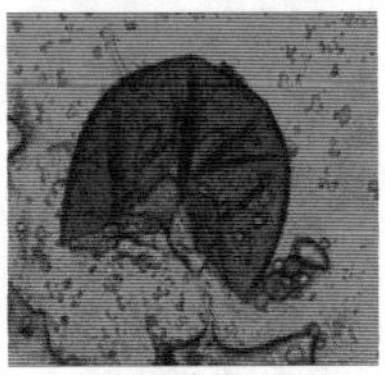 *Alfisols 1*	Globose	White	Smooth	-	Change from white to orange-brown
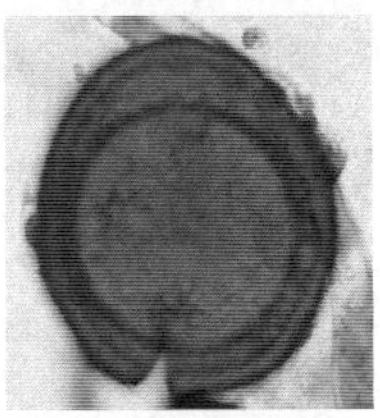 *Alfisols 2*	Subglobose	Red-orange	Smooth	cicatrix	-
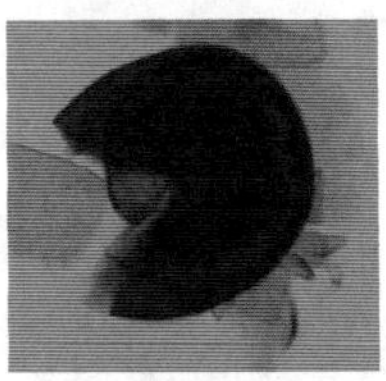 *Alfisols 2*	Globose	Dark orange brown	Smooth	-	A change color inside

According to table 3, *Acaulospora* sp. were existed at all type of soils. Andisols 1, Andisols 2 and Alfisols 1 had one species of *Acaulospora* sp., while Alfisols 2 had two species of *Acaulospora* sp. The characteristics of the spores are colors consisted of sub hyalin, white, golden brown, red-orange and dark orange-brown; the majority had globose shape but there was subglobose to irregular shapes, and had a smooth surface. Each spore had special characters including cicatrix, subtending hyphae and reaction with Melzer reagent. However, the main approaches of spores included in *Acaulospora* sp. were the thick spore walls and separating flexible inner spore walls when pressed slightly on prepared glass [21].

Table 5. Morphology of *Gigaspora* sp. from Alfisols 1 and Alfisols 2

	Shape	Colour	Spores surface texture	Special characteristics	Reaction with Melzer
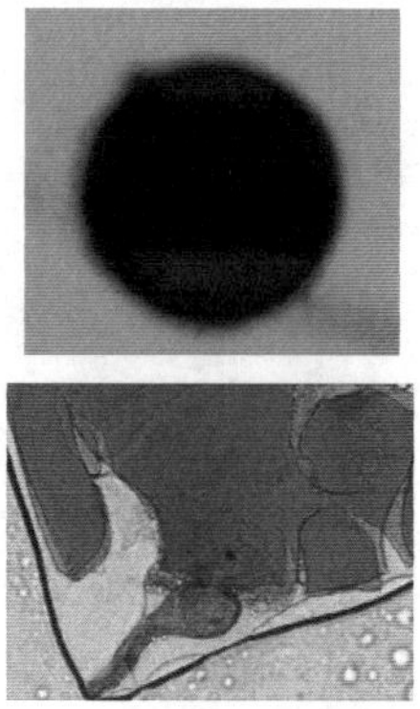	Globose	Dark Orange	Smooth	Swelling hyphal tip	-

The last identified mycorrhizal spores were *Gigaspora* sp.(Table 4), which was only found in Alfisol soils. These soils had low soil pH and very low P availability. The characteristic is dark orange color, globose shape, and smooth surface. The main approaches of spores included in *Gigaspora* sp. were the swelling hyphal tip [21].

4. Conclusion

There are three genera of mycorrhizal spores include *Glomus* sp., *Acaulospora* sp. and *Gigaspora* sp. with special characteristics at each altitude and soil type. *Glomus* sp. and *Acaulospore* sp. were found as a common in all areas, but *Gigaspora* sp. was only found in acidic Alfisol soils. Rhizosphere soil of stevia from Alfisols 2 indicated the highest mycorrhizal infectivity at 56% and diversity consisted of *Glomus* sp., *Acaulospora* sp. and *Gigaspora* sp., however, Alfisols 2 also indicated the lowest mycorrhizal spore density which was only 39 spores per 100 gram soil.

The altitude, soil pH, and P availability positively correlated with mycorrhiza spore density, where the increase of altitude soil pH, and P availability also increased the mycorrhizal spore density. Mycorrhizal infectivity negatively correlated with C/N ratio, indicated by the decrease of C/N ratio significantly increased mycorrhizal infectivity.

References

[1] Mandal S, Evelin H, Giri B and Singh V P 2013 *J. Applied Soil Ecology* **72** pp 187-94
[2] Püschel D, Rydlová J, Sudová R, Gryndler M and Vosátka M 2011 *J. Plant Nutr. Soil Sci.* **174** pp 664–72
[3] Gianinazzi S, Gollotte A, Binet M N, van Tuinen D, Redecker D and Wipf D 2010 *J. Mycorrhiza* **20** pp 519-30
[4] BPS (Badan Pusat Statistik) Kabupaten Karangayar *http://karanganyarkab.bps.go.id* 2009 [Online] Available: http://karanganyarkab.bps.go.id/website/pdf_publikasi/Karanganyar-Dalam-Angka-2009.pdf [Accessed:15-09-2017]
[5] BPS (Badan Pusat Statistik) Kabupaten Karangayar *http://karanganyarkab.go.id* 2015 [Online] Available: http://karanganyarkab.go.id/wp-content/uploads/2015/04/Karanganyar-Dalam-angka-2011.pdf [Accessed: 15-09-2017].
[6] BPS (Badan Pusat Statistik) Kabupaten Karangayar. *http://karanganyarkab.go.id* 2015 [Online] Available: http://karanganyarkab.go.id/wp-content/uploads/2015/04/Karanganyar-Dalam-Angka-2013.pdf [Accessed: 15-09-2017].
[7] Midmore D J and Rank A H 2002 *RIRDC Pub* No. W02/22, p. 55.

[8] Ramesh K, Singh V and Megeji NW 2006 *Advance in Agronomy Pub* **89** pp 137-177
[9] Gregersen S, Jeppesen P B, Holst J J and Hermansen K 2004 *J.Metabolism* **53(1)** 73-6
[10] Kinghorn A D 2002 *England Medical and aromatic plants/industrial profiles* **19** pp 1-17
[11] Rosenblueth M and Martinez-Romero E 2006 *MPMI* **19** pp 827-37
[12] Berruti A, Lumini E, Balestrini R and Bianciotto V 2016 *J.Frontiers in Microbiology* **6**
[13] Khalvati M, Bartha B, Dupigny A and Schröder P 2010 *J. Soil Sediments* **10** pp 54-64
[14] Zhao R, Guo W, Bi N, Guo J, Wang L, Zhao J and Zhang J 2015 *J.Applied Soil Ecology* **88** pp 41-9
[15] Yano K and Takaki M 2005 *J.Soil Biol Biochem* **37** **pp** 1569-72
[16] Van der Heijden M G A, Klironomos J N, Ursic M, Moutoglis P, Streitwolf-Engel R, Boller T, Wiemken A and Sanders I R 1998 *J.Nature* **396** pp 69-72
[17] Bellgard S E and William S E 2011 *J.Diversity* **3** pp 8-90
[18] Olsen S R, Cole C V, Watanabe F S and Dean L A 1954 *J.USDA cir.* **939** pp 242-6
[19] Phillips J M and Hayman D S 1970 *Trans. Br. Mycol. Soc.* **5** pp 158-61
[20] Gerdermann J W and Nicolson T H 1963 *Trans. Br. Mycol. Soc.* **46** pp 235-44
[21] West Virginia University [International culture collection of (vesicular) arbuscular mycorrhizal fungi] http://invam.wvu.edu 2017 [Online] Available: http://fungi.invam.wvu.edu/the-fungi/species-description.html [Accessed: 05-08-2017]
[22] Dumbrell AJ, Nelson M, Helgason T, Dytha C and Fitter A H 2010 *ISME J.* **4** pp 337-45
[23] Wu H B, Guo Z T and Peng C H 2001 *J.Quaternatry Sciences* **21(4)** pp 366-76
[24] Gai J P, Tian H, Yang F Y, Christie P, Li X L and Klironomos J N 2012 *Int.J.Soil Biology-Pedobiologia* **55** pp 145-51
[25] Guo Y J, Ni Y, Raman H, Wilson B A L, Ash G J, Wang A S and Li G D 2012 *J.Plant soil* **351** pp 389-403
[26] Shi Z, Wang F, Zhang K and Chen Y 2014 *Can J Microbiol* **60** pp 811-8

Land use, climate parameters and water quality changes at surroundings of Code River, Indonesia

Muryanto[1,4], Suntoro[2], T Gunawan[3] and P Setyono[2]

[1]Doctoral program of Environmental Science Department of Environmental Science, Sebelas Maret University, Jl. Ir. Sutami 36A, Kentingan, Surakarta, 57126 Indonesia
[2]Department of Environmental Science, Sebelas Maret University, Surakarta, Indonesia
[3]Department of Geography, Gadjah Mada University, Yogyakarta, Indonesia
[4]Corresponding author: muryanto_65@yahoo.com

Abstract. Regional development of an area has the potential of adverse impact on land use, vegetation, or green space. The reduction of green open space is known to contribute to global warming. According to the Intergovernmental Panel on Climate Change (IPCC), global warming has become a serious and significant phenomenon in human life. It affects not only ecological environment but also social and cultural environment. Global warming is a rise in global annual temperature due to, one of which, greenhouse gases. The purpose of this research is to determine the effects of land use change on water pollution and climate parameters at Code river. The results showed that Code River is experiencing land use conversion. Rice field was the most extensively reduced land use, by 467.496 ha. Meanwhile, the other land uses, namely plantation, grass, and forest, were reduced by 111.475 ha, 31.218 ha, and 1.307 ha, respectively. The least converted land use was bushed, whose decreased 0.403 ha. The land use conversion in the study area deteriorated the water quality of river , as proven by the increasing trend of COD and BOD from 2012 to 2016. The COD from 2012 to 2016 was 14, 16.6, 18.7, 22.5, and 22.8 ppm, respectively. Meanwhile, the BOD from the same observation years was 6, 7.2, 8.9, 9.3, and 10.3 ppm, respectively.

1. Introduction

Regional development in an area has a potential adverse impact on land use, vegetation, or green space. Land use change is the conversion of one land use into another, which suits the objectives of human activities in fulfilling their needs. Therefore, the conversion includes a shift of the land's function. Along with the expansion of housing construction, the function of green space is increasingly turned into a built-up area or, in other words, the extent of green space constantly diminishes. The availability of green space plays a major role in the arrangement of the integrated area along Code River, particularly to compensate buildings density. Land use conversion also disturbs the hydrological cycle. In the middle of this cycle, water is retained and stored (temporarily) in rivers, lakes or reservoirs and soil, providing a chance for humans and other creatures to utilize it [1].

Open green space plays a role in provide shade from the warming environment and as a medium for the people along the upper Code River to socialize and interact with nature. The actualization of green space is an initial effort to raise awareness of media that promote environmental harmonization especially regarding the decreasing availability of clean and healthy air supply due to environmental degradation along the river.

Content from this work may be used under the terms of the Creative Commons Attribution 3.0 licence. Any further distribution of this work must maintain attribution to the author(s) and the title of the work, journal citation and DOI.
Published under licence by IOP Publishing Ltd

According to the Intergovernmental Panel on Climate Change [4], global warming has become a serious and significant phenomenon in human life. It affects not only ecological environment but also social and cultural environment. Global warming is a rise in global annual temperature due to, one of which, greenhouse effect. In general, the presence of green space along Code River is expected to reduce global warming and air pollution level in the Special Region of Yogyakarta. Therefore, an awareness of green space preservation and function restoration, i.e., as an absorber of global warming, becomes necessary. Such awareness can at least start from the process of 'thinking together' with the local people in the study area as the smallest scope. This process is expected to contribute a greater thought to the other regions around Code River to preserve and organize the open green space as part of urban open space, which plays a major role in the sustainability of the living beings around it.

Relying on the potential of green space, the concept of the spatial planning in the study area is expected to harmonize the existence of vegetation, as well as the ecological environment, with the development of regional infrastructure. The harmonization helps achieve the objectives of sustainable development and overcome the current global warming problems. The purpose of this research is to determine the effects of land use change on water pollution and climate parameters at Code river

2. Methods

This research took place at Code Watershed in the Special Region of Yogyakarta. The secondary data was acquired from related institutions, while the primary data was obtained from interviews with the major figures in the community. The interviews aimed to dig information of the geographic condition and topography along Code River and to collect data on settlement areas. The research data includes land use map, climatological data (i.e., rainfall and air temperature) in 1976-2005 [2], and the water quality of Code River in 2012-2016 [3]. These data were analyzed using a descriptive method. The determinants of land use conversion consist of physical and biological factors, economic considerations, and institutional factors. The physical factor includes suitability to the physical properties of an area, such as geology, soil, water, climate, and population, while the biological factor consists of flora and fauna. As for the economic consideration, it represents profitability, market condition, and transportation. The institutional factor is the role of government in regulating and organizing an area.

The land use in the study area is divided into nine classes, as presented in Figure 1. Figure 1 shows that the dominant land use in 2012 and 2017 is settlements. Referring to the administrative boundaries, almost all of the land in Yogyakarta City (at the center of the Special Region of Yogyakarta) is used for settlements. The study area is located in the tropics, specifically in the south of the equator and in the western part of Indonesia. Consequently, it is strongly influenced by the east-southeast and west-northwest monsoonal climate. The rainfall in the Special Region of Yogyakarta is thereby categorized as having a tropical rain climate.

Rainfall events are recorded by rain gauges, which are installed at rain stations in predefined locations. Therefore, to identify the rainfall events in the study area, the research used a set of rainfall data obtained from various rain stations. In addition to rainfall data, this research used temperature data from a weather station owned by the Indonesian Agency for Meteorology, Climatology, and Geophysics. This station is located in Adi Sucipto International Airport at an elevation of 350 m above sea level. It records both daily rainfall and temperature. The methods for identifying land use change by analyzing land use maps. The water sampling used water sampler, by taking river water at the middle and left part of the river then mixed as a composite. Sampling is done every year for 5 years at the same point. Laboratory analysis of using spectrophotometric method, BOD analysis using Winkler method (volumetric titration), and COD analysis with potassium permanganate, with spectrophotometric method.

3. Results and discussion

3.1 Land use change in Code Watershed

The analysis of land use change not only concerns on the conversion of forest to non-forest (settlement) land use, but it also includes the analysis of each land use change. From the perspective of urban development, the rapid growth of elite settlements (public housing) and various trading sites is the main factor of land use change.

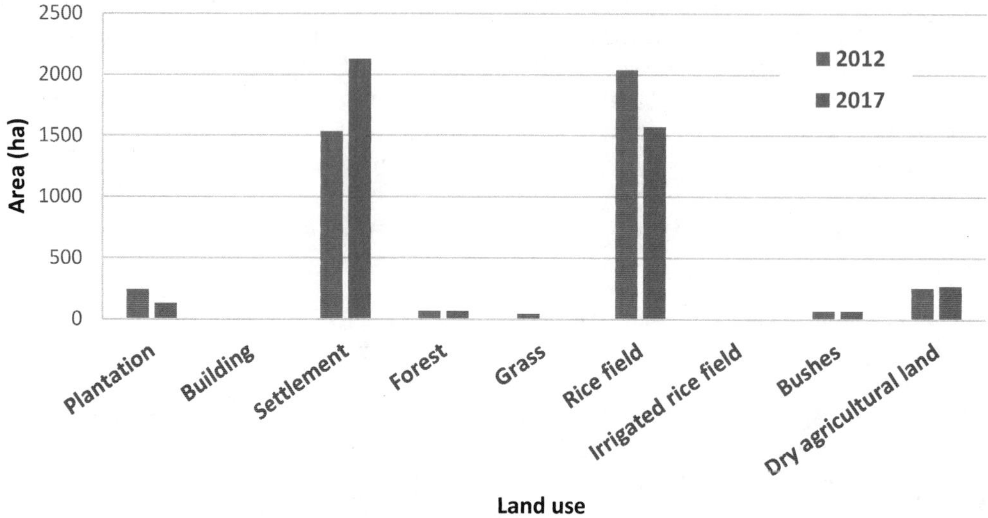

Figure 1. The Land Utilization in Code Watershed in 2012 and 2017

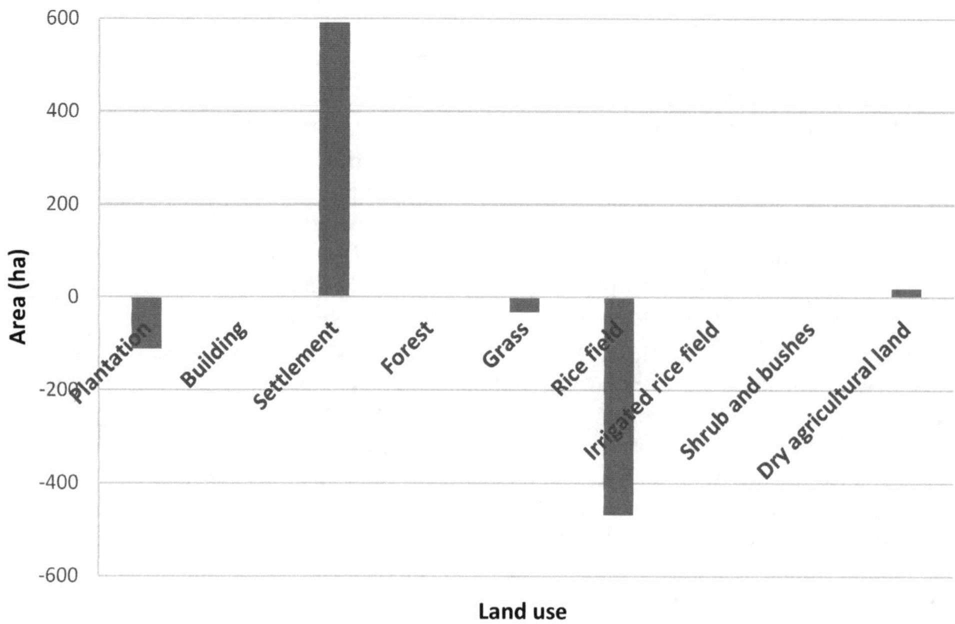

Figure2. Land Use Change in the Study Area in the Last Five Years (2012-2017)

The extent of land use change, as presented in Figure 2, was calculated from the analysis of land use maps in 2012 and 2017. The negative (-) sign presents diminished extent for each type of land use. Settlements are the land use increased extensively in the study area, i.e., from 1,535.505 ha in 2012 to 2,127.088 ha in 2017. In these years, the settlement area has increased by 591.583 ha or 118.316 ha per year. The extensive land use conversion to settlement in 2012-2017 was caused by population growth. On the contrary, rice field continuously decreased, from 2012 to 2017 shrank by 467.496 ha or 93.4992 ha per year. Furthermore, within five years, the areas of some other land uses also became smaller, namely plantation area (decreased by 11.475 ha or 22.295 ha per year) and grass (31.218 ha or 6.2436 ha per year), forest (1.307 ha or 0.2614 ha per year). As for shrubs and bushes experienced the smallest reduction, i.e., 0.403 ha or 0.10075 ha per year. The constantly diminishing open green space leads to climate change.

3.2 Rainfall and air temperature changes

3.2.1 Rainfall. The largest source of groundwater in the world is rainwater, which makes regional precipitation the determinant of groundwater potential in an area. When a high rainfall is complemented with highly permeable aquifer and suitable land use, the groundwater resource becomes highly reliable. The data consists of monthly rainfall, i.e., from January to December, in 30 years, as presented in Figure 3.

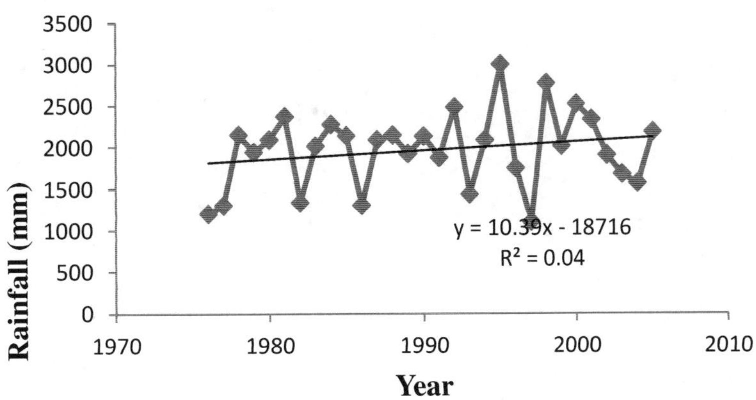

Figure 3. The fluctuation of rainfall in 1976-2005

3.2.2 Air Temperature. Such conversion decreases the open green space and hence, raises the air temperature. The air temperature records in 30 years (1976-2005) are depicted in Figure 4.

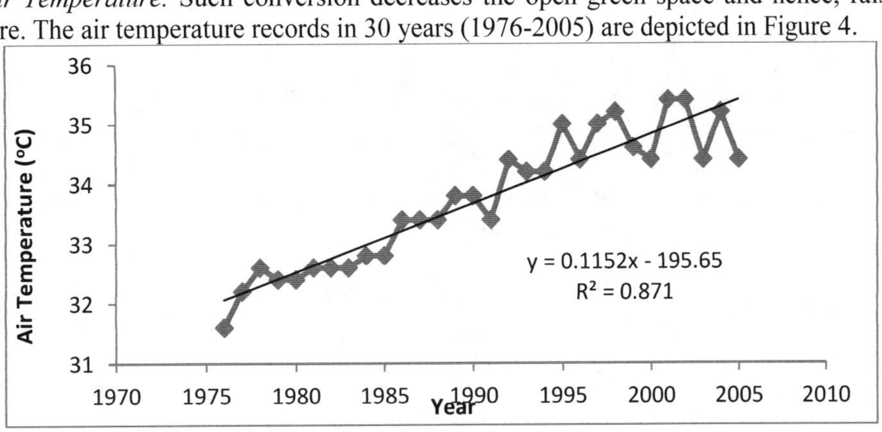

Figure 4. Air temperature rise in 1976-2005

The annual air temperature rise in the study area is caused by the reduction of open green space. A temperature rise is observable from 30-year time series data. Figure 4 shows that the annual temperature increase in the study area is 0.115^{0}C.

3.3 Water quality of Code river

The land use change from agricultural areas to settlements indicates an expansion of residential areas, or in other words the population density also increases. Population growth increases the number of household wastes disposed into the river. Consequently, the concentration of domestic waste in the river also increases, which deteriorates the water quality of Code River. The analysis results of the water quality data in the study area (Sayidan Bridge, Gondokusuman) in 2012-2016 are presented in Figure 5.

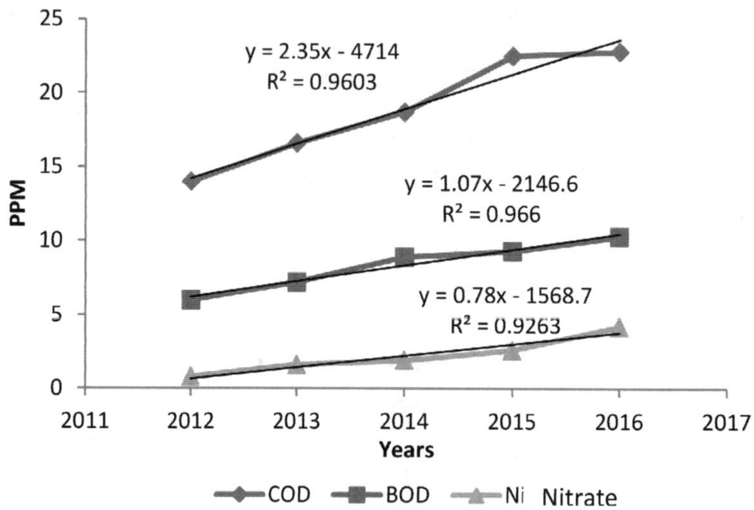

Figure 5. Increasing pollution in Code river from 2012 to 2016

Figure 5 shows an increase in the concentrations of COD, BOD, and Nitrate at Code River. The deterioration of river water quality in the study area is indicated by the annual increase of COD (i.e., 2.35 ppm), BOD (i.e., 1.07 ppm), and Nitrate (i.e., 0.78 ppm), all of which are caused by the conversion of many types of land use to settlements. The expansion of residential area is likely followed by population growth as well as the constantly increasing volume of waste disposal to the river. In this case, a larger settlement leads to the deterioration of river water quality.

Considering those changes, an environmental management dealing with global warming and climate change become necessary. The preservation of open green space and the cultural movement of tree planting significantly increase the vegetation cover in an area. Areas with critical land degradation can adopt the following management strategies: planting perennials, using natural energy source, and forbidding logging to maintain water storage and filtration (especially during continuous rain events). Bare land is also susceptible to landslides because the absence of tree roots makes the soil easily eroded and transported by water.

4. Conclusion

The land use conversion along Code River is quite extensive. In five years (2012-2017), many types of land use were converted to settlements(591.583 ha or 118.3166 ha per year). This extensive conversion is caused by the population growth. The mostly decreased land use was rice fields (467.596 ha), followed by plantation (111.475 ha), grass (31.218 ha), forest (1.307 ha), and shrubs and bushes (0.403 ha) as the least affected land use. The land use conversion in the study area reduced the extent of open green space but widened the residential area. Consequently, it causes a greater volume of domestic wastes, air temperature rise, and the deterioration of river water quality, as indicated by an increase in the concentrations of COD (2.35 ppm per year), BOD (1.07 ppm per year), and Nitrate (0.78 ppm per year). An environmental management dealing with global warming and climate change become necessary.

References

[1] Asdak C 1995 *Hidrology and Management Watershed* (Yogyakarta: Gadjah Mada University Press)

[2] Meteorology Climatology and Geophysics Council 2005 *Adisucipto Meteorological Station Monitoring Data* (Yogyakarta: Meteorology Climatology and Geophysics Council)

[3] Environmental agency. 2016 *Monitoring Data of the Environment Agency of the Special Province of Yogyakarta* (Yogyakarta: Environmental Agency)

[4] Bogner J, Pipatti R, Hashimoto S, Diaz C, Mareckova K, Diaz L, Kjeldsen P, Monni S, Faaij A, Gao Q, Zhang T, Ahmed M A, Sutamihardja R T M and Gregory R. 2008 *Mitigation of global greenhouse gas emissions from waste: conclusions and strategies from the Intergovernmental Panel on Climate Change (IPCC) Fourth Assessment Report. Working Group III (Mitigation),* Los Angeles

The effect of nano-silica fertilizer concentration and rice hull ash doses on soybean (*Glycine max* (L.) Merrill) growth and yield

T Suciaty[1,2], D Purnomo[3], A T Sakya[3] and Supriyadi[4]

[1]Doctoral Program of Agricultural Science Graduate School, Universitas Sebelas Maret, Jl. Ir. Sutami 36A, Kentingan, Surakarta, Central Java 57126 Indonesia
[2]Departemen Agrotechnology, Faculty of Agriculture, Universitas Swadaya Gunung, Jl. Pemuda No.32, Sunyaragi, Kesambi, Cirebon, West Java 45132 Indonesia
[2]Department of Agrotechnology, Universitas Sebelas Maret, Jl. Ir. Sutami 36A, Kentingan, Surakarta, Central Java 57126 Indonesia
[3]Department of Soil Science, Universitas Sebelas Maret, Jl. Ir. Sutami 36A, Kentingan, Surakarta, Central Java 57126 Indonesia

Abstract. Agriculture is facing a number of challenges included limited water supply, low nutrient use efficiency, etc affected by climate change. Nano-silica is a product of nanotechnology, the frontier technologies to enhance crop productivity under climate change threats. The purpose of the research was to investigate the effects of nano silica concentration and rice hull ash on growth and yield of soybean. The experiment was conducted at Gagasari village, Cirebon, West Java from March until June 2017. The treatments were arranged by using factorial completely randomized block design with two factors. The first factor was a concentration of nano silica fertilizer consisted of four levels i.e., 0, 1.75, 2.5, and 3.75 $ml.l^{-1}$. The second factor was doses of rice hull ash consisted of four levels i.e., 0, 1, 2, and 3 $ton.ha^{-1}$. Each treatment combinations was repeated three times. The result showed that concentration of nano silica individually affected the number of leaves and number of branches, NAR and RGR, productive branches at 21, 30-45, and 35 daps, respectively. It also affected the seed dry weight $plant^{-1}$ and $plot^{-1}$. Meanwhile, doses of rice hull ash affected LAI, NAR, and RGR, 15-30, and 30-45 dap, respectively. Dry seed weight $plot^{-1}$ was also affected by doses of rice hull ash. There was an interaction effect between nano-silica concentration and doses of rice hull ash on number $pods.plant^{-1}$. Combinations of 2.5 $ml.l^{-1}$ nano-silica and 3 $ton.ha^{-1}$ of rice hull ash gave the highest number $pods.plant^{-1}$.

1. Introduction

Climate change is a serious threat to the agricultural sector and potentially make a new problem for food production sustainability and agricultural production systems in general. Climate change affects agricultural sectors such as natural resource, agricultural infrastructure, agricultural production systems, food security, and farmers welfare, and society [1].

Climate change scales down agricultural production especially food crops (rice, maize, and soybean). Soybean (*Glycine max* (L.) Merrill) is an essential food crop after rice and maize in Indonesia, due to it contains healthy substances. Soybean acts as a source of vegetable protein which is very important, in order to build up the society nutrition, it is also relatively cheap compared to the animal protein source [2].

Content from this work may be used under the terms of the Creative Commons Attribution 3.0 licence. Any further distribution of this work must maintain attribution to the author(s) and the title of the work, journal citation and DOI.
Published under licence by IOP Publishing Ltd

According to Kastono [3], new domestic soybean production is able to meet the needs of about 30% and at least 70% must be imported. Numerous efforts to strengthen the soybean production should be noted that the plant is able to produce high potency. Additional elements such as Si can be applied as a support plant growth since several studies explain that Si in the soil can boost the phosphorus availability. Si generates plant stronger and not easy to fall. The collapse causes a plant production declining, thus fertilizing Si is considered to multiply crop production [4].

The Si application greatly affects the rice plant growth, impacts on the plants stems strengthening, plant protection from pests, root reinforcement, and others [5]. The super-small sized fertilizer employment has several advantages, such as, more reactive, directly reach the target, and only needed in small amounts [6]. With these advantages, the nano fertilizer is expected to be a breakthrough technology to intensify agricultural production that is continuous and environmentally friendly [7].

The potassium responsibility in the body hull ash is to strengthen the plant leaves and flowers do not fall and stimulate the root hairs formation [8]. It purposes that can enhance soil physical and chemical properties. In addition, it serves to loosen the soil, thus can simplify roots nutrients absorption [9].

According to Fitiyani dan Haryanti [10] research stated that different rates of concentration nano silica fertilization treatment gave an effect on plant height, leaf number, and root length. Nano silica fertilizer 75% of concentration or by dissolving 3.75 ml nano-silica in 1.5 liters of water gave the highest growth of tomato (*Solanum lycopersicum*) var. Round. According to Melati et al [11], a combination of 5 ton of manure with 2 $ton.ha^{-1}$ of rice hull ash can provide the soybeans highest yield. It also donates the same effect with 150 $kg.ha^{-1}$ of KCl. The purpose of this study was to analyze the effect of nano silica fertilizer concentration and dosage of rice hull ash on soybean (*Glycine max* L. Merrill) growth and yield.

2. Methodology

The experiment was conducted in the Gagasari village, Gebang, Cirebon. with ± 2 meters of altitude above sea level on March until June 2017. Soybean varieties which exploited in this experiment were Grobogan. The experiment was arranged by utilizing two factors RCBD. The first factor was nano silica concentration that consisted of four levels i.e., 0, 1.25, 2.50, and 3.75 $ml.l^{-1}$. The second factor was rice hull ash dose that existed of four levels i.e., 0, 1, 2, and 3 $ton.ha^{-1}$. Each treatment combination repeated three times so there were 48 plots. The variable observation includes plant growth and yield.

3. Results and discussion

3.1. Growth components.

3.1.1 Plant height, number of leaves, and stem diameter. Based on the analysis, it showed that the effect of nano-silica fertilize concentration employment and rice hull ash on soybean crop did not affect plant height and stem diameter at 14, 21, and 28 daps, however, it significantly affected the trifoliate leaf number at age 14 and 21 daps.

According to Indradewa et al [12], the plant metabolism ability was associated with the plant age. At the age of 14-21 dap was still in the plant growth phase was slow, whereas at 28 dap due to nutrients absorption from fertilizers nano-silica and rice hull ash content provided together, thus providing the same effect on plant height and stem diameter.

There was no interaction between of nano-silica fertilizer concentration and doses of rice hull ash on the number of leaves trifoliate, but at age 21 dap its treatment gave individually affected on the leaves trifoliate number at 21 and 28 dap, it is in line with the opinion Yukamgo and Yuwono [13], that Si is quite profitable biochemical and chemical-poss process for plant growth and Makarim et al [14] assert that Si in the plant causes the plant roots stronger so the nutrient absorption was more intensive.

Table 1. Effect of nano-silica fertilizer concentration and doses of rice hull ash on plant height, number of trifoliate leaves and stem diameter at 14, 21, 28 daps on soybean plants.

Treatment	Plant height (cm)			Number of trifoliate leaves			Stem diameter		
	14 dap	21 dap	28 dap	14 dap	21 dap	28 dap	14 dap	21 dap	28 dap
	concentration nano-silica Fertilizer (mll^{-1})								
0	10.72a	15.86a	26.46a	1.77a	4.46a	12.84a	2.43a	3.33a	5.16a
1.25	10.31a	15.75a	23.58a	1.77a	5.05bc	13.68a	2.40a	3.40a	5.10a
2.50	10.68a	15.83a	25.43a	1.73a	4.89b	13.62a	2.32a	3.36a	4.89a
3.75	10.38a	15.49a	24.87a	1.75a	5.40c	13.77a	2.33a	3.29a	4.99a
	doses of rice hull ask ($ton.ha^{-1}$)								
0	10.44a	14.45a	24.78a	1.71a	4.87ab	12.73ab	2.39a	3.35a	4.90a
1	10.30a	15.94a	25.37a	1.80a	4.96ab	13.47ab	2.32a	3.30a	4.98a
2	10.60a	15.79a	25.19a	1.69a	5.20b	14.82b	2.35a	3.37a	5.12a
3	10.75a	15.7a	24.99a	1.82a	4.78a	12.89a	2.41a	3.35a	5.13a

Description: The average number followed by the same letter in the same column is not significantly different shows according to Duncan's multiple range test at 5% level.

3.1.2 Number of branches and LAI. The analysis result variations portrayed that there was no interaction between nano-silica fertilizer concentration and rice hull ash dose on the number of branches and LAI. It presented an independent effect on the number of branches by 21 and 28 dap and LAI at 28 daps.

Table 2. Effect of nano-silica fertilizer concentration and doses of rice hull ash to total branch at 21 and 28 dap and LAI at 14, 21 and 28 daps on soybean plants.

Treatment	Number of branches		LAI		
	21 dap	28 dap	15 dap	30 dap	45 dap
	concentration nano-silica fertilizer (mll^{-1})				
0	1.49a	3.91ab	0.0839a	0.3826a	1.5099a
1.25	2.28b	4.29b	0.0888a	0.4164a	1.6389ab
2.50	1.93b	3.66a	0.0866a	0.4188a	1.7687ab
3.75	1.95b	4.07ab	0.0652a	0.3404a	2.1019b
	doses of rice hull ash ($ton.ha^{-1}$)				
0	2.03a	3.93a	0.0701a	0.3826a	1.5632a
1	1.98a	4.04a	0.1000b	0.4248a	1.6288a
2	1.88a	3.98a	0.0770ab	0.4068a	1.4839a
3	1.76a	3.97a	0.0775ab	0.3399a	2.3436b

Description: The average number followed by the same letter in the same column is not significantly different shows According to Duncan's Multiple Range Test at 5% level.

Table 2 shows that the treatment of nano-silica fertilizer concentration demonstrated a significant effect on the of branches number 21 dap compared with controls, whereas at 28 dap treatment of nano-silica fertilizer concentration 0 $ml.l^{-1}$, 1.25 $ml.l^{-1}$, and 3.75 $ml.l^{-1}$ were not significantly different yet it opposite performed with the treatment of 2.50 $ml.l^{-1}$. This was due to the treatment of S_1 has the most reasonably optimal dose, so it can be used and the maximum available as a nutrients stimulator by plants, beneficial Si nutrients to support plant growth [14].

A large number of soybean crop branches were affected by the leaves number because the soybeans branch is grown over the books petiole, but the branches number will be less than the leaves number. This was due to the growth of the books leaves will be followed by the branches and flowers development, or one of its branches or flowers [15].

Table 2 displays that there was no interaction between the treatment of nano-silica fertilizer concentration and doses rice hull ash to the LAI. It had a significant effect on LAI at 45 daps, the higher of its value was followed by the LAI's. This is in line with the results Sumida [16], which states that the supply of Si is quite capable for providing a good crop, because it helps more upright stems and leaves, thus increasing the photosynthesis rate. Similarly, the doses of rice hull ash 3 ton.ha^{-1} were independently significantly different from the other dosage of rice hull ash. Rice hull ash 3 ton.ha^{-1} provided sufficient Si into the soil, so it can be used for plant growth.

LAI was illustrative of the leaf surface ratio of the land area which was occupied by the plant. Leaf area escalated with increasing age of the plant, therefore, the leaves number into lots [6], the leaf area increment was expected of the nutrients absorption effects larger than mutual shade effect inflation between leaf.

3.1.3 NAR, RGR, and total productive branch. Results analysis variation showed that the treatment of nano-silica fertilizer concentration and doses of rice hull ash did not provide interaction against the NAR, RGR at age 15-30 dap and 30-45 dap and the productive branches number at 35 daps. It had an independently significant effect.

Table 3. Effect of concentration nano-silica fertilizer and doses of rice hull ash to NAR, RGR of ages 15-30 dap and 30-45 dap and total productive branch on soybean plants.

Treatment	NAR (x10^{-6}) gcm^{-2}day^{-1}		RGR (x10^{-4}) gday^{-1}		Number of productive branches
	15-30 dap	30-45 dap	15-30 dap	30-45 dap	35 dap
concentration nano-silica fertilizer (mll^{-1})					
0	1,152.58a	1,664.77ab	1,201.23a	1,228.73a	4.02b
1.25	1,139.50a	1,447.80a	1,243.92a	1.185.94a	4.33c
2.50	1,089.33a	1,5884.45a	1,329.45a	1,230.47a	3.68a
3.75	1,256.50a	2,380.53b	1,385.43a	1,518.29b	3.88ab
doses of rice hull ash (ton.ha^{-1})					
0	1.591.41b	1,664.77a	1,431.16a	1,096.29a	3.80a
1	869.41a	1,725.51a	1,174.22a	1,295.57ab	3.94a
2	1,072.03a	1,884.55a	1,296 .20a	1,327.56ab	4.08a
3	1,105.08ab	1,947.45a	1,258.44a	1,444.02b	4.09a

Description: The average number followed by the same letter in the same column showed no significant according to Duncan's Multiple Range Test at 5% level.

From Table 3 above portrays that treatment of nano-silica fertilizer concentration 3.75 ml.l^{-1} were significantly different from other treatments both to the net assimilation rate (NAR) is 2,380.53x10^{-6} g.cm^{-}2.day^{-1} and relative growth rate (RGR) of 1,518.29x10^{-4} g.day^{-1} at the age of 30-45 dap, the same as is shown by its treatment of 3.75 ml.l^{-1} of the LAI with the highest LAI values 2.1019 (Table 2). LAI is the net assimilation unity leaf area and time result. It can be concluded that it can enhance the LAI, NAR, and RGR. This is in accordance with the opinion of Gardner et al [17], that the plant growth rate is affected by the NAR and LAI. High NAR and optimum LAI will boost its rate. It is the independently significant effect on the productive branches number, nano-silica fertilizer treatments concentration 1.25 ml.l^{-1} to provide the best effect on the productive branches number at age 35 dap is 4.33. It is suspected that Si accumulated in the leaves which make it moves upright and stretching well into the leaf surface consequently get more sunlight, so a more optimal photosynthesis. More photosynthate result, it will be distributed and stored in the plant's vegetative organ of such as roots, stems, and leaves as a food reserve for plant growth [18].

The dose of rice hull ash independently significant effect on the NAR aged 13-30 dap, while at the age of 30-45 dap were not significant. Its 3 ton.ha^{-1} provides the best effect against the RGR to amounted RGR to 1,444.02x10^{-4} g.day^{-1}, but the dose of rice hull ash did not significantly affect the number of productive branches.

3.2 Yield component

3.2.1 The number of pods per plant. Results of analysis of variance showed interaction effect between concentration nano-silica fertilizer and dose rice hull ash to the number of pods per plant.

Table 4. Effect of nano-silica fertilizer concentration and doses rice hull ash on contents number of pods per plant.

Dose of rice hull ash (ton.ha^{-1})	Concentrations of nano-silica fertilizer (ml.l^{-1})			
	(0)	(1.25)	(2.50)	(3.75)
0	59.23a	63.07a	65.17a	75.50a
	A	A	A	B
1	55.20a	60.60a	61.17a	77.40a
	A	A	A	B
2	53.20a	70.57b	67.07a	77.60a
	A	B	B	C
3	74.17b	72.63c	**79.37b**	69.70a
	A	A	**B**	A

Description: Lowercase for columns, capitals of the line. The average number followed by the same letter in the same row and column shows no significant according to Duncan's Multiple Range Test at 5% level.

At 0 mll^{-1} and 3.75 mll^{-1} of nano-silica concentration level combined with a level of dose husk ash began to 0-3 ton ha^{-1} did not significantly affect the pods per plant number, while at 1.25 mll^{-1} of nano-silica and 2.50 mll^{-1} combined with rice hull ash dose level of 0-3 ton ha^{-1} significantly influence the pods per plant number, its concentration extent 1.25 ml.l^{-1} in combination with rice hull ash dose of 2 ton ha^{-1} and 3 ton ha^{-1} served the best effect on the pods number i.e., 70.57 and 72.63, respectively.

Combination treatment of nano-silica fertilizer concentration 2.50 ml.l^{-1} and rice hull ash with the dose of 3 ton.ha^{-1} indicated the highest (79.37) of pods number. According to Kusuma et al [9] rice hull ash presented the plant growth significant effect, especially the growth in the leaves number and pods number escalation, and described by several studies that Si can multiply the availability of P and reduce the toxic metals activity, one of the functions P is for helps the flowers and fruit composition.

3.2.2 Weight dry seeds per plant, the weight of 100 grains dry beans and dry seed weight per plot. The Results variation analysis, there was no interaction between the treatment of nano-silica fertilizer concentration and doses and rice hull ash to the dry seed weight per plant, the weight of 100 seed, dry weight of seeds and dry weight seeds per plot, but there was an independent effect of each variable observation.

From Table 5 it can be seen that the treatment of nano-silica fertilizer concentration did not significantly affect the weight of dry seeds per plant and weight of 100 of dry beans grains, while it was independently instant effect on the dry seeds weight per plot, treatment 2.50 ml.l^{-1} provided the best effect with dry grain weight per plot was 950.17 g.plot^{-1}. It is assumed that Si fertilizing can boost the uptake of P and phosphorus needed by plants in cell division and as energy in every plant metabolic process [19].

Table 5. Effect of nano-silica fertilizer concentration and doses rice hull ash on weight dry seeds per plant, weight of 100 grains and weight dry seeds per plot

Treatment	Dry seeds per plant (g)	Weight of 100 seeds (g)	Dry seed per plot (g)
Concentration nano-silica fertilizer ($ml.l^{-1}$)			
0	26.93a	21.41a	870.83a
1.25	24.99a	21.24a	867.83a
2.50	30.89a	21.75a	950.17b
3.75	27.03a	21.82a	857.08a
Doses of rice hull ask ($ton.ha^{-1}$)			
0	26.02a	21.57a	878.67a
1	25.77a	21.36a	838.92a
2	28.07ab	21.43a	880.50a
3	29.98b	21.84a	947.83b

Description: The average number followed by the same letter in the same column showed no significant according to Duncan's Multiple Range Test at 5% level.

The treatment dose rice hull ash was significantly affected the dry seed weight per plant and per plot independently, while against the weight of 100 grains of dry beans were not significantly different. Rice hull ash dose of 3 $ton.ha^{-1}$ served the best effect on the dry seed weight (29.98 $g.plant^{-1}$) and seed weight per plot (947.83 g). It is suspected that the rice hull ash can replace or the potassium substitution content in soil under soybean cultivation. The rice hull ash provision can increase the content of K highly supportive crop production. The more the number or the dose is given to the soil, the better [20].

4. Conclusion

In the treatment of nano-silica fertilizer concentration of 2.5 $ml.l^{-1}$ portrayed the best result in the dry seed weight per plot of 950.17 g, equivalent to 1.58 $ton.ha^{-1}$ and doses of rice hull ash 3 $ton.ha^{-1}$ provided the best results on a dry seed weight per plot of 947.83 g, equivalent to 1.57 $ton.ha^{-1}$.

References

[1] Wiyono S "Climate change and pest and plant disease" *http://www.repository.ipb.ac.id* 2007 [Online] Available: http://repository.ipb.ac.id/handle/123456789/30756 [Accessed:28-Sept-2017]
[2] Sudaryanto T and Swastika D K S "Soybean economy in Indonesia" *http://balitkabi.litbangpertanian.go.id* 2016 [Online] Available: http://balitkabi.litbang.pertanian.go.id/wp-content/uploads/2016/03/dele_1.tahlim-1.pdf [Accessed: 5-Agt-2017]
[3] Kastono D 2005 Respons on growth and yield of black soybean in the usage of organic fertilizer and biopesticide of slam weed (*Chromolaena odorata*) *J. Agricultural Science* **12(2)**103-116.
[4] Rosmarkam A and Yuwono N W 2002 *Soil Fertility Science* (Yogyakarta: Kanisius) pp 86-8
[5] Ma J F and Takahashi E 2002 *Soil, Fertilizer, and Plant Silicon Research in Japan* (Amsterdam: Elsevier B.V.) pp 49-59
[6] Rochman N and Jumarman 2008 Opportunity and strategy for developing nanotechnology in Indonesia *J. Industrial Research* **2(1)** 56-63.
[7] Syahri R, Djayadi, Sumarni T and Nugroho A 2016 Effect green manure (*Crotalaria juncea* L.) and silica nano fertilizer concentration on growth and yield sugarcane after 9 months *J. Produksi Tanaman* **4(1)** 73-81
[8] Kiswondo S 2011 Use of husk ash and ZA fertilizer on the growth and yield of tomato plants

(*Lycopersicum esculentum* Mill.) *Embryo* **8(1)** 9-17
[9] Kusuma A H, Izzati M and Saptaningsih E 2008 The effect of adding charcoal and hull ash with different proportions to the permeability and porosity of clay as well as growth green bean (*Vigna radiata* L). *Bulletin Agronomy and Physiology* **21(1)** 1-9
[10] Fitiyani H P dan Haryanti S 2016 The effect of using nano-silica fertilizer on growth tomato plant (*Solanum lycopersicum*) var. Round *Bulletin Anatomi dan Fisiologi* **24(1)**34-41
[11] Melati A, Aisah A and Rianawati D 2008 The application of organic manure and its residue for vegetable soybean production *Bul. Agron.* **36(3)** 204-13
[12] Indradewa D, Sastrowidodo S, Notohadisuwarno S and Prabowo H 2004 Nitrogen metabolism of soybean under saturated soil culture *J. Agr. Sci.* **4(2)** 68-75
[13] Yukamgo E and Yuwono N W 2007 The role of silicon as a useful element in sugarcane crops *J. Soil Science and the Environment* **7(2)** 103-16
[14] Makarim A K, Suhartatik E and Kartohardjono A 2007 Silicon: Important nutrition on rice production systems *J. Food Crop Scince* **2(2)** 195-204
[15] Bakhtiar, Hidayat T, Jufri Y and Safriati S 2014 The performance of growth and yield of soybean varieties in Aceh Besar *J. Floratek* **9** 46-52
[16] Sumida H. 1992. Silicon supplying capacity of paddy soils and characteristics of silicon uptake by rice plants in cool regions in Japan. *Bulletin Tohoku Agric.* **85** 1-46
[17] Gardner F P, Pearce R B and Mitchell R B 1991 *Physiology of Crop Plants* (Jakarta: Universitas Indonesia Press) p 428
[18] Putri F M, Suedy S W A and Darmanti S "The effects of nano-silica fertilizer on the number of stomata, chlorophyll content and growth of black rice (*Oryza sativa* L. Cv. *Japonica*)" *www.ejournal.undip.ac.id* 2017 [Online] Available: http://www.ejournal.undip.ac.id/indek.php/baf/index (Accessed:1-Feb-2017)
[19] Zulputra, Wawan dan Nelvia 2014 Response of upland rice (*Oryza sativa* L.) on the application of silicate and phosphate fertilizer on ultisol *J. Agroteknologi* **4(2)** 1-10.
[20] Baon J B and Sugiyanto 2011 Soil chemical properties as affected by plant-derived ash to replace potassium fertilizer and its conversion value *J. Coffee and Cacao* **27(2)** 98-108.

Soil degradation level under particular annual rainfall at Jenawi District– Karanganyar, Indonesia

A Herawati[1], Suntoro[1], H Widijanto[1], I Pusponegoro[2], N R Sutopo.[1], and Mujiyo[1,3]

[1]Department of Soil Science, Faculty of Agriculture, Universitas Sebelas Maret, Jl. Ir. Sutami No.36A, Jebres, Surakarta, Central Java 57126
[2]Department of Environmental Science, Graduate School, Universitas Sebelas Maret, Jl. Ir. Sutami No.36A, Jebres, Surakarta, Central Java 57126
[3]Corresponding author: mujiyo@staff.uns.ac.id

Abstract. The study of the climatic elements such as rainfall is vital for the sustainable development of agriculture at a region. The aims of the study were to evaluate the soil degradation based on the annual rainfall and to determine the key factors which responsible for the soil degradation at in Jenawi Sub-District. The mapping of soil degradation potency is an identification of initial soil condition to discover the potential of the land degradation. The mapping was done by overlaying the map of soil, slope, rainfall and land use with the standard procedures to obtain the value and status of Soil Degradation Potency (SDP). The result showed that SDP in Jenawi District categorized in very low (SDP I) 0.00 ha (0.00%); low (SDP II) 109.01 ha (2.57%); moderate (SDP III) 1,935.92 ha (45.63%); high (SDP IV) 1,959.54 ha (46.19%) and very high (SDP V) 238.08 ha (5.61%). The rainfall is the factor which has the strong correlation with the SDP ($r = 0.65$, $P < 0.01$, $n = 306$). The changes in the rainfall as the impact of climate change need to be anticipated to minimize soil degradation. The result can be adapted to the rainfall changes in various ways based on local soil-land characteristics.

1. Introduction

Global warming is currently one of the world's most challenging issues. The effect of anthropogenic changes to the land and atmosphere on climate change is being pursued as a dynamic multi-disciplinary problem. Increases in atmospheric greenhouse gases, *i.e.* water vapor, carbon dioxide, methane, nitrous oxide and ozone are considered the major reason for global warming. The land is a part of the resources that important on the biomass production, so its sustainability must be maintained to keep the function, and stay sustain for life that goes both above and below the earth surface. The importance of land is not only for its function as an anchorage for plants and the supplier of important resources and footholds, but also as a part of the ecosystem [1]. The process of land to produce biomass cannot be separated from the existences of the soil damage, which caused by several factors both natural and human activities. The soil damage may result in the decrease of land productivity. Rain is one of the main factors in triggering soil erosion and thus leads to the soil damage. The soil eroded by the effects of rainwater, especially on cropland will cause the decreasing on the quality of physical, chemical, and biological properties of the soil; also reduce the nutrients and organic matter, as well as decline the productivity.

In recent decades the impact of global climate change has also affected the increase of annual rainfall as the result from rising air temperature due to global warming. That will lead to more water

Content from this work may be used under the terms of the Creative Commons Attribution 3.0 licence. Any further distribution of this work must maintain attribution to the author(s) and the title of the work, journal citation and DOI.
Published under licence by IOP Publishing Ltd

evaporating into the air, so that the potential for heavy rainfall will also highly increase [2]. According to the Regulation of the Indonesian Minister of Environment No. 7 of 2006 who stated that the higher the rainfall the higher potential of soil damage. The aims of the study were to evaluate the soil degradation based on the annual rainfall and to determine the key factors which responsible for the soil degradation at in Jenawi Sub-District. This result will be an information to identify areas with high rainfall and need more attention and anticipation to prevent land damage.

2. Materials and Methods

The research was conducted in Jenawi district, Karanganyar regency which consists of 9 villages namely Anggrasmanis, Balong, Gumeng, Jenawi, Lempong, Menjing, Seloromo, Sidomukti, and Trengguli. The materials used were thematic maps, i.e.: land map, rainfall 1: 50,000 [3], slope [4], and land use 1: 25,000 [4]. The tools used included a set of survey tools and ground investigations, computers and software ArcView GIS 3.3 [5]. The land maps were made with transect method survey followed by a soil profile investigation and then completed with the analysis of soil samples in the laboratory [6][7][8].

The mapping of land damage potential based on the land units from the uniformity resulted from overlaying rainfall, slope, and land use map using ArcView GIS 3.3. The overlay land units were scored based on the multiplication of the soil weight and the land rating (See Table 1), the slope (See Table 2), the rainfall (See Table 3), and the land use (See Table 4). The scoring results were then classified to obtain several classes of the land damage potential (See Table 5). The type of the land use under study is an effective area for biomass production in the form of forests, rice fields, gardens, shrubs, grasslands, and moorlands. The mapping of potential land damage following standard procedures in [9][10].

The correlations between factors of the potential soil degradation were determined using the correlation analysis and the Pearson's correlation coefficient was calculated as well. All statistical analysis was performed with SPSS Statistics 17.0 software.

Table 1. The assessment of SDP based on the soil types

Soil types	SDP	Scores	Status	Weight score (score x status)
Vertisols	Very low	2	1	2
Oxisols	Low	2	2	4
Alfisols, Mollisols, Ultisols	Moderate	2	3	6
Inceptisols, Entisols, Histosols	High	2	4	8
Spodosols, Andisols	Very high	2	5	10

Table 2. The assessment of SDP based on the land slope

Slope (%)	SDP	Scores	Status	Weight score (score x status)
1-8	Very low	3	1	3
9-15	Low	3	2	6
16-25	Moderate	3	3	9
26-40	High	3	4	12
>40	Very high	3	5	15

Table 3. The assessment of SDP based on the rainfall

Rainfall (mm/year)	SDP	Scores	Status	Weight score (score x status)
< 1.000	Very low	3	1	3
1.000-2.000	Low	3	2	6
2.000-3.000	Moderate	3	3	9
3.000-4.000	High	3	4	12
> 4.000	Very high	3	5	15

Table 4. The assessment of SDP based on the land use

Land use	SDP	Scores	Status	Weight score (score x status)
Natural forest, paddy field, pure fertile reed	Very low	2	1	2
Mix farm, shrub, savanna	Low	2	2	4
Production forest, dry land	Moderate	2	3	6
Dry land (annual crop)	High	2	4	8
Open field	Very high	2	5	10

Table 5. The SDP classes based on the final score

Symbol	SDP	Weighting score (final score)
SDP I	Very low	< 15
SDP II	Low	15 – 24
SDP III	Moderate	25 – 34
SDP IV	High	35 – 44
SDP V	Very high	45 – 50

3. Results and Discussion

The dominant annual rainfall was 3,000 - 4,000 mm/year with the status of 4 (high damaged potential) in the north-central area (Jenawi Village, Trengguli, Sidomukti, Balong, south-east Lempong, Menjing and Seloromo), and eastern part of Gumeng. The rainfall which was > 4,000 mm/year is categorized in the status of 5 (very high potential damage) which is located in Anggrasmanis Village and the west of Gumeng. The rainfall in a small part of the northern area of Seloromo, Menjing, and Lempong villages was 2,000 - 3,000 mm/year which is categorized in the status of 3/moderate (potentially damaged).

In affecting the soil damage, both rainfall intensity and pattern play an important role. The different patterns of the rainfall intensity are: (1) increase; (2) increase then decrease; (3) decrease, and (4) decrease then increase. All the patterns contributed different effects on soil damage [11]. The increase rainfall intensity resulted in the highest sediment, total runoff and sediment concentrations.

The soil type at research site consists of 3 types, namely Alfisols, Inceptisols, and Andisols. According to Table 1, Alfisols with status 3(moderately damage potential) (3) spread in the middle-north of research site. Inceptisols with high potentially damaged status (4) distributed in the north and the middle-south. Andisols with a very high potential damage status (5) are located in the southern area which is on the hillside to the top of Mount Lawu. The slope varies from 0 to 40%, and the top area of Mount Lawu has a slope > 40% with a very high damaged potential status (5). Below the top of Mount Lawu the central part-the south area has a slope of 16 - 40% with potential damage status between 3 -5 (moderate-high). The central-north area has a slope of 0-25% with a potentially damaged status between 1 - 3 (very low - moderate). Land use type which is effective for biomass production are: (1) forest, the area was 552.07 ha with the status of potential damage 1 (very low); (2) garden, the total area is 877,22 ha with the status of potential damage 2 (low); (3) grass, with total area 1.38 ha with the status of potential damaged 2 (low); (4) paddy field has an area of 544.94 ha with the status of potential damage 1 (very low); (5) the shrubs have an area of 1086.87 ha with the status of potential

IOP Conf. Series: Earth and Environmental Science **129** (2018) 012010 doi:10.1088/1755-1315/129/1/012010

damage 2 (low); and (6) the moor has an area of 1,183.55 ha with potentially damaged status of 4 (high).

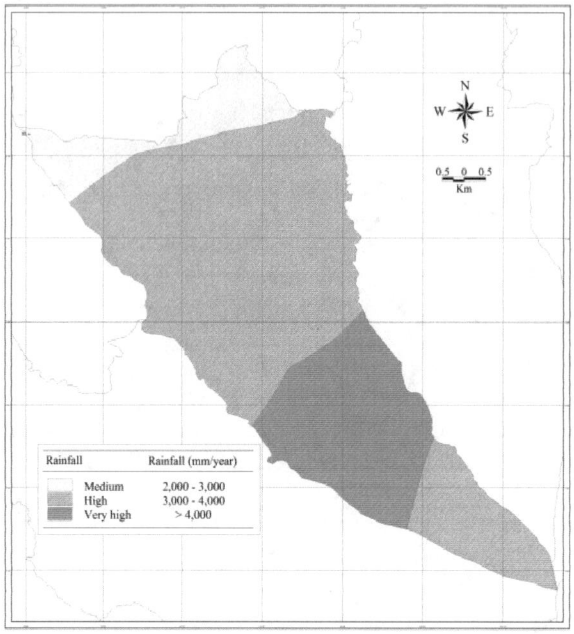

Figure 1. Map of annual rainfall, Jenawi District

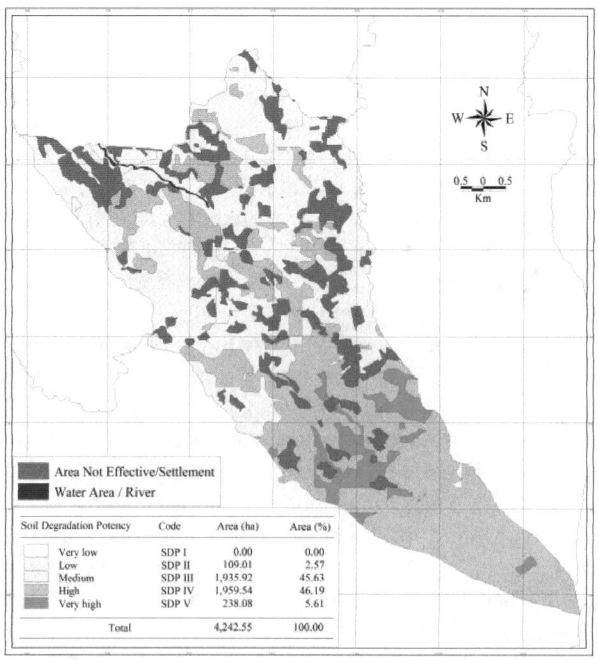

Figure 2. Map of soil degradation potency, Jenawi District

Table 6. The Area with the SDP status in Jenawi

No.	Soil Degradation Potency	Area (ha)	Area (%)
1.	Very low	0.00	0.00
2.	Low	109.01	2.57
3.	Moderate	1,935.92	45.63
4.	High	1,959.54	46.19
5.	Very High	238.08	5.61
	Total	**4,242.55**	**100.00**

The mapping result of the soil damage potential (See Figure 2) shows that most of the areas in the study site are categorized in the high damaged potential with the total area of 1,959.54 ha (46.19%) which is dominant in Gumeng and Anggrasmanis. The medium damage potential with the total area of 1,935.92 ha (45.63%) is distributed throughout the study site and small part in Gumeng. The very high damaged potential status of the total area of 238.08 ha (5.61%) is located in Gumeng, Anggrasmanis, and a small part at Jenawi. While the rest of the total area which is 109.01 ha (2.57%) is categorized as low damaged potential status located in Balong, Lempong, Seloromo, Trengguli, Sidomukti and Menjing. The area with very low damage potential was not found in the study site.

The correlation test showed that slope is the most affected factor on soil damage ($r = 0.76$, $P < 0.01$, $n = 306$) as it causes the soil erosion. The steeper the slope, the more soil splashes by rainfall [12]. If the steepness of the slope increases twice, then the erosion will rise 2 - 2.5 times. The increasing slope will raise the soil loss [13][14]. This is also supported by the results of [15] who stated that the loss of soil will be higher if the slope becomes steeper, especially in uncultivated soil. The second factor affecting the soil damage potential is rainfall ($r = 0.65$, $P < 0.01$, $n = 306$) as it can increase the surface runoff, soil loss and soil particle displacement [12]. The rain which directly falls on the soil surface and the flow of the surface will accelerate the erosion as the water is not infiltrated into the soil [16].

The third factor which affects the soil damage potential is soil type ($r = 0.61$, $P < 0.01$, $n = 306$) as the soil damage occurs due to rainfall in certain soil conditions [17]. The soil properties such as texture, organic matter and calcium content will affect the extent of soil damage through its effect on the surface flow [18]. The fourth factor is the type of land use ($r = 0.39$, $P < 0.01$, $n = 306$). Land use will also promote the surface flow which then leads to soil erosion [19]. The open agricultural system causes big erosion as when the surface flows, most of the C- organic are eroded [20]. On the other hand, the forest land has a higher C-organic content resulted by the accumulation of organic matter on the bottom layer of the forest [21]. Paddy field has a low SDP as it commonly lies on the flat area and the conservation method by the farmer resulting both of paddy bund and terrace maintenance [22].

Rainfall is one of the important factors that affect the soil damage potential. As shown in Figure 1 and Figure 2, there is a similar trend between the rainfall levels and the soil damage potential. The areas which have high potential damage are those with high annual rainfall (> 4,000 mm/year). The lower rainfall found in the upper and middle areas, the less soil damage occurred. The increasing rainfall intensity will rise soil loss, and thus increase the soil damage potential [13][14]. The changes and differences in the rainfall within a region can cause different levels of soil damage potential. Rain is an external factor of the land damage that cannot be controlled globally. The changes in rain properties are affected by global climate conditions, which also effects the environment and organism. The anticipation on land damage due to rain properties changes is urgent, and shall be supported by observing the landforms such as the slope, soil type, and land use. The conservation-based land management can be implemented to minimize the potential of land damage, especially those caused by rainfall.

This study needs a further study to identify the actual soil degradation through detail verification in the field and soil samples analyzed in the laboratory. Similarly with soil water regimes, soil surface roughness, tortuosity, and soil management practices also contribute to the surface flows and soil erosion [23][24][25]. Therefore, the adaptation to the climate change, particularly to rainfall can be carried out in various ways to increase water use efficiency; reduce evaporation, surface runoff, seepage and filtration losses; increase water storage capacity and available moisture range of soils [26].

4. Conclusion

The soil damage potential (SDP) at Jenawi district can be categorized to several classes, namely: very low (SDP I) 0.00 ha (0,00%); low (SDP II) 109.01 ha (2.57%); medium (SDP III) 1,935.92 ha (45.63%); high (SDP IV) 1,959.54 ha (46.19%); and very high (SDP V) 238.08 ha (5.61%). Annual rainfall strongly affected the SDP. The higher the annual rainfall, the more soil degradation occurs. The changes in rainfall properties as resulting by climate change phenomenon need an anticipation to minimize soil degradation based on local soil-land characteristics.

Acknowledgements
This research was supported by Sebelas Maret University Surakarta under the PNBP UNS 2017 research grant. We would like to thank Nur Aida Suheri and Citra Recha Sari for paper preparation assistance.

References

[1] Prasetyo H and Thohiron M 2012 *J.PAL* **4(1)** 63
[2] Quaile I 2013 "Floods the impact of global climate change" [Online] Available: *http://p.dw.com/p/18k6E*
[3] Puslittanak 1994 Map of Soil Type and Agroclimate (Bogor: Indonesian Soil Research Institute)
[4] Bakosurtanal 2010 Indonesia Earth's Surface Map (Jakarta: National Land Survey Coordination Agency)
[5] ESRI 2000 *ArcView 3.3 Desktop Help* (United Kingdom: Environmental Systems Research Institute, Inc.)
[6] FAO 2006 *Guidelines for soil description. 4th Edition* (Rome: Food and Agriculture Organization of The United Nations) p 97
[7] Eviati and Sulaeman 2009 *Technical manual 2nd edition: Chemical analysis of soil, plant, water, and fertilizer* (Bogor: Indonesian Soil Research Institute)
[8] Soil Survey Staff 2014 *Keys to soil taxonomy, 12th edition* (Washington: USDA-Natural Resources Conservation Service) p 360
[9] Anonim 2000 Government Regulation No. 150 (Jakarta: President of Republic Indonesia)
[10] Anonim 2006 Regulation of the State Minister of Environment Number 7 (Jakarta: State Minister of Environment)
[11] Mohamadi M A and Kavian A 2015 *J. International Soil and Water Conservation Research* **3** 273-81
[12] Arsyad S 2010 *Soil and Water Conservation Second Edition* (Bogor: IPB Press)
[13] Ziadat F M and Taimeh A Y 2013 *J. Land Degradation and Development* **24** 582-90
[14] Shen H, Zheng F, Wen L, Han Y and Hu W 2016 *J. Soil and Tillage Research* **155** 429-36
[15] Sun W, Shao Q, Liu J and Zhai J 2014 *J. Catena* **121** 151-63
[16] Widiatinigsih A, Suntoro and Mujiyo 2017 *Land degradation mapping in Keduang Sub-Watershed (Case Study In Jatipurno District). Proceeding National Seminar The Utilization of Geospatial Information for Synergy Improvement of Environmental Management. ISBN 978-602-73302-1-4* (Surakarta: Graduate School of PKLH UNS)
[17] Sinoga J D R and Diaz A R 2010 *J. Geomorphology* **118** 359-68
[18] Vaezi A R, Bahrami H A, Sadheqi S H R and Mahdian M H 2010 *J. Hydrol. Earth Syst. Sci.*

Discuss **7** 2577–607
[19] Fang F N, Shi Z H, Li L, Guo Z L, Liu Q J and Ai L 2012 *J. Catena* **99** 1–8
[20] A Monde, N Sinukaban, K Murtilaksono, N Pandjaitan 2008 *Jurnal Agroland* **15(1)** 22- 6
[21] Hairiah K, Sulistyani H, Suprayogo D, Widianto, Purnomosidhi P, Widodo R H and van Noordwijk M 2006 *J. Forest Ecology and Management* **224** 45-7
[22] Mujiyo, Rahayu, Ustiatik R and Anggrahini DS 2016 *J. Sains Tanah – Journal of Soil Science and Agroclimatology* **13(1)** 25-30
[23] Römkens M J M, Heling K and Prasad S N 2001 *J. Catena* **46** 103–23
[24] Bramorski J, De Maria I C, Lamos e S R and Crestana S 2012 *J. R. Bras. Ci. Solo* **36** 1291-97
[25] Biddoccu M, Opsi F and Cavallo E 2014 *J. Soil Science and Plant Nutrition* **60** 92–9
[26] Várallyay G 2010 *J. Agronomy Research* **8 (2)** 385–96.

Study on rhizobium interaction with osmoprotectant rhizobacteria for improving mung bean yield

Y Maryani[1,2,3], Sudadi[1], W S Dewi[1] and A Yunus[1]

[1] Doctoral Program in Agricultural Science, Graduate School, Sebelas Maret University, Surakarta, Indonesia
[2] Faculty of Agriculture, Universitas Sarjanawiyata Tamansiswa, Jl. Batikan no. 7 Yogyakarta, Indonesia
[3]Corresponding author: ym_ust@yahoo.com

Abstract. Gunungkidul has calcareous soil with limitations including calcareous stone, mostly hilly terrain, and shallow cultivated layer. Furthermore, nowadays we face the disadvantages climates such as long dry seasons, a short rainy season and high temperatures caused by climate change. Climate change leads to irregular rainwater availability for microbes and crops. Research in this field is currently needed as climate change affected directly on crop production, while we need to find the strategy to keep high productivity of the plant. This research aimed to determine the ability of osmoprotectant rhizobacteria and rhizobium to support mung bean yield. Osmoprotectant rhizobacteria were isolated and screened from the calcareous soil in Gunungkidul with disadvantageous climates such as a long dry season, a short rainy season and high temperature. This research was arranged in Completely Randomized Design. The result showed that osmoprotectant rhizobacteria isolate of strain Al24-k and Ver5-k can produce 9.6306 mg g^{-1} cell of glycine betaine in a soil density 1.7667 x 10^7 CFU g^{-1} and 11.4870 mg g^{-1} cell of glycine betaine in a soil density 1.9667 x 10^7 CFU g^{-1}. Inoculation of isolates osmoprotectant rhizobacteria can support mung bean yield. Osmoprotectant rhizobacteria isolate did not effect rhizobium in mung bean rhizosphere.

1. Introduction

Gunungkidul has limitations by having calcareous soil including calcareous stone, deep hardpan layer, mostly hilly terrain, and shallow cultivated layer. Furthermore, we face the climate drawbacks such as long dry seasons, short rainy season, and high temperatures which caused by climate change nowadays. It leads to irregular rainwater availability for microbial and crops. Microbes are abundant in the soil, it was necessary to investigate the interactions between microbes in soil and plants to boost crop production. Rhizobium always lives contiguous rhizobacteria in the rhizosphere. It always helps mung bean to synthesis nodule at mung bean root. This research aimed to study the ability of osmoprotectant rhizobacteria isolates and rhizobium to support mung bean yield. Calcareous rhizobacteria isolates were A124-k and Ver5-k strains. It can produce osmoprotectant as glycine betaine.

Mung bean is the third most important food plant in Indonesia after soybean and peanut. Mung bean production is low. It is not capable of fulfilling the need and productivities of 1.44 ton h^{-1} which is low potential yield [1].

Mung bean always lives with rhizobium symbiotically by nature. Rhizobium helps mung bean to tie up Nitrogen in the air. There are many other bacteria in it. Some of the bacteria are free living. The

Content from this work may be used under the terms of the Creative Commons Attribution 3.0 licence. Any further distribution of this work must maintain attribution to the author(s) and the title of the work, journal citation and DOI.
Published under licence by IOP Publishing Ltd

rhizobium is always living with many other rhizobacteria in rhizosphere's mung bean. The rhizobacteria do not affect rhizobium growth and activity [2].

Shokley *et al.* [3] explained that inoculation rhizobium and *Arbuscular mycorrhiza* at legume could boost the agroforestry plant prosperity. It is at nonlegume did not influence plant growth. It can escalate legume and nonlegume plant development.

The rhizobacteria can help to maintain water and nutrient. Supporting nutrition for a plant such as, tie up nitrogen, siderophore, decomposed organic, antagonist activity to plant disease and pest, and plant growth which promoting rhizobacteria [4].

2. Material and methods

The research was employing 2 factorial experiments in *Completely Randomized Design*. The first-treatment factors were non inoculated rhizobium (R0) and inoculated rhizobium (R1). The second-factor treatment was inoculation of osmoprotectant rhizobacteria isolates which was consisting of non inoculation of rhizobacteria isolates (I0), inoculation osmoprotectant rhizobacteria isolate Al24-k (I1), and inoculation osmoprotectant rhizobacteria isolate Ver5-k (I2).

This experiment used a sterile insectisol soil media. Mung bean seeds were washed by using sterile water. Mung bean seeds were submerged in a solution which containing osmoprotectant rhizobacteria isolates with 10^6 of density during 30 minutes. Seeds were plated in sterile media.

The observed variable in this research was **Rhizobacteria density in mung bean rhizosphere.** It was performed by plating method which according to [2]. Rhizobacteria density was observed by six weeks after planting age. Rhizobacteria density was observed by soil dilution method and measured in Colony Form Unit (CFU) [2]. **Glycine betaine production, the** procedure for determining glycine betaine adopted the periodide method which was developed by [4] [5]. **Mung bean yield,** mung bean yield include dry grain weight, 100-grain weight per mung bean plant. **Analysis of variance** (ANOVA), using statistical analysis system (SAS) was performed on the data (glycine betaine) and the treatment means were compared using Duncan's test at $p \leq 0.05$.

3. Results and Discussions

Table 1. The density of osmoprotectant rhizobacteria isolates in mung bean's rhizosphere and glycine betaine production.

Treatment	Osmoprotectant rhizobacteria isolates in mung bean's rhizosphere	
	Density	Glycine betaine
Inoculation of Rhizobium		
Non inoculated of Rhizobium	1.4556×10^7 p	6.8503 p
Inoculated of Rhizobium	1.6237×10^7 p	7.1402 p
Inoculation of osmoprotectant rhizobacteria :		
Non inoculated rhizobacteria	0 c	0.0000 c
Inoculated rhizobacteria Al24-k	1.7559×10^7 a	9.7006 b
Inoculated rhizobacteri Ver5-k	1.8279×10^7 a	10.8670 a

Information: Numbers followed by the same lower case letters show nonsignificant difference based on Duncan's test $p \leq 0.05$.

Table 1shows the glycine betaine density and production of osmoprotectant rhizobacteria isolates in mung bean's rhizosphere of non inoculated rhizobium were same as inoculated rhizobium. This indicates that rhizobium did not affect the osmoprotectant rhizobacteria isolates activity,

Table 1, displays osmoprotectant rhizobacteria isolates ver5-k density in mung bean's rhizosphere has the same density of osmoprotectant rhizobacteria isolates Al24-k. Glycine betaine production of osmoprotectant rhizobacteria isolates Ver5-k was higher than glycine betaine production of osmoprotectant rhizobacteria isolates Al24-k. This illustrates that osmoprotectant rhizobacteria isolate Ver5-k have more activity than osmoprotectant rhizobacteria isolates Al24-k.

Table 2. Dry weight plant, dry grain weight, 100-grain weight per mung bean plant in six weeks after planting.

Treatment	Variable		
	Plant dry weight (g)	Dry grain weight (g)	100-grain weight (g)
Inoculation of rhizobium:			
Non inoculated rhizobium	1.6419 b	49.8701 b	4.70 b
Inoculated rhizobium	1.7337 a	51.7899 a	5.25 a

Information: Numbers followed by the same lower case letters show nonsignificant difference based on Duncan's test $p \leq 0.05$.

Table 2, Inoculation of Rhizobium in mung bean shows significant effects on dry plant weight, dry grain weight, 100-grain weight per plant. Inoculation of Rhizobium in mung bean displayed dry plant weight, dry grain weight, 100-grain weight per plant was higher than non inoculated rhizobium [6]. Inoculation rhizobium and *Arbuscular mycorrhiza* at legume can escalate agroforestry plant growth. Inoculation rhizobium at nonlegume cannot boost plant growth. Rao [2] demonstrated that rhizobium only helps legume to synthesis nodule root and gives hand to tie up nitrogen in the air. Perruod and LeRudulier [8] justified that nodule synthesis and activity of external and indigenous rhizobium illustrated significant effects. Inoculated rhizobium of legume improved disease and pest plant endurance, and also raised plant growth and yield.

Table 3. plant dry weight, dry grain weight, 100-grain weight per plant mung bean six weeks after planting.

Treatment	Variable		
	Plant dry weight (g)	Dry grain weight (g)	100-grain weight (g)
Inoculation of osmoprotectant rhizobacteria:			
Non inoculated rhizobacteria	1.5661 b	43.2029 b	4.77 b
Inoculated rhizobacteria Al24-k	1.7544 a	54.9857 a	5.04 a
Inoculated rhizobacteri Ver5-k	1.7428 a	54.3014 a	5.11 a

Information: Numbers followed by the same lower case letters show nonsignificant difference based on Duncan's test $p \leq 0.05$.

Table 3 inoculated osmoprotectant rhizobacteria isolates for mung bean demonstrates significant effects on plant growth and yield mung bean. Inoculated osmoprotectant rhizobacteria isolates of strain Al24-k and Ver5-k on mung bean gave higher plant growth and yield than non inoculated osmoprotectant rhizobacteria isolate. Rhizobacteria supported water for plant growth and yield through many activities synthesis glycine betaine [4]. Nicolaus *et al.* [6] explained that organic osmolyte was produced by a microorganism to accumulate in the cell. By employing this microorganism adaptation mechanism has wide range adaptation.

Salmonella typhimurium, Escherichia coli, Klebsiella pneumonia, Azospirillum amadaerahnse, A. lipoferum, A. brasilense and *A. halopraeferens* were bacteria that produce osmolyte under osmotic stress [8][9][10]. Osmolite can motivate water potential. It can cause water to flow into the cell, so cell turgidity l can be maintained. According to [11] *Rhizobium meliloti, ectothiorhodospira halochloris* secretes glycine betain. Glycine betaine was osmoprotectant.

4. Conclusion

Osmoprotectant rhizobacteria were isolating and screening from the calcareous soil of Gunungkidul with climates disadvantageous such as, long dry season, short rainy season, and high temperature which are caused by the climate change. Inoculation of isolates osmoprotectant rhizobacteria can support mung bean yield. Osmoprotectant rhizobacteria isolate did not affect rhizobium in mung bean rhizosphere.

References

[1] Widyawati, Harjasa T and Taufik T T 2016 Organik fertilizer application on yield green bean (*Vigna radiata* L) varieties N ultisol *Jurnal Kultivasi* **15(3)** pp 159-163

[2] Rao N S S 1994 *Soil microorganism, and plant growth* Indonesia University Press Jakarta.

[3] Shokley F W, R McGrow and H F Gareet 2004 Growth and nutrient concentration of two native forage legumes inoculated with rhizobium and mycorrhiza in Missouri USA *Agroforestry System* **60** pp 137 – 142

[4] Maryani Y 2010 Khemotaksis rhizobacteria osmotolerant pada rhizosphere mug bean (*Vigna radiata* L) *Biota* **15 (3)** pp 486 - 493

[5] Barak A J and Tuna D J 1981 Determination of choline, phosphorylcholine and *betain Methods of Enzymol* **72** pp 287 – 292

[6] Nicolaus, Lanzotti B V, Trincone A, DeRosa M, Grant W D and Gambacarta A 1989 Glycine-betaine and polar lipid composition in Halophytic archaebacteria in response to growth in different salt concentration *FEMS Microbiol Leff* **59** pp 157 – 160

[7] Suki K 2011 Rhizobiol measures to evade host defense strategy and indigenous threats to present symbiotic nitrogen fixation: a focus two legume-rhizobium model system *Cellular and Molecular Life Sciences* **68** pp 1327 – 1339

[8] Perruod B D and LeRudulier 1985 Glycine-betaine Transport in E. coli: Osmotic *Modulation J Bacteriol* **161** pp 391 – 401

[9] Smith L T, Pocard J A, Bernard T and LeRudulier D 1988 Osmotic control of glycine betain biosynthesis and degradation in Rhizobium meliloti. *J. Bacteriol* **170** pp 3142 – 3149

[10] Hartmann A 1988 Ecophysiological aspects of growth and nitrogen fixation in Azospirillum sp *Plant and Soil* **110** pp 225 – 238

[11] Tschichholz I and Truper H G 1990 Fate of compatible solutes during dilution stress in Ectothiorhodospira halochloris *FEMS Microbiol Ecol* **73** pp 181 -186

IOP Conf. Series: Earth and Environmental Science **129** (2018) 012012 doi:10.1088/1755-1315/129/1/012012

Graphical approach to assess the soil fertility evaluation model validity for rice (case study: southern area of Merapi Mountain, Indonesia)

E A Julianto[1,2], W A Suntoro[3], W S Dewi[3] and Partoyo[2]

[1]Doctoral Program of Agricultural Science, Graduate School, Sebelas Maret University, Surakarta Jl. Ir. Sutami No. 36A Surakarta, Indonesia
[2]Dept. Of Agrotechnology, Universitas Pembangunan Nasional "Veteran" Yogyakarta, Jl. SWK 104 Lingkar Utara Condong Catur Yogyakarta, Indonesia
[3]Dept. of Soil Science, Fac. of Agriculture, Sebelas Maret University, Jl. Ir. Sutami No. 36A Surakarta, Indonesia

Abstract. Climate change has been reported to exacerbate land resources degradation including soil fertility decline. The appropriate validity use on soil fertility evaluation could reduce the risk of climate change effect on plant cultivation. This study aims to assess the validity of a Soil Fertility Evaluation Model using a graphical approach. The models evaluated were the Indonesian Soil Research Center (PPT) version model, the FAO Unesco version model, and the Kyuma version model. Each model was then correlated with rice production (dry grain weight/GKP). The goodness of fit of each model can be tested to evaluate the quality and validity of a model, as well as the regression coefficient (R^2). This research used the Eviews 9 programme by a graphical approach. The results obtained three curves, namely actual, fitted, and residual curves. If the actual and fitted curves are widely apart or irregular, this means that the quality of the model is not good, or there are many other factors that are still not included in the model (large residual) and conversely. Indeed, if the actual and fitted curves show exactly the same shape, it means that all factors have already been included in the model. Modification of the standard soil fertility evaluation models can improve the quality and validity of a model.

1. Introduction

Climate change is known to cause diminishing land resources including soil fertility deterioration. Soil as the source of life, is the most vital and valuable natural resource which is not renewable quickly. Soil fertility is a dynamic natural property that can change the impact of natural and human-derived factors [1]. Soil, being the natural medium for plant growth has a direct impact on yield and quality of crops growing on it. Measurement of the fertility of an agricultural soil informs a lot of about the potential productivity. Fortunately, producers can control fertility by managing the plant's nutritional status [2], and good management on fertility requires a good fertility evaluation model. Speaking of models, one imagines something intricate, complex, and difficult, and thus this research is an attempt to bridge those knowledge gap. Any input from a farming operation is aimed at stepping up the welfare of the farming community. Choosing an evaluation model highly correlated with crop productivity is a rational step for efficiency and effectiveness in planning soil fertility management. This is an important effort to ensure that a set of recommendations on fertilizing from the research results can be restricted in terms of its area in the form of a map based on the homogeneity of the

Content from this work may be used under the terms of the Creative Commons Attribution 3.0 licence. Any further distribution of this work must maintain attribution to the author(s) and the title of the work, journal citation and DOI.
Published under licence by IOP Publishing Ltd

IOP Conf. Series: Earth and Environmental Science **129** (2018) 012012 doi:10.1088/1755-1315/129/1/012012

biophysical and socio-economic characteristics of a farming operation [3]. The soil fertility evaluation models evaluated were the PPT (Indonesian Soil Research Center) version model, the FAO Unesco version model, and the Kyuma version model. Each model was correlated with wet field rice production (dry grain weight). There are several indicators that can be used/seen to infer that the quality of a model is good. In the language of statistics, it is to test a model's goodness of fit. After fitting a regression model, it is important to determine whether all the necessary model assumptions are valid before performing inference. If there are any violations, subsequent inferential procedures may be invalid resulting in faulty conclusions. Therefore, it is crucial to perform appropriate model diagnostics. In constructing our regression models, we assumed that the response y to the explanatory variables were linear in the parameters and that the errors were independent and identically distributed normal [4]. Model diagnostic procedures to analyze the quality of a model can be seen from the regression coefficient (R^2) and a graphical approach. The objective of this research is therefore to assess the validity of a soil fertility evaluation model using a graphical approach in the southern area of Merapi Mountain, Indonesia.

2. Research method

The research was conducted on the southern slope of Mount. Merapi (Cangkringan, Pakem, Turi, Tempel, and Seyegan subdistricts) of Sleman Regency, the Province of the Special Region of Yogyakarta, Indonesia. Research stations were scattered on the southern slope of Merapi Mountain (37 experiment stations) which were planted with rice (*Oryza sativa*) of the Ciherang variety. The research was carried out from April 2015 to April 2016. The observation was done by taking soil fertility samples in points (37 stations) determined through stratified purpose sampling. The main basis used was a wet field land (at least half-irrigated) on land map units and its area. Each point of observation was in the fieldwork map whose geographical position (coordinates) had been determined, then traced in the field, and its sample of soil fertility was taken. The soil samples were then analyzed in the laboratory to ascertain the nutrient status covering C-org, N-total, pH H_2O, pH KC1, P_2O_5, K_2O, P-Bray, Morgan K_2O, Ca, Mg, K, Na, Ca-total, Mg-total, Cation Exchange Capacity (CEC), Si, Texture, Alkali Saturation, P-retention, and P-Olsen. As for the component of crop production, the observation was carried out based on the dry weight grain. The statistical analyses used Eviews 9. Regression and correlation analyses was employed on soil fertility parameters in respect with crop production. Regression was done in single and multiple ways. By using Eviews 9, a good quality model could be seen from indicators R^2 and a graphical approach. A high R^2 did not always show a good quality model. From the graphical approach, a graphic with three curves was obtained, namely data regression curve, the actual data, and residual data. If the data regression curve and the actual data coincide with each other, the model is considered to have good quality as there are only a few external factors that affect Y.

3. Results and discussion

3.1. Pre-indicators to assess the validity of a soil fertility evaluation model

The results of regression analysis of standard and modified PPT, respectively, are presented in Table 1. Table 1 shows that on the CEC log the significant prob at a reliability level of 99%, on the regression coefficient of CEC if the CEC increased by 1%, the unit (dry grain weight) of rice production would increase by 2.04%, with the assumption that the other factors remained constant (cateris paribus). The total prob. in standard PPT =0.0469 means that the reliability level=99.9531, whereas in the total prob. of modified PPT =0.0034 means that the reliability level= 99.9966. This shows that modified PPT is more reliable.

Table 1. Regression results of standard PPT and modified PPT.

Standard PPT			Modified PPT		
Variable	Coefficient	Prob.	Variable	Coefficient	Prob.
C	-11277.41	0.0283	C	-4.97	0.1842
LOG(CEC)	2.04	0.0006	LOG(CEC)	1.36	0.0024
LOG(CORG)	1.82	0.0009	LOG(CORG)	1.34	0.0010
LOG(KB)	1.14	0.0454	LOG(KB)	6.95	0.0871
LOG(P2O5)	3.84	0.4511	LOG(BRAY1)^2	1.03	0.0000
LOG(K2O)	4.74	0.4167	LOG(K2O)	3.42	0.4193

When compared with a higher R^2 ex: $R^2=0.82$, the total prob. that affects significantly can be greater although per a particular parameter e.g. from 2.04 to 1.3, this is not applicable in Table 1 Equation as shown in Table 1 on the left is :

$$\text{GKP} = -11277.41 + 2041.68*\text{LOG(CEC)} + 1820.78*\text{LOG(CORG)} + 1138.23*\text{LOG(KB)} + 383.96*\text{LOG(P2O5)} + 474.10*\text{LOG(K2O)} \quad (1)$$

Regression coefficient (R^2) = 0.603

While equation in Table 1 on the right is :

$$\text{GKPLOG} = 1.08 + 0.02*\text{LOG(CEC)}\hat{}2 + 0.09*\text{LOG(CORG)} + 0.05*\text{LOG(KB)}+0.01*\text{LOG(BRAY1)}\hat{}2+ 0.03*\text{LOG(K2O)} \quad (2)$$

Regression coefficient (R^2) = 0.82

The results of regression analysis of standard and modified FAO, respectively, are presented in Table 2. Table 2. shows that total prob. in standard FAO=0.0469 means the reliability level=99.9531, whereas the total prob. of modified PPT means the reliability level=99.9966 so that it can be inferred that the modified PPT was more reliable. In the standard FAO, there were 5 significant parameters while in modified FAO there were 6 significant parameters. In terms of its R^2 the modified FAO $R^2=0.903$ whereas in the standard FAO $R^2=0.82$. The equation as shown in Table 2 on the left is :

$$\text{GKP}=-16098.05+1695.53*\text{LOG(CEC)}+1663.19*\text{LOG(CORG)}+367.94*\text{LOG(K)}+323.07*\text{LOG(OLSENP2O5)}+513.78*\text{LOG(P-RETENTION)}+1581.70*\text{LOG(CA1)}+940.08*\text{LOG(MG1)}-98.71*\text{LOG(K2O)}+ 306.29*\text{LOG(P2O5)} \quad (3)$$

While equation in Table 2 on the right is :

$$\text{GKPLN}=3.37+0.001*\text{(LN)(CEC)}\hat{}2+0.11*\text{(LN)(CORG)}\hat{}2+1.27*\text{(LN)K}+0.001*\text{(LN)BRAY1}-0.0001*\text{(LN)RETENSIP}+2.56\text{e-}05*\text{(LN)CA1}+0.0003*\text{(LN)MG1}-0.0007*\text{(LN)MORGANK2O} + 0.0004*\text{(LN)P2O5} \quad (4)$$

Table 2. Regression results of standard FAO and modified FAO.

Standard FAO			Modified FAO		
Variable	Coefficient	Prob.	Variable	Coefficient	Prob.
C	-16098.05	0.0032	C	-11160.00	0.0235
LOG(CEC)	1.70	0.0007	LOG(CEC)	1.51	0.0005
LOG(CORG)	1.66	0.0001	LOG(CORG)	1.46	0.0001
LOG(K)	1.37	0.0004	LOG(K)	1.10	0.0022
LOG(OLSENP2O5)	-3.23	0.3133	LOG(BRAY1)	4.05	0.0158
LOG(RETENSIP)	5.14	0.2176	LOG(RETENSIP)	1.48	0.7065
LOG(CA1)	1.58	0.0002	LOG(CA1)	1.14	0.0047
LOG(MG1)	9.40	0.0239	LOG(MG1)	7.72	0.0386
LOG(K2O)	-9.87	0.8302	LOG(MORGANK2O)	-1.98	0.3871
LOG(P2O5)	3.06	0.4171	LOG(P2O5)	1.19	0.7338

The results of regression analysis of standard and modified Kyuma, respectively, are presented in Table 3.

Table 3. Regression results of standard Kyuma and modified Kyuma.

Standard Kyuma			Modified Kyuma		
Variable	Coefficient	Prob.	Variable	Coefficient	Prob.
C	1.98	0.5788	C	0.97	0.0012
LOG(CORG)	1.26	0.0022	LOG(CORG)	0.09	0.0005
LOG(N)	7.23	0.1956	LOG(N)	0.05	0.1167
LOG(P2O5)	-1.45	0.7034	LOG(P2O5)	-0.01	0.5770
LOG(BRAY1)	5.34	0.0018	LOG(BRAY1)	0.03	0.0025
LOG(CAMG)	-3.43	0.3654	LOG(CAMG)	-0.02	0.2680
LOG(K)	4.93	0.1674	LOG(K)	0.04	0.0650
LOG(CEC)	1.35	0.0047	LOG(CEC)	0.08	0.0032
LOG(SI)	2.53	0.0967	LOG(SI)	0.06	0.3746
LOG(PASIR)	1.26	0.1241	LOG(PASIR)	0.09	0.0390
			LOG(KB)	0.05	0.0228
			LOG(RETENSIP)	0.008	0.7139

Table 3 shows that total prob. in standard Kyuma=0.0087 means that the reliability level=99.9913, whereas in the total prob. of modified Kyuma 0.0062 means that the reliability level=99.9938 so that it can be inferred that modified PPT was more reliable. In terms of the high R^2 ex: R^2=0.896 there was a more significant prob. (5 parameters) whereas in standard Kyuma there were 3 parameters. In term of its R^2 the modified Kyuma=0.896 whereas the standard Kyuma R^2=0.84. The equation as shown in Table 3 on the left is :

GKP=1984.29+1261.61*LOG(CORG)+723.08*LOG(N)-144.59*LOG(P2O5)+
533.72*LOG(BRAY1)-343.19*LOG(CAMG)+492.60*LOG(K)+1350.66*LOG(CEC)+
2533.54*LOG(SI) + 1258.80*LOG(PASIR) .... (5)

While equation in Table 3 on the right is :

GKP(LN)=3.09+0.21*(LN)CORG+1.16*(LN)N-0.0002*(LN)P2O5+0.002*(LN)BRAY1-
0.01*(LN)CAMG+0.61*(LN)K+ 0.018*(LN)CEC + 0.29*(LN)SI + 0.003*(LN)PASIR
+ 2.15e-05*(LN)KB + 9.56e-05*(LN)P-RETENTION .... (6)

3.2. Analysis of model quality with graphical approach

The results of regression analysis of standard and modified PPTis presented in Table 4.

Table 4. Regression results of standard PPT and modified PPT

Evaluation Model	R^2	Prob. (F-statistics)
PPT/SRC LOGARITHM (LOG 10) Standard PPT (Fig. 1 left)	0.473	0.000007
GKP LOG, LOG (CEC)^2 and LOG (BRAY1)^2 (Fig. 1 right)	0.815	0.000000

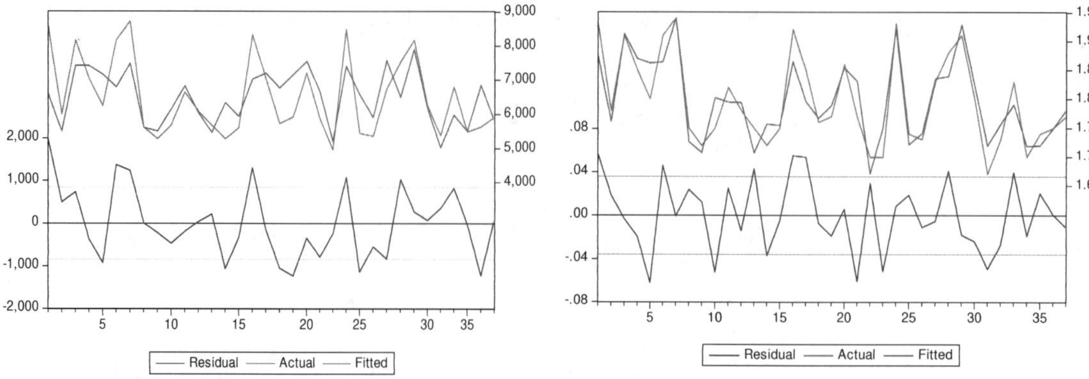

Figure 1. Quality of PPT model using graphical approach.

Two-dimensional plots of residuals versus fitted values or predictors are traditionally used to assess lack of fit of a regression model. The general idea is that if the model is correct then the sample residuals should appear independent of the predictors, with allowance for typically negligible dependence caused by substituting estimates for parameters [5]. Table 4 shows that the high regression coefficient (0.815) is followed by simultaneous influence/less Prob (F-statistic) (greater significance) at 0.000000. Conversely, the low/small regression coefficient (0.473) is followed by simultaneous influence/greater Prob (F-statistic) (less significance) at 0.000007. this is confirmed in Figure 1 where the actual and fitted curves are distantly spaced (wide apart) or irregular, indicating that the quality of the model is not good, or there are many other factors that have not been included in the model (larger residual curves) as shown in Figure 1 on the left. Conversely, if the actual and fitted curves coincide with each other the quality of the model is good, or other factors that have not been included in the model are few (small residual curves). Indeed if the actual and fitted curves became just one curve it means that there are no other factors that have not been included in the model, as seen in Figure 1 on the right.

Table 5. Regression results of standard FAO and modified FAO.

Evaluation Model	R^2	Prob. (F-statistics)
FAO		
Standard FAO (Fig. 2 left)	0.440	0.000019
*GKP*2 (*LN*) (*CEC*)^2 and (*LN*) (*CORG*)^2 (Fig. 2 right)	0.903	0.000000

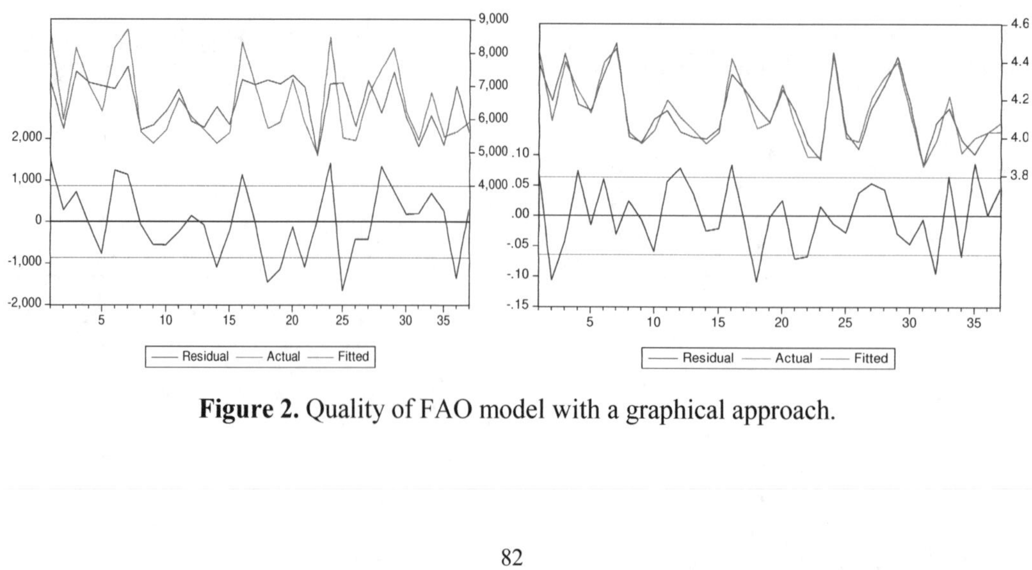

Figure 2. Quality of FAO model with a graphical approach.

Table 5 shows that the higher regression coefficient(R^2= 0.903) is followed by simultaneous influence Prob (F-statistic), with greater significance at 0.000000. Conversely, the low regression coefficient (R^2) (0.440) is followed by simultaneous influence/more Prob (F-statistic) (less significance) at 0.000019. This is to confirm Figure 2, where the actual and fitted curves are distantly spaced (wide apart) or irregular indicating that the quality of the model is not good, or there are still many other factors that have not been included in the model (large residual curve) as shown in Figure 2 on the left. Conversely, if the actual and fitted curves coincide with each other, the quality of the model is good, or other factors that have not been included in the model are few (small residual curve); indeed, if the actual and fitted curves become just one curve, it means that there are no other factors that have not been included in the model, as shown in Figure 2 on the right.

Table 6. Regression results of standard Kyuma and modified Kyuma.

Evaluation Model	R^2	Prob. (F-statistics)
Kyuma		
OM (Fig. 3 left)	0.122	0.042645
GKP (LN) and without coefficient (+ (LN) P-RETENTION and (LN) KB (Fig. 3 right)	0.896	0.000000

Table 6 shows that the high regression coefficient (R^2= 0.896 is followed by simultaneous influence Prob (F-statistic) (greater significance) at 0.000000. Conversely, the low regression coefficient (R^2= 0.122 is followed by simultaneous influence/more Prob (F-statistic) (less significance) at 0.042645. This is to confirm Figure 3, where the actual and fitted curves are distantly spaced (apart) or irregular indicating that the quality of the model is not good, or there are still many other factors that have not been included in the model (large residual curve) as shown in Figure 3 on the left. Conversely, if the actual and fitted curves coincide with each other, the quality of the model is good, or other factors that have not been included in the model are few (small residual curve); indeed, if the actual and fitted curves become just one curve, it means that there are no other factors that have not been included in the model, as shown in Figure 3 on the right.

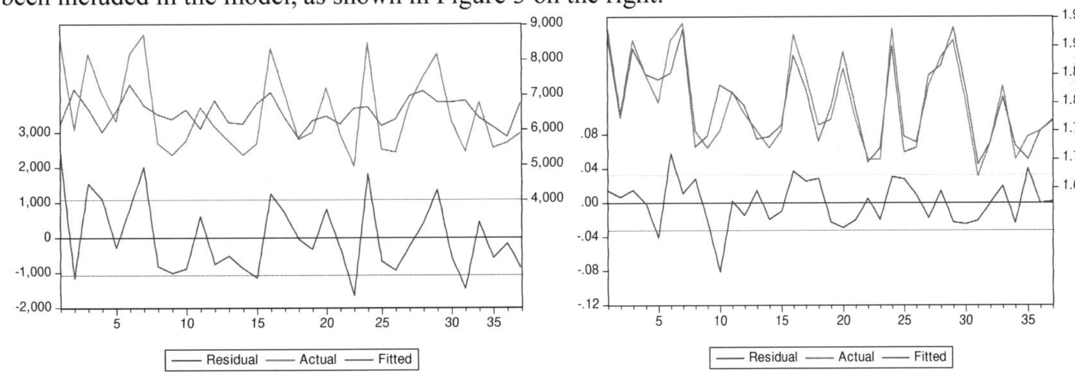

Figure 3. Quality of Kyuma model with graphical approach

4. Conclusion

A high regression coefficient (R^2) is followed by simultaneous influence/less Prob (F-statistic) (greater significance) and there are more parameters involved. If the actual and fitted curves are distantly spaced (wide apart) or irregular, the quality of the model is not good, or there are many other factors that have not been included in the model (large residual curve). Conversely, if the actual and fitted curves coincide with each other, the quality of the model is good, or other factors that have not been included in the model are few (small residual curve); indeed, if the actual and fitted curves become just one curve, it means that there are no other factors that have not been included in the model. Modification of the standard Soil Fertility Evaluation can improve the quality of an evaluation model.

IOP Conf. Series: Earth and Environmental Science **129** (2018) 012012 doi:10.1088/1755-1315/129/1/012012

The resulted model by graphical approaching could be used to evaluate soil fertility, so it can reduce the risk of climate change effect on plant cultivation.

Acknowledgement
Dedicated to Ministry of Research, Ministry of Technology and Higher Education, Republic of Indonesia, for funding this study.

References
[1] Kavitha C and Sujatha M P 2015 Evaluation of soil fertility status in various agroecosystems of Thrissur District, Kerala, India. *Int. J. Agric. Crop Sci.* **8** pp 328–38
[2] Flynn R, Shane T B and Baker RD 2004 *Sampling for Plant Tissue Analysis Guide A-123* (New Mexico: College of Agriculture and Home Economics New Mexico State University) p 92
[3] Dobermann A *et al.* 2003 Estimating indigenous nutrient supplies for site-specific nutrient management in irrigated rice *Agronomy Journal* **95** pp 924-35
[4] "Model Diagnostics for Regression" *www.stat.columbia.edu* 2017 [Online] Available: www.stat.columbia.edu/~martin/W2024/R7.pdf [Accessed: 27-08-2017]
[5] Pardoe I and Cook R D "Graphical Method for Assessing the Fit of the Logistic Regression Model" *http://citeseerx.ist.psu.edu* 2002 [Online] Available: http://citeseerx.ist.psu.edu/viewdoc/download?doi=10.1.1.138.4256&rep=rep1&type=pdf [Accessed: 27-08-2017]

IOP Conf. Series: Earth and Environmental Science **129** (2018) 012013 doi:10.1088/1755-1315/129/1/012013

The effect of pricing level to the loss of welfare costs (case study: Indonesia region II water company)

B Rosalina E W K[1,3], E Gravitiani[2], M Raharjo[2] and T Mulyaningsih[2]

[1] Magister Dept of Economics and Development Studies, Graduate School, Universitas Sebelas Maret, Jl. Ir. Sutami No.36A, Kentingan, Surakarta, 57126, Indonesia
[2] Dept. of Economy Development, Universitas Sebelas Maret, Surakarta, Indonesia
[3] Corresponding author: rosalinabernadeta@gmail.com

Abstract. Climate change makes the water balance composition being unstable, both quality and quantity. As a company which responsible for water management, Regional Drinking Water Company (abbreviated as PDAM) is often unable to solve the problem. Welfare costs are indicators to evaluate the economic efficiency. This study aims to calculate the welfare cost of the people lost due to the price determination of PDAM Indonesia in region II with deadweight loss (DWL) approach, so it can provide information to pricing regulator, pricing decision makers and for coIDRorate management. DWL is a loss of economic efficiency that can occur when equilibrium for a good or a service is not achieved, caused by monopoly pricing of artificial scarcity, an externality, a tax or subsidy, or a binding price ceiling or price floor such as a minimum wage. Results showed that the pricing rules set by PDAM yielded different DWL, depending on margin set by the company DWL PDAM ranges between IDR 260,485.66/M^3 to IDR 127,486,709.86/M^3 which is actually shared to improve the welfare of customers, other communities, and PDAM itself. Data analysis used PDAM performance in 2015 that have not Good CoIDRorate Governance Management and Efficiency.

1. Introduction

The 1945 Constitution of article 33, paragraph 3 explains that the earth, water, and all the natural wealth contained therein are controlled by the State and should be fully utilized for the welfare of the society. Based on the regulation, the water resources management reference is fully controlled by the government. Its concern is to emphasize that all water resources can be maximally supervised and equally profit-sharing to society.

Climate change generates the water balance composition being imbalanced, both quality and quantity [1]. Water is a basic human necessity, so it is a very profitable business, almost 99% of the world's population needs water. The government becomes a decision maker to solve this problem by establishing a Local Water Company (abbreviated as PDAM) in every region to reach all the people's needs. It has large market potentials but based on data by 2010, Indonesian which using its facilities as drinking water were only 14%, and by 20% for clean water purposes [2]. It's minor product utilization because is still bad and not ready for consumption. It should be able to contribute needs which is ready to be consumed and processed or at least clean, but in reality, there are still many enterprises whose its quality is below the clean water standard. Its production which has not maximized, due to the public confidence level to eat up becomes low.

PDAMs have spent a lot of money for water sources, reservoirs, treatment, pipelines, and distribution. The funds are purely from local or governmental finances, as well as foreign debt or aid. The companies have high debt trends and low production cost efficiency [3]. It is caused by the

Content from this work may be used under the terms of the Creative Commons Attribution 3.0 licence. Any further distribution of this work must maintain attribution to the author(s) and the title of the work, journal citation and DOI.
Published under licence by IOP Publishing Ltd

number of consumer requests who are still small, so it cannot cover production costs incurred by the corporation. The small number of requests in addition to the low level of public confidence to this firm, as well as it is located in every city or district, while the number of people in those locations is different. Regions that have more communities are more likely to be those clients than others who do not have. Its supply capability differences and the public demand potential for water cause different tariffs. It has high production costs and a low number prospects will have a high rate as well. The price escalation can cause a loss of people's welfare, as they have to lose more money to meet water needs. It can be measured by the Dead Weight Loss (DWL) approach, i.e. the loss of economic efficiency in monetary measure if the applied cost to the public is not in accordance with the marginal price.

The government as the holder of PDAM price regulator should know the economic factors, including the water price elasticity, in order to determine the balanced price. Its performance in 2015 portrayed that 368 companies in Indonesia, only 53% had healthy indicators, 27% were less healthy, and 20% were sick, one of the main factor, due to its high and inefficient production costs and high margin requirements [4], which was provoking a DWL value roses up.

The purpose of this paper is to analyze the economic welfare of the people who are lost due to the increase of water supply tariff in PDAM region II Indonesia, which includes the sick PDAM. This analysis approach was performed by calculations from DWL with different levels of demand elasticity assumptions. It employed the demand levels assumptions from different price levels.

2. Methodology

The method which exploited to determine the lost welfare was by calculating deadweight loss approach. It was employed by different approaches to the demand elasticity. The calculation formula of deadweight loss from several companies in the same year is [5]:

$$\text{DWL}i = \frac{(|Q1-Qo|*|P-MC|)}{2} \tag{1}$$

Where Q1 is the predicted amount of consumed water when the price is equal to MC, Qo is the amount of consumed water at the current price (P * Qo). From equation (Q1-Qo) can be calculated by the formula:

$$\Delta\text{Q} = \frac{(MC-P)}{P} * \varepsilon * \text{Qo} \tag{2}$$

Where ε is the level of demand elasticity. Its water demand has not accomplished by many researchers, therefore in this study operating by different elasticity levels were 0.2, 0.5, and 0.8.

3. Sample data

Observations were conducted by utilizing PDAM data in Indonesia Region II that have the less healthy condition. Data was taken from the performance of 12 diseases in 2015 [6]. It was picked up from the annual report by 2015 which was created by Development Agency of Development of Drinking Water Supply System to see how much loss of society welfare level.

The marginal data of PDAM disease in region II shows a price difference variation between tariff and marginal price ranged between IDR 64.50/m^3 up to IDR 693/m^3. This margin was high enough, therefore the losing welfare chances became great. As an example of Table 1 above that if Kediri's applied marginal rates to the tariff, then public consumption of PDAM water production will increase from 2,587,769.00 m^3 to 2,943,020.40 m^3, 417,020.40 m^3 of increasing amount. The predicted climb was quite large, it shows how much the people lost welfare.

4. Result

The results of this study will examine the marginal differences and tariff margins which were employed by PDAMs to change the water consumption demand of public water companies with different elasticity approach to bridge all possible conditions, i.e. -0.2, -0.5, and -0.8 of elasticity. This analysis will look at how enterprises have lower tariffs than marginal prices and it have higher tariff than marginal prices, with all consumers record who are already their clients.

Table 1. Data on profit, initial consumption, predicted consumption, and deadweight loss from 0.8 elasticity.

Performance of PDAM Indonesia region II					
No	**Region**	**Margin**	**Water consumption**	**Predicted water consumption**	**Deadweight loss**
		(IDR/m^3)	**(m^3)**	**(m^3)**	**(IDR)**
1	Sukabumi City	173.81	3,793,069.00	3,900,387.67	9,326,666.08
2	Banjar City	693.00	2,534,611.43	2,872,072.67	116,930,317.51
3	Banjarnegara District	94.00	1,617,937.00	1,676,828.03	2,767,878.28
5	Pacitan District	380.24	2,966,239.00	3,160,522.96	36,937,051.57
7	Blitar District	671.12	2,559,640.00	2,939,561.28	127,486,709.86
8	Kediri District	472.20	2,526,000.00	2,943,020.40	98,459,077.55
9	Pamekasan District	64.50	2,587,769.00	2,620,078.48	1,041,942.62
10	Lamongan District	439.03	3,871,066.00	4,206,684.93	73,672,848.50
11	Blitar City	381.23	88,258.00	97,639.44	1,788,262.42
12	Ponorogo District	109.19	2,963,620.00	3,055,569.92	5,020,011.14
	Average	**347.83**	**2,550,820.94**	**2,747,236.58**	**47,343,076.55**

Source: Annual report elaboration of PDAM performance in 2015.

This calculation analysis can also be used to compare PDAM, for example, PDAM Blitar and Pamekasan Regencies cooperation. Both businesses have the same 2,500,000 m^3 of gross water consumption, but PDAM Blitar margin was very high at IDR 671.12/m^3, while Pamekasan Regency has a low margin of IDR 64.50/m^3. Huge gap difference made Blitar district firm lost more prosperity than Pamekasan regency, so Blitar deadweight loss was bigger than Pamekasan district.

Table 2. The total deadweight loss in the different level of elasticities.

Performance of PDAM Indonesia region II				
		DWL		
No	**Region**	**-0.2**	**-0.5**	**-0.8**
			(IDR/M3)	
1	Sukabumi City	2,331,666.52	5,829,166.30	9,326,666.08
2	Banjar City	29,232,579.38	73,081,448.44	116,930,317.51
3	Banjarnegara District	691,969.57	1,729,923.93	2,767,878.28
5	Pacitan District	9,234,262.89	23,085,657.23	36,937,051.57
7	Blitar District	31,871,677.47	79,679,193.66	127,486,709.86
8	Kediri District	24,614,769.39	61,536,923.47	98,459,077.55
9	Pamekasan District	260,485.66	651,214.14	1,041,942.62
10	Lamongan District	18,418,212.13	46,045,530.31	73,672,848.50
11	Blitar City	447,065.61	1,117,664.01	1,788,262.42
12	Ponorogo District	1,255,002.78	3,137,506.96	5,020,011.14
	Average	**11,835,769.14**	**29,589,422.85**	**47,343,076.55**

Table 3 displays the deadweight loss overall number from different levels of elasticity. Sukabumi city at elasticity 0.2, has IDR 2,331,666.52 of deadweight loss and at elasticity 0.8 was IDR 9,326,666.08 which exactly 4 times bigger. The total deadweight loss was IDR 11,835,769.14 to IDR 47,343,076.55 with the highest deadweight loss was Blitar district. Based on table 3, it can be seen that in an average of PDAM wasted welfare value with IDR 260,485.66 to IDR 127,486,709.86 of deadweight loss approach level.

The Results portrayed that a low-level efficiency, there was expected the government role to improve PDAM performance. By developing, companies expected performance can help by solving uneven water necessary due to the climate change impact.

5. Discussion and policy implications

The regulation of tariff formation of PDAM water supply should be taken seriously by the government because, according to the 1945 Act, water is the natural resource which becomes state responsibility and to fulfill the people's welfare. The first policy that must be accomplished by the government is to control the productivity and working pattern of the companies, in order to produce water production quality to increase customer trust in PDAM therefore, people hope who has not yet become PDAM members, will join later. The second policy that must be done was to control the price tag, so the prevailing tariff was a reasonable amount and in accordance with the ability, level to pay the community. Welfare loss can be seen from the deadweight loss value, if it can be managed properly, it can be utilized to improve the customer's welfare themselves, other communities, and the company itself. By improving its performance, it can solve and manage uneven water necessary due to the climate change impact

6. Conclusion

By improving PDAM performance, it can solve and manage uneven water necessary due to the climate change impact. Loss welfare analysis with deadweight loss approach indicates that freedom of price tag determination by each company without central government control would cause price level variation. Low or negative margin prices have a positive impact on society as their welfare rises, however, for enterprises would experience continuous and welfare losses, while those high margins will cause massive losses to society and greater deadweight loss. This paper wishes to show the firms with the same amount of public consumption, rate, and margins have unequal characteristics and need further research on conduct and performance.

References

[1] Trenberth K E 2011 Water Cycles and Climate Change; Global Environmental Change *Preprint* HB_GlobalEnvChange_30

[2] Prihatin R B 2013 Clean water problem in urban areas (In) *Info Singkat* Vol V No. 07/I /P3DI/April/2013

[3] The Foundation Of Indonesian Consumer Institutions – YLKI "Review of Debt Implications on the Performance of Regional Drinking Water Companies (PDAMs) in Indonesia" (In) *http://kruha.org* 2004 [Online] Available: http://kruha.org/oneMODUL/oneDocument/1300268639.pdf [Accessed: 10-09-2017]

[4] Ministry of Public Works and Public Housing of the Directorate General of Human Settlements "Handling Drinking Water Field" (In) *http://ciptakarya.pu.go.id* 2015 [Online] Available: http://ciptakarya.pu.go.id/dok/ebook/konsinyasi/Volume%202.%20Penanganan%20Bidang %20Air%20Minum.pdf [Accessed: 10-09-2017]

[5] Porcher S, Maziotis A and Senante M 2017 The welfare costs of nonmarginal water pricing: evidence from the water only companies in England And Wales *Urban Water Journal* **14(9)** pp 1-10

[6] Indonesian Agency for System Development Support of Regional Drinking Water Company - Indonesian Ministry of Public "Works PDAM Performance 2015: Java Island 2015" (In) *http://www.bppspam.com* 2015 [Online] Available: http://www.bppspam.com/index.php?option=com_docman&task=cat_view&gid=51&Itemid =108

IOP Conf. Series: Earth and Environmental Science **129** (2018) 012014 doi:10.1088/1755-1315/129/1/012014

Study on osmoprotectant rhizobacteria to improve mung bean growth under drought stress

Y Maryani[1,2,3], Sudadi[1], W S Dewi[1] and A Yunus[1]

[1] Doctoral Program in Agricultural Science, Graduate School, Sebelas Maret University, Jl. Ir. Sutami 36A Kentingan, Surakarta, Central Java 57126 Indonesia
[2] Faculty of Agriculture, Universitas Sarjanawiyata Tamansiswa, Jl. Batikan No 06 Tahunan, Umbulharjo, Yogyakarta 55167 Indonesia
[3] Corresponding author: ym_ust@yahoo.com

Abstract. Climate change leads to irregular rainwater availability for crops and thus enhances drought stress. Furthermore, nowadays we face climate disadvantages such as long dry season, short rainy season and high air temperature caused by climate change. This research aimed at studying the ability of osmoprotectant rhizobacteria isolates to support mung bean growth under drought stress. The rhizobacteria were isolated from mung bean's rhizosphere. The results showed that isolates of strain Al24-k and Ver5-k produced glycine betaine 9.6306 mg g^{-1} cell, 1.7667 x 10^7 CFU g^{-1} soil and 11.4870 mg g^{-1} cell, 1.9667 x 10^7 CFU g^{-1} soil. The isolated rhizobacteria from mung bean's rhizosphere under field capacity of soil moisture produced glycine betaine 6.8000 mg g^{-1} cell, 1.2556 x 10^7 CFU g^{-1} soil. Under 75% field capacity of soil moisture, isolates produced glycine betaine of 6.4059 mg g^{-1} cell, 1.3111 x 10^7 CFU g^{-1} soil, while under 50% from field capacity, the isolates produced glycine betaine of 7.4108 mg g^{-1} cell, 1.6667 x 10^7 CFU g^{-1} soil. The osmoprotectant rhizobacteria improved the resilience of mung bean to drought stress.

1. Introduction

Gunungkidul regency is occupied by calcareous soils. The calcareous soil has the limitations which majorly dominated by hilly terrain, deep hardpan layer, calcareous stone and shallow cultivation layer. Furthermore, nowadays we are facing the climate disadvantages such as long dry seasons, short rainy season and high air temperatures due to climate change. Climate change leads to irregular rainfall and thus enhances drought stress on crops because the water availability is insecure. This research aimed to study the ability of osmoprotectant rhizobacteria isolates to support mung bean growth under drought stress. Strains A124-k and Ver5-k are calcareous rhizobacteria isolates, which can produce osmoprotectant as glycine betaine and tolerant to osmotic stress.

The rhizobacteria isolates are capable to produce the osmoprotectant compound as glycine betain, can help other organisms secreting glycine betaine into the environment. Glycine betaine is an osmolyte synthesized by rhizobacteria to protect it from osmotic stress such as dry soil, which also indicator of *E. coli* osmotic stress during rhizobacteria colony forming [1].

Mung bean is the third most important commodity in Indonesia after soybean and peanut, but the production is low and not sufficient to meet the needs. Productivity is very low, only of 1.162 ton ha^{-1} [2]. The efforts to mung bean production can be done by field extension, including to marginal land such as dry land. Mung bean can be cultivated on dry land with technology support called inoculation of osmoprotectant rhizobacteria, the plant is tolerant to drought stress.

Content from this work may be used under the terms of the Creative Commons Attribution 3.0 licence. Any further distribution of this work must maintain attribution to the author(s) and the title of the work, journal citation and DOI.
Published under licence by IOP Publishing Ltd

2. Material and methods

The research was designed with 2 factor in the *Completely Randomized Design*. The first factor was the inoculation of osmoprotectant rhizobacteria consisted of three levels, i.e. (1) inoculation of osmoprotectant rhizobacteria isolates consisting of non inoculation of rhizobacteria isolates (I0), (2) inoculation osmoprotectant rhizobacteria isolate Al24-k (I1), and (1) inoculation osmoprotectant rhizobacteria isolate Ver5-k (I2). The second factor was drought stress consisting of three levels, i.e. (1) field capacity (C1), (2) 75% field capacity (C2), and (3) 50% field capacity (C3).

This experiment used to steril insectisol soil media. Mung bean seeds were washed using sterile water. And then submerged in a solution containing osmoprotectant rhizobacteria isolates with 10^6 density during 30 minutes. And then planted in sterile media.

Variable observed was **Rhizobacteria density in mung bean rhizosphere,** using serial dilution plating technique according to [3]. Rhizobacteria density was observed at six weeks after planting, by soil dilution method and in Coloni Form Unit (CFU). **Glycine betaine content** was investigated by adopting the periodide method developed by Barak and Tuma [4] and Nicolaus [5]. **Mung bean growth parameter,** including plant height, leaf number, leaf width, and biomass were measured using the volumetric method. **Analysis of variance** (ANOVA) and Duncan's advanced test (DMRT) test at $p \leq 0.05$ was performed to the analyzed data.

3. Results and Discussions

Table 1. Osmoprotectant rhizobacteria isolate density mung bean's rhizosphere and glycine betain production.

Treatment	Osmoprotectant rhizobacteria isolates in mung bean's rhizosphere			
	Density		Glycine betain	
Inoculation of osmoprotectant rhizobacteria :				
Non inoculated rhizobacteria	0	c	0.0000	c
Inoculated rhizobacteria Al24-k	1.7667×10^7	b	9.6306	b
Inoculated rhizobacteri Ver5-k	1.9667×10^7	a	11.4870	a
Soil moisture :				
Field capacity	1.2556×10^7	p	6.8000	q
75 % field capacity	1.3111×10^7	p	6.9059	p
50 % field capacity	1.6667×10^7	p	7.4108	p

*Mean followed by the same letters is non significantly different at $\alpha = 5$ %.

Table 1, shows the density of osmoprotectant rhizobacteria isolates ver5-k in mung bean rhizosphere was higher than that at osmoprotectant rhizobacteria isolates Al24-k. It also can be seen in table 1 that glycine betain production of osmoprotectant rhizobacteria isolates Ver5-k is higher than that at osmoprotectant rhizobacteria isolates Al24-k. This indicates that osmoprotectant rhizobacteria isolate Ver5-k had more activity and ability than that of osmoprotectant rhizobacteria isolates Al24-k.

It performed in table 1 that, soil moisture treatment showed no significant effects on osmoprotectant rhizobacteria isolates density. but significantly effected the glycine betaine production. Soil moisture 75% field capacity and 50% field capacity increased glycine betaine production compared in that of field capacity. It indicates that drought stress can increase the activity of osmoprotectant rhizobacteria isolates to produce glycine betain.

Table 2 shows that inoculation of osmoprotectant rhizobacteria isolates in mung bean rhizosphere showed that strain Al24-k and Ver5-k higher plant height, leaf number and leaf width than non inoculated osmoprotectant rhizobacteria isolates.

Table 2 also displays that moisture influenced plant height, leaf number, and leaf width, where 75% and 50% of field capacity decreased plant height, leaf number, and leaf width.

Table 2. Plant height, leaf number and leaf width per plant six weeks after planting.

Treatment	Variable		
	Plant height	Leaf number	Leaf width (cm^2)
Inoculation of osmoprotectant rhizobacteria:			
Non inoculated rhizobacteria	16.37 b	16.7 b	188.679 b
Inoculated rhizobacteria Al24-k	16.78 ab	17.7 a	215.479 a
Inoculated rhizobacteri Ver5-k	17.00 a	17.6 a	221.169 a
Soil moisture:			
Field capacity	17.24 a	17.9 a	226.376 a
75 % field capacity	16.69 ab	17.2 b	205.931 b
50 % field capacity	16.23 b	16.9 b	193.020 c

*Mean followed by the same letters is non significantly different at $\alpha = 5$ %.

Table 3. Dry weight per plant six weeks after planting.

Treatment	Dry weight (g)
Non-in_field capacity (control)	1.613 cd
Non-in_75% field capacity	1.520 d
Non-in_50% field capacity	1.180 e
Al24-k_ field capacity	1.883 a
Al24-k_75% field capacity	1.743 bc
Al24-k_50% field capacity	1.630 cd
Ver5-k_ field capacity	1.786 ab
Ver5-k_75% field capacity	1.776 ab
Ver5-k_50% field capacity	1.620 cd

*Mean followed by the same letters is non significantly different at $\alpha = 5$ %.

Table 3 performs that, the inoculation between osmoprotectant rhizobacteria isolates and moisture soil effected the biomass. The interaction of inoculation osmoprotectant rhizobacteria isolates Al24-k with field capacity soil moisture; Ver5-k at field capacity soil moisture; Al24-k at 75% field capacity soil moisture; Ver5-k at75% field capacity soil moisture showed higher plant growth than at non inoculation in field capacity (control). This indicates that osmoprotectant rhizobacteria isolate Ver5-k and Al24-k increased glycine betaine production along with the decrease of soil moisture.

The results also show that inoculation of osmoprotectant rhizobacteria isolates Al 24-k and Ver5-k improved mung bean tolerance on drought stress. According to Truper and Galinski [6], the microorganism under osmotic stress have two adaptation mechanism i.e. KCl type and organic osmolyte type, respectively. According to Smith [7], organic osmolyte was produced by a microorganism and accumulated in cell and blocked the cellular process, so resulting in the compatible solute. Microorganism with this adaptation mechanism has wide range adaptation, as isolate A82 was able to help upland to growth well under 40 % field capacity soil moisture [8]. Isolate A82 in osmotic stress condition produced glycine betain. The isolation and screening of soil in Rajasthan desert, India obtained 24 bacteria isolates which resistant to osmotic stress [9], namely *Bacillus sp* 45.8 %, *Corynebacterium sp* 41.7 % and *Acinetobacter sp*, *Aeromonas sp*, *Staphylococcus sp* group 2.5 %. *Bacillus megaterium* BOFC15 rhizobacteria increased plant growth and help the plant to be tolerated for drought stress [10] because *Bacillus megaterium* produced spermidine (Spd) which is polyamine type. Spermidine can increase plant dry weight and improve the architecture of plant root, as well as increase plant adaptation to drought stress.

4. Conclusion

Inoculation of osmoprotectant rhizobacteria which isolated and screened from the calcareous soil of Gunungkidul supported mung bean growth under field capacity soil moisture and improved the resilience of mung bean to drought stress. This is important for face the recent climate disadvantages affected by climate change.

Acknowledgement
The authors are indebted to the ministry of research DIKTI Department, Indonesia for the Doctor Grant research funding of Research DIKTI.

References

[1] Yuwono T, Lolita E S, Ikhsan A, Ngadiman, Sovitri M, Mursyanti E and Soedarsono J 1997 Studies on drought of rhizobacteria and their potential use as growth – promoting agent for gogo rice *Annual Reports of Icbiotech* **20** pp 667-80
[2] Wibowo W 2016 The Effect of doses and concentration fertilizer to growth and yield mug bean (*Vigna radiata* L.) var. Vima-1 Faculty of Agriculture Sarjanawiyata Tamansiswa University Yogyakarta
[3] Yoetono, Soedarsono J, Hartadi S, Kabirun S, Soesanto and darmosuwito S 1972 *Basic of Microbiology for College* (Yogyakarta: Department of Microbiology Faculty of Agriculture Gadjah Mada University) p 122
[4] Barak A J and Tuma D J 1981 Determination of choline, phosphorylcholine and betain *Methods of Enzymol* **72** pp 287-92
[5] Nicolaus B, Lanzotti V, Trincone A, DeRosa M, Grant W D and Gambacarta A 1989 Glycine-betaine and polar lipid composition in halophitic archaebacteria in response to growth in different salt concentration *FEMS Microbiol Leff* **59** pp 157-60
[6] Truper H G and Galinski E A 1990 Biosynthesis and Fate of compatible solutes in extremely Halophillic Phototrophic Eubacterium *FEMS Mirobiol Rev* **75** pp 247-54
[7] Smith L T, Pocard J A, Bernard T and LeRudulier D 1988 Osmotic control of glycine betain biosynthesis and degradation in Rhizobium meliloti *J. Bacteriol* **170** pp 3142-9
[8] Yuwono T, Handayani D and Soedarsono J 2005 The role of osmotolerant rhizobacteria in rice growth under different drought conditions *Australian Journal of Agriculture Research* **56(7)** pp 715-21
[9] Sharma R, Manda R, Gupta S, Kumar S and Kumar V 2013 Isolation and characterization of osmotolerant rhizobacteria from desert of Western Rajasthan, India *Revista de Biologia Tropical* **61(4)** pp 1551-62
[10] Zhou C, Ma Z, Zhu L, Xiao X, Xie Y, Zhu J and Wang J 2016 Rhizobacterial strain Bacillus megaterium BOFC15 induces cellular polyamine changes that improve plant growth and drought resistance International *Journal of Molecular Science* **17(6)** pp 976-81

IOP Conf. Series: Earth and Environmental Science **129** (2018) 012015 doi:10.1088/1755-1315/129/1/012015

Local adaptive capacity as an alternative approach in dealing with hydrometeorological risk at Depok Peri-Urban City

I S Fitrinitia[1,2], P Junadi[1], E Sutanto[1], D A Nugroho[1], A Zubair[1] and E Suyanti[1]

[1] Urban Development Studies: School of Strategic and Global, Universitas Indonesia, Salemba Raya Street Number 4, Central Jakarta, Indonesia.
[2] Corresponding author: irenesondang20@gmail.com

Abstract. Located in a tropical area, cities in Indonesia are vulnerable to hydrometeorological risks such as flood and landslide and thus become prone to the climate change effects. Moreover, peri-urban cities had double burden as the consequences of main city spill over and also lack of urban facilities in overcoming the disaster. In another perspective, the city has many alternative resources to recover, so its create urban resiliency. Depok city becomes a case study of this research regarding with its development following the impact of Jakarta growth. This research purposes to capture how the local city dwellers could anticipate and adaptive with flood and landslide with their own mitigation version. Through mix method and spatial analysis using GIS techniques, it derives the two comparison approach, the normative and alternative that had been done by the city dwellers. It uses a spatial analysis to have a big picture of Depok and its environmental changing. It also divided into 4 local group of communities as a representative of city dwellers regarding the characteristic of a settlement with their level of risk. The result found type or characteristic of settlement which influenced the local adaptive capacity, from the establishment of infrastructure, health fulfillment and social livelihood with different kind of methods.

1. Introduction

Cities as a melting pot of population and innovation growth become a place where the frequency of disasters are escalated and have been exposed to several kinds of environment risk [1]. City and urbanization are two inseparable incidents as it becomes a base for urban development. Rapid unplanned urbanization, population pressure with unstable and unclear policies makes urban areas at great risk [2]. Meanwhile, since 2011, Indonesia has entered a period of development, called urban millennia, a situation where the number of urban residents is greater than in rural areas. Major cities in Indonesia have a strategic role in regional development as a service, collection and distribution node, which has a backward relationship with its small and hinterland cities as well as forward relationships with other big cities [3]. Unfortunately, the growth rate of cities in Indonesia is relatively uneven and uncontrolled. Examples such as development in some cities, it is not likely to see the characteristics of the city itself consequently the physical burden that must be borne by the city even further increases. If the situation is not immediately addressed, then the magnitude of potential disasters and weak growth process will slow the pace of development of cities in Indonesia.

In other side, cities in Indonesian archipelagic have many challenging issues regarding the environmental depletion because of its location. Regarding the tropical area, it seems very common with dynamic climate and environmental change. Environmental changing at the archipelagic and tropical area is different with the mainland at others continental. As the consequences, 62 % of 497 cities in Indonesia have a high-risk level of disaster [4].

Content from this work may be used under the terms of the Creative Commons Attribution 3.0 licence. Any further distribution of this work must maintain attribution to the author(s) and the title of the work, journal citation and DOI.
Published under licence by IOP Publishing Ltd

Moreover, besides the negative effects from the urban growth, the city also become "one-stop solution" which are ideas and technology grow in cities become a solution for disaster mitigation. At the national level, the government takes a major role in coping the disaster at-risk area. It includes how to mainstreaming disaster issues at provincial, city and municipalities levels. However, government past paradigm was a top-down approach which is a service provider and distribution of development outcomes. Especially at the provincial level, since there is regional autonomy so province government takes a lead for the local program. This kind of participation seldom encountered problems that causing project-oriented, passive benefits and increasing inequality [5].

In order to reduce disaster risk, we couldn't only stand to a government program, meanwhile local capacity at society or community level already adaptive with the risks itself with their version. The focus of this research is to figure out how the local group of communities could identify their own capacity as local adaptive dealing with environmental risk surround their neighborhood. Focusing on hydro-meteorological risks determined with 2 kinds of hazards such as flood and landslide. As a study case, Depok is a peri-urban received pressure of Jakarta growth and development. Since 2007, Depok experienced a devastating flood disaster that resulted in huge losses. Flood disaster in Depok is caused by several factors such as heavy rain that can cause river water flows, piles of garbage along the river and the depths of which have started shallow and filled with garbage. By knowing the local adaptive, its aims to understand there are alternative approach from communities as city dwellers in coping with flood and landslide risks. So it can't depend on government action only, but also gathering the local community's participation.

2. Materials and method

2.1. Materials

Rising urban populations will have implications for increased urban activity and space needs. The increasing need for urban space faced with land constraints leads people to place unsuitable areas for their activities and settle. Poor people tend to make settlement enclaves in restricted locations for housing because they're economically less affordable, for example on riverbanks. Slums are scattered in several parts of the city. Studies show that vulnerable populations inhabited slums due to proximity to urban economic opportunities, despite disregarding the safety of disasters and the environment [6].

In addition to the emergence of slum settlements, environmental degradation is also a consequence of urban development. Increasing the number of people imbalanced with the availability of space to live and doing activities makes vulnerable groups occupy areas prone to disaster. Bad habits and lack of basic infrastructure services for housing, food, water, sanitation and energy cause environmental problems, such as littering, using rivers for bathing, washing, cooking, and others. Many found the river to be a garbage dump for residents living on the banks of the river [3].

The disaster threats experienced by middle cities in Indonesia are hydro-meteorological disasters (i.e floods, landslides, droughts and heat waves), rather than the geophysical disasters (i. e., earthquakes and volcanic eruptions), that has been trending upwards in recent decades. This rising trend of climate-related hazards suggests a possible connection between these hazards and in turn disasters on the one side and climate change on the other [7].

Hydrometeorological disaster risk reduction is not only done with the construction and management of facilities and infrastructures, in accordance with National Law Nu. 26, 2007 on Spatial Planning, Republic of Indonesia if it located in disaster-prone areas requires spatial planning based on disaster mitigation as an effort to improve the safety and comfort of life and preserve the environment [8]. Floods are indirectly related to so massive land use changes. Land use will be given for making land use monitoring system [9]. It is understood that floods occur because of the unplanned rapid urbanization, change in land use and poor watershed management mainly in floodplains become important issues for consideration as the flood causes [10].

Based on the identification of potential disaster variables and historical data related to the hydro-meteorological disaster that threat Depok is flood and landslide [11]. The increasing number of impervious land surfaces from Depok urban development, as well as the increasing utilization of Depok's water resources, put significant pressure on water resources and hydrological systems in Depok city. Assuming it will be including increased overland flow, erosive flow, instability increment river morphology, and changes in river biodiversity [12]. The pressure on this hydrological system is exacerbated by weather and climate trends worsening due to climate change. The impact of this phenomenon is a very high variability in terms of availability from time to time and spread unevenly perceived as floods and droughts. Intensive anthropogenic activities in Depok City and inadequate infrastructure management have consequences that exacerbate the above impacts, water pollution and landslide disaster.

Another consequence that is sometimes not in the spotlight when a hydro-meteorological disaster occurs is how well prepared the community. One way to look at it can be through the level of social vulnerability of Depok society. Based on previous research [11], areas in Depok with low social vulnerability are located in some districts located in the suburbs. In addition, in the region in some indicators also shows are in low grade. Broadly speaking, it can be said that the more towards the center of the city, the level of social vulnerability is increasing.

2.2. Research method

The method used in this research is mixed-method. Quantitative methods used are a statistical analysis of tabular data and spatial data with the help of geographic information system software (GIS). Qualitative method used is an interview with respondent, Shared learning dialogue (SLD) through focus group discussion (FGD) with the related institution and social element.

Data collection in this study is divided into three, according to categories of data to be collected. Secondary data were obtained by literature study and data from related institutions. Primary data will be obtained by field survey including in-depth observation and observation, ground check for secondary data validation, and interviews with respondents and shared learning dialogue (SLD). In the following data classification table, data collection is classified based on three types of data, including health data, basic infrastructure data and social data.

Communities were divided into 4 local group as a representative of city dwellers regarding the characteristic of settlement: 1) slum with high risk, 2) slum with low risk, 3) formal residential with high risk and 4) formal residential with low risk. Each local group is spread 2 different questionnaires. 1 questionnaire intended for ordinary resident and other questionnaires for stakeholder's/community leaders/ RT/ RW heads. The number of questionnaires distributed was 30 in each local group. The questionnaire consists of several closed questions (with options) and the rest is an open question. Questionnaires were processed and analyzed statistically.

3. Results and discussions

3.1. Hydrometeorological risk in Depok

Floods are overflow water that exceeds the capacity of rivers in their canals that are usually preceded by high rainfall or high flow water from upstream areas. Due to its vast territory, flood in different areas manifests themselves in different types and with various characteristics [13]. Flood and inundation problems are caused by high rainfall, modified or disturbed river morphological conditions, poor drainage systems, and other external factors such as high river sedimentation and landslides.One of the causes of high rainfall is impacted by the changing of climate surround.

The flooded districts spread throughout Depok City and are dominated in the east of the city, as shown in Figure 1. Some areas prone to flooding were passed by large streams such as Ciliwung and Pesangerahan River. Besides that, the absence of a lake is one of the factors that make the region prone to flooding. According to Figure 1, the flood-prone areas are 126.4 Hectares. Sukmajaya District became the most potentially affected by floods with an estimated total area of 32.8 Hectares. In addition, flood-prone areas in Depok City tend to be located at an altitude of fewer than 100 meters which includes a declivous area.

Figure 1. Flood area in Depok City in 2015 [11]

Landslide in Depok City occurs due to several factors from the physical condition, natural factors, and human activity. Factors caused by humans can be in the form of piles of garbage and the accumulation of material caused by the expansion of settlements so that piling up the soil around the valley leads to overloaded land, and farms activity on the slopes.

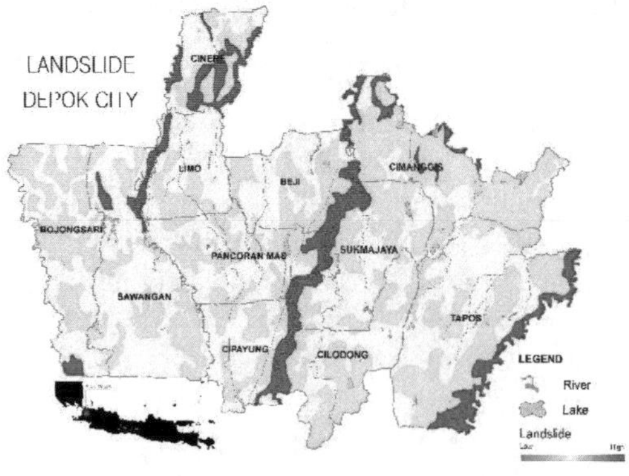

Figure 2. Depok City's landslide area in 2015 [11].

The landslide-prone area in Depok City has a slope between 8-15% which means that the slope condition is quite steep and has a steep hill or valley. This region is often found along a large river flow that has a width of more than 10 m. Steeper the slope, the higher the landslide potential, but this can't be separated from the type of rock and soil type in the area. Figure 2 shows that areas prone to landslides reach 1.807 Hectares. The most vulnerable areas of landslides are Tapos and Cinere Districts with a total of 7 vulnerable villages. This region has a fairly steep slope along the stream.

3.2. Normative approach regarding risk at Depok

Instead of climate change management, it is more common with disaster management while climate change effects include in it. Although, the policies and regulations on the disaster in Depok City are still overlapping. This is due to the absence of agencies that focus on disaster management affairs. Unlike other cities such as Jakarta that already have Regional Board for Disaster Management (abbreviated as BPBD DKI Jakarta), the Disaster Management Regional Board Depok needs to be completed because the level of disaster threat is increasing every year.

Recently, disaster-related policies are downgraded into several plans and programs can be identified through urban planning regulation (RDTR Depok City in City Regulation No.1 Year 2015). The plans and programs are as follows:

Table 1. Flood and landslide management program.

Item	Program
Flooded Area Plan	Normalization of drainage channels
	Building new channels to the river
	Building embankments
Landslide Area Plan	Protecting and building of river/lake walls
	Normalization of riverbank and lake with mud dredging and develop jogging track
	Building an evacuation road
	Reforestation by planting landslide prevention plants along riverbank
	Limit the utilization of space around riverbank

Source: Urban planning of Depok City.

Several programs can also be identified through a series of Focus Group Discussions (FGD) with the Local Government of Depok City and field findings.

Regarding the early warning system, there is no standard scheme regarding early warning system in Depok City. Society has its own way to know when and where the disaster will happen. For example, some people living near the river basin said that until now there is no early warning from the government, they actually have an early warning system via messages on mobile phones. Communities are inevitably forced to adapt to their homes. The importance of communication between households also makes it easier for information about disaster events.

Meanwhile, disaster preparedness in Depok City is still very low include climate change effect management than the other cities. This condition is reflected by the inadequate condition of infrastructure and the lack of government's role in tackling disasters. This research believes that humanitarian interventions should as soon as possible include efforts to help affected people rebuild a disaster management and develop the capacity of local communities to recover if a new disaster would occur. With this approach, vulnerable people in hazard-prone areas allow building resilience by themselves. Therefore, strengthening the understanding of resilience at the community level becomes an important factor.

3.3. Alternative Approach in 4 Categories Areas

The inadequate condition of infrastructure and the lack of government's role in tackling disasters, leads people in Depok to manage their environment by their own “style” due to their capacity. Each community has local adaptive in managing floods and environmental degradation.

Based on the interview results, the coverage of the health center with house location is important. Access to local public health services center called *puskesmas* is a problem because there are differences in health care systems such as providing treatment to citizens KIS card (*Kartu Indonesia Sehat*) and so on. Some residents also stated to seek treatment tailored to the availability of facilities in the hospital or *puskesmas*. In addition, the public also often get regular health information counseling from *puskesmas* and government. This issue gives some impacts either from health aspects or material loss. Some efforts to overcome these problems are still done individually and communally. The absence of a maximum strategy with the local government to prevent the coming of the disaster.

The fulfillment of water needs in Depok City according to 4 research areas tend to give the same picture with the report given by the government. 90.25% of people use groundwater for toilet and 72% for cooking. Fulfilling the need for a drink is still dominated by gallon water as much as 58.75%. This illustrates that most people in Depok are still dependent on the groundwater used every day and it takes an additional cost to buy clean water. Waste management in 4 research areas tended to be done individually. As with garbage management, many of the 66.04% of the community don't have trash cans and dispose of the waste by burning or dumped it into the river. 31.13% of them have done communal management or have a good garbage disposal system [3]

Some forms of community role in tackling disasters can be seen from the level of the role of formal leaders in the scope of RT, RW, sub-district and local government. In addition, in interviews, some roles of informal and community leaders are also present in responding to and tackling when a hydrometeorological disaster occurs. Some of the roles of these institutions have been summarized and can be seen from the table below.

Table 2. The roles of the institution.

Institution	Forms of role
RT	Population data collection, *fogging*, evacuate during a disaster, community services program
RW	Community services program, *fogging*, food aid.
Sub District	Logistic assistance
Puskesmas	Health counseling and immunization, aid post during disaster
City Government	Logistic aids, health counseling, medical aid, flood embankment
NGO	Cleaning the trash in the river, incoming flood's information, evacuation
Mosque	Flood's announcement and temporary shelter
Other Party	Help in terms of borrowing money and medication.

Hydrometeorological hazards directly provide the various impacts that must be anticipated by the community. Based on interviews, some forms of adaptation and community mitigation of floods, landslides and droughts are as follows:

Table 3. Forms of adaptation and mitigation.

Hazard	Forms of adaptation and mitigation	
	Short term	Long-term
Floods	Refuge to a higher place	Elevate the foundation of the house
	Cleaning the trash in the river	Raised the terrace of the house / added floor of the house
		Build stilts house
		Build a water channel
Land Slide	Home improvement	Build a cemented foundation of the house
	Refuge to relatives/neighbors	Moving to lower area
Drought	Buy a water/gallon	Deepening the depth of the well
	Ask for water to relatives/neighbors	Relocating the water

4. Conclusion

The current local capacity in adapting to disaster and climate change impact in Depok city is still at the community level. Lack of preventive action been to force people to adapt individually and communally as to elevate the house, create a stronger and taller foundation and create waterways. The presence of a strongly linked familial relationship is also one of the adaptation and mitigation factors running as long as the community is struggling with disasters. In addition, the distance factor between the residence of a relative or family becomes important when the post-disaster mitigation period occurs.

The stronger the social interaction in the community, the stronger the level of adaptation to disasters including climate change effects. Factors of spatial closeness and the existence of reciprocal relationships facilitate the community in the process of adaptation and mitigation. It is also closely related to community resilience to disasters occurring in the environment.

Acknowledgment

The authors gratefully acknowledge that the present research is supported by Universitas Indonesia's Publication Grant for Thesis and Dissertation (Hibah PITTA UI).

References

[1] Rahman R, Shaw A, Surjan and G A Parvin "Urban Disasters and Resilience in Asia" *http://nikhildubey.com* 2016 [Online] Available: nikhildubey.com/library/download/asin=0128021691&type=sream [Accessed: 10-07-2017]

[2] Kita S M 2017 Urban vulnerability, disaster risk reduction and resettlement in Mzuzu city *Malawi International Journal of Disaster Risk Reduction* **22** pp 66-158

[3] Soemabrata J 2016 *Measuring the Risk of Hydro-Meteorological Hazard on Slum Village from Health Aspect, Citizens Adaptation Capacity, and Availability of Basic Infrastructure* (In) (Depok: Universitas Indonesia) pp 1-34

[4] State Ministry for Development Planning 2014 *Disaster Index Reduction in Indonesia* (In) (Jakarta: National Board for Disaster Management) pp 2-5

[5] Tampi D M 2016 *Bottom Up Approach. KANCIL, The Strive of Local Informal Communities* (In) (Jakarta: Urban Development Studies Universitas Indonesia) pp 2-12

[6] Lall S V and Deichmann U 2009 *Density and Disasters: Economics of Urban Hazard Risk* (Washington DC: Policy Research Working Paper 5161 World Bank) 1-4

[7] Thomas V 2014 Confronting climate-related disasters in asia and pacific (Ger) *Jahrbuch für Wirtschaftswissenschaften Review of Economics* **2** pp 31-121

[8] Amri M R *et al.* 2016 *Indonesian Disaster Risk* (In) (Jakarta: Indonesian National Board for Disaster Management) pp 2-34

[9] Djaja K *et al.* 2017 The integration of geographic information system (GIS) and global navigation satellite system-real time kinematic (GNSS-RTK) for land use monitoring *International Journal of Geomate* **13** pp 4-31

[10] Ibrahim N F *et al.* 2017 Identification of vulnerable areas to floods in Kelantan River Sub-Basons by using flood vulnerability Index *International Journal of Geomate* **12** pp 14-107

[11] Soemabrata J 2017 *Risk Mapping Studies of Hydro-meteorological Hazard in Depok Middle City* (Brisbane: Third International Conference on Science, Engineering & Environment) pp 1-7

[12] Komínková D 2012 The urban stream syndrome – a mini-Review *The Open Environmental and Biological Monitoring Journal* **5** pp 9-24

[13] Guihui Z, Liu S, Han C and Huang W 2014 Urban flood mapping for jiaxing city based on hydrodynamic modeling and GIS Analysis *Journal of Coastal Research* **1** pp 75-168

Genetic diversity of pigeon pea (*Cajanus cajan* (l.) Millsp.) based on molecular characterization using randomly amplified polymorphic DNA (RAPD) markers

N Khoiriyah[1], E Yuniastuti[2], and D Purnomo[2]

[1]Agronomy Department, Graduate School Program, Sebelas Maret University, Jl. Ir. Sutami 36A, Kentingan, Surakarta, Central Java 57126 Indonesia
[2]Agrotechnology Department, Agriculture Faculty, Sebelas Maret University, Surakarta, Central Java 57126 Indonesia

Abstract. Pigeon pea (*Cajanus cajan* (L.) Millsp.) is an annual leguminous crop (perennial) which has advantages over other local leguminous crops as drought resistant, hold collapsed and strong pods. The research on drought resistance plant is very important to adapt to climate change adverse impact to support food security. The potential of pigeon pie has not been supported by accurate data. To explore the potential of pigeon pea, it is necessary to record the important properties by characterization, one of which is molecular. Increasing genetic diversity can be done through mutation which widely used gamma ray for the induction. The purpose of this study was to identify the genetic diversity of pigeon pea of black, white and brown seeds type resulted by gamma-ray irradiation with a wavelength of 100, 200 and 300 grays by using RAPD method. The experiment resulted 14 bands, 12 of them are polymorphic bands and 2 of them are monomorphic with size varied from 300bp to 1.3kbp. The dendrogram showed from 30 accessions are divided into two main clusters, B shows clear genetical divergence from other clusters and some others split randomly. The range of similarity coefficient is from 0.43 to 1.00

1. Introduction

Climate change becomes one of the problems which quite influence on agriculture activities. It is clearly visible to notice an Indonesian climate change cycle. It switches off rainy and dry which influence on plant metabolism growth process and development especially in photosynthesis, transpiration, and respiration speed that is determined by plant production level.

To anticipate the climate change, a plant that tolerant toward this condition become necessary to develop. Pigeon pea has several advantages compared other leguminous, such as tolerant toward dryness source, resistant to collapse, pod not easy to crack, and suitable for any kinds of soil. *C. cajan* can live in dry area condition because it can adapt to the long rooting system [1]. Therefore, Pigeon pea plant has the potential for being developed in dry areas and rather barren, that is an area that cannot grow well soy [2].

Pigeon pea (*Cajanus cajan* (L.) Millsp.) is plant that contains good nutrition sources for family (every 100 gram seed contain 57 – 59% of carbohydrate, 14 – 30% of protein, 1 -9% of fat, and high vitamin A, B complex, and C); can be created for veggie friends eat, such made for soybean cake or cooked with milk squeezed from coconut; all parts of plant (root, steam, leaf, flower, and seed) can be established as natural medicine for treating many kinds of disease; it also can use as wood fire; its

Content from this work may be used under the terms of the Creative Commons Attribution 3.0 licence. Any further distribution of this work must maintain attribution to the author(s) and the title of the work, journal citation and DOI.
Published under licence by IOP Publishing Ltd

ability for binding nitrogen from air from root nodule that contain rhizobium that can help to repaire the soil fertility: and often used for reforestation plant especially in dry climate area [3].

Pigeon pea potential commodities are not supported yet by an accurate data such as peanuts, soy, and etc, unlike Pigeon pea genotype. Therefore, digging abilities, it needs an urgent characteristic data collection by operating characterization. Its function was for qualifying the plant diversity that has in the field as morphology characteristic diversity, agronomy, physiology, as well as molecular markers.

Utilizing expected molecular markers can contribute a high accuracy character description and enough on individual genetic diversity appearance, both at the species level or as well as distant relatives. RAPD is one of the molecular techniques for employing certain markers to study genetic diversity. Its basis analysis is to utilize a PCR machine which capable of amplifying DNA sequences in vitro. This technique involves single primary attachment with random nucleotide sequences [4].

The RAPD method is based on a homologous DNA sequence with the oligonucleotide primer sequence [5]. It has been used for numerous applications in plant molecular genetics research despite having disadvantages of poor reproducibility and not generally being associated with gene regions [6]. With the simplest and fastest detection technology, have been successfully employed for the determination of intraspecies genetic diversity in several grain legumes [7]. These include *Cajanus cajan* [2,8], *Vigna radiata* [9], *Phaseolus* sp. [10], *Glycine* sp. [11], and *Pisum* sp. [12]. Genetic variability of pigeon pea has been studied using several genetic markers such as RAPD [13,14,15], Amplified Fragment Length Polymorphism (AFLP) [16], and Restriction Fragment Length Polymorphism (RFLP) [17,18].

The application of gamma-ray plant recovery has huge advantages in developing new mutant varieties. 64% of 1585 varieties which were released since 1985 have been developed by practicing gamma-ray [19]. The x-ray and gamma-ray induction mutation were the most utilization for developing mutant varieties [20]. The irradiation gamma ray manipulating was one of increasing alternation to escalate genetic varieties of *C.cajan*. Mutation enlarges the genetic diversity opportunity of plant population. The relative relation among accession can provide information about the special character of each accession formed group. The information of among accession relatives can be used in plant breeding for determining its potentials which can be developed more.

The purpose of this study was to identify the genetic diversity of pigeon pea of black, white, and brown seeds type which was resulted by gamma-ray irradiation with a wavelength of 100, 200 and 300 grays by using RAPD markers.

2. Materials and Methods

Research materials were *C. cajan* leaf sample. It consists of brown guided seed type, white, and black which was radiated by gamma light with long of wave 100, 200, and 300 grays (table 1). The sampling method was carried out in accordance with the DNA sampling guide in the field.

Table 1. Sample collection of *C. cajan*

Population	Accession Name	Population	Accession Name	Population	Accession Name
Black	H0	White	P0	Brown	C0
Black	H1.1	White	P1.1	Brown	C1.1
Black	H1.2	White	P1.2	Brown	C1.2
Black	H1.3	White	P1.3	Brown	C1.3
Black	H2.1	White	P2.1	Brown	C2.1
Black	H2.2	White	P2.2	Brown	C2.2
Black	H2.3	White	P2.3	Brown	C2.3
Black	H3.1	White	P3.1	Brown	C3.1
Black	H3.2	White	P3.2	Brown	C3.2
Black	H3.3	White	P3.3	Brown	C3.3
Sum	**10**		**10**		**10**

2.1 Extraction and DNA Isolation

Extraction and DNA Isolation genome *C. Cajan* were based CTAB method that modified. The quantity of every DNA isolation result was measured by Fluorometer (Shimadzu UV1201), while its quality looked on electrophoresis 1,0%. DNA extraction result which producing a good enough quantity and quality DNA, and was continued by Polymerase Chain Reaction (PCR).

2.2 PCR Optimization and DNA Amplification

DNA genome *C. Cajan* was amplified by using a random primer that consists of 6 bases (decamer) (Operon Tech). PCR Optimization was done for getting PCR optimal condition. Several variables such as primer concentration, concentration DNA mold, and primer adhesive were exploited for PCR learning and trial for getting an optimal PCR product. To choose the primer will be used on RAPD analysis, every population *C. Cajan* represented by one individual that took at random and amplified using 6 primers from Operon Tech. (OPA 4, OPA5, OPB 17, OPC1, OPD 3, and OPD 5). Primer gave ribbons accretive amplification and clearly also producing polymorphic DNA ribbons of which was taken to amplify all of *C.cajan* DNA sample. DNA amplify was done based on Williams method [4] that modified. Furthermore, PCR was accomplished on total volume 10 µl every PCR reaction which consists of 5 µl PCR *mix* Go *Taq polymerase*, 0.25 µl 100 µMprimer (*Sigma-prologue*), 2.5 µl DNA sample (*template*), and 2.25 *nuclease-free water.*

The amplification DNA was operated using PCR equipment BIO-RAD system t100-Thermal Cycler. pre-denaturation did 94°C during 1 minute, then followed 45 cycles with temperature and time every cycle was denaturation on temperature 94°C during 30 seconds, annealing on temperature 37°C during 30 seconds, and elongation on 72°C during 1 minute 30 second, which followed by post elongation on 72°C during 7 minutes.

PCR DNA result then electrophoresed using 1.0 % (b/v) agarose that was added *florosafe* DNA *stain* as coloring, on TBE *buffer* (that consist from 0.45 M Tris-HCl pH 8, 0.45 M *Boric acid,* 20 mM EDTA) by voltage 100 volt during 45 minutes. The result was visualized by applying UV light.

2.3 Data Analysis

Base on data analysis which results by having or nothing DNA band. DNA profile ribbons were translated into Byner data utilizing provisions value 0 for nothing ribbons and 1 for having DNA ribbons on 1 position equal to a variety of *C. cajan* that compared. Analysis of colony (*cluster analysis)* done using program NTSYSpc (*Numerical Taxonomy and Multivariate Análisis System*) version 2.02i by method *Unweight Pair Group Method Arithmetic* (UPGMA) function SIMQUAL (*Similarity Qualitative*) [21]. Genetic similarity matric was counted based Dice coefficient.

3. Results and Discussion

3.1 Profile RAPD

The total DNA genome amplification results using two random primers (OPA 5 and OPB 17) at 30 accessions of *C. cajan* produced PCR outputs that could be read and scored, so the results could be analyzed (Fig. 1,2,3,4,5,6). The 2 primers produced 87.5% of polymorphism with OPA 5 and 83.3% of polymorphism with OPB 17 (Table 3). The experiment resulted 14 bands, 12 of them were polymorphic bands and 2 of them were monomorphic. The band's size was varied from 300bp to 1.3kbp. This result was RAPD markers which used to have a high level of polymorphism (> 50% polymorphic bands). Present study obtained 85.7% polymorphic band with an average 7 band per primer which was comparable to earlier studies on pigeon pea using RAPD marker [8,22]. The band's number which was generated after DNA amplification with PCR was highly dependent on how the primer recognizes the complementary DNA sequence in the DNA template (DNA template) used [23].

3.2 Genetic Diversity

Genetic similarities was measured through data analysis on the 14 RAPD markers revealed varying degrees of genetic relatedness among black, white, and brown seeds type and the genetic diversity which resulted by gamma-ray irradiation with a wavelength of 100, 200, and 300 grays. Dice similarity coefficients were between the *C. cajan* accession from 0.89 to 0.99. The value of this inequality shows that changes were not large but can be distinguished [24].

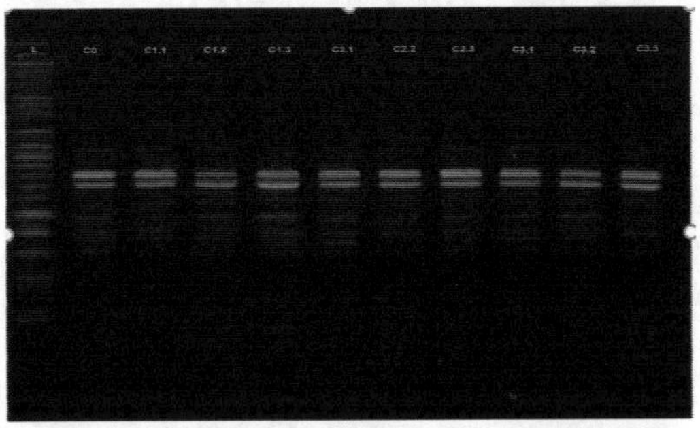

Figure 1. Amplification patterns of different pigeon pea produced brown seeds type accession by primer OPA 5

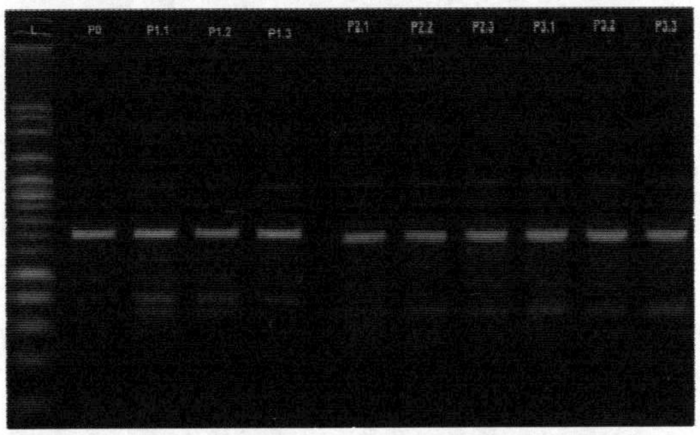

Figure 2. Amplification patterns of different pigeon pea produced white seeds type accession by primer OPA 5

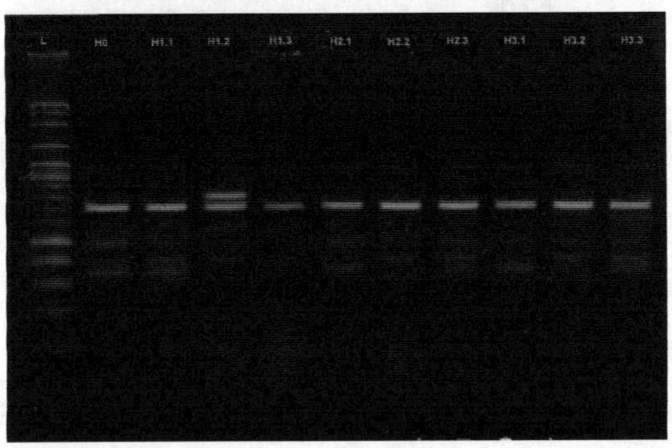

Figure 3. Amplification patterns of different pigeon pea produced black seeds type accession by primer OPA 5

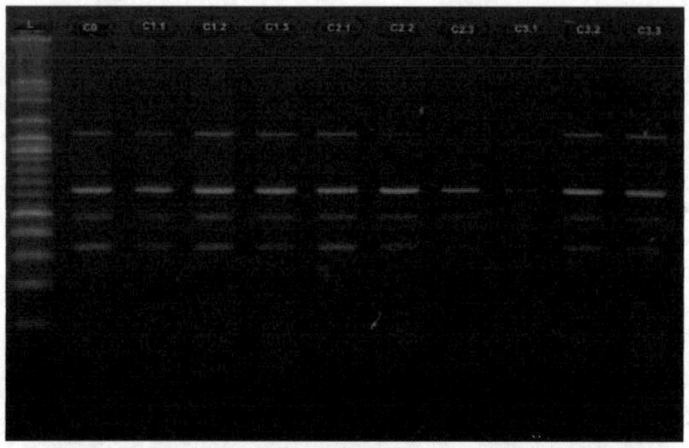

Figure 4. Amplification patterns of different pigeon pea produced brown seeds type accession by primer OPB 17

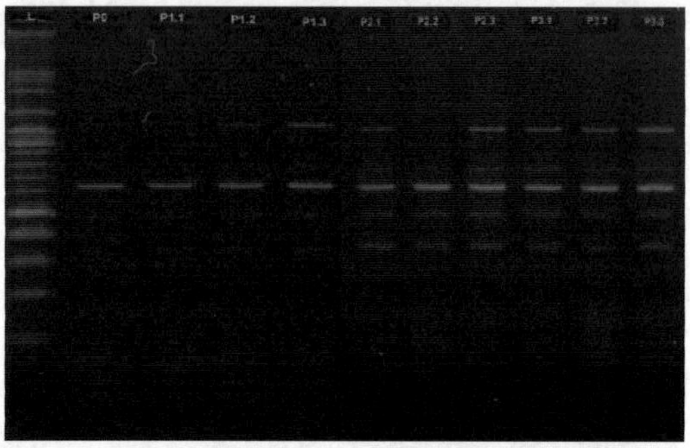

Figure 5. Amplification patterns of different pigeon pea produced white seeds type accession by primer OPB 17

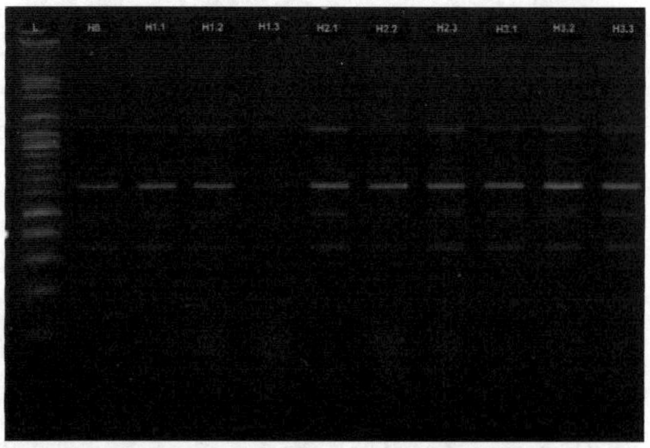

Figure 6. Amplification patterns of different pigeon pea produced black seeds type accession by primer OPB 17

3.3 Analysis cluster between accession

The phylogenetic dendrogram results obtained using NTsys software pc 2.2 from 27 accessions of *C. cajan* was divided into two main groups. Distance similarity coefficients were achieve 0.43 to 1.00 by splitting into at random.

Cluster analysis of genetic similarity in 30 accessions of *C. cajan* shows accessions separation into two main clusters, namely (A) similarity coefficient 0.69 and (B) similarity coefficients of 0.43. A cluster consists of two sub-cluster, namely sub-cluster which consists of C (which consists of E and F). Cluster F split randomly. While the cluster D consists of C31 (fig 7).

It shows the genetic diversity exists between individuals and populations. The first prediction was polymorphic loci which used in the high polymorphism analysis that have been selected. Due to genetic mutations in gamma-ray irradiance with different wavelengths was resulting in considerable genetic diversity. The same was also reported by Fox [25] by using isozyme. The high rates effect of genetic inequality because of several things. The first prediction was that the polymorphic loci which were used in this analysis have been selected which has a high polymorphism.

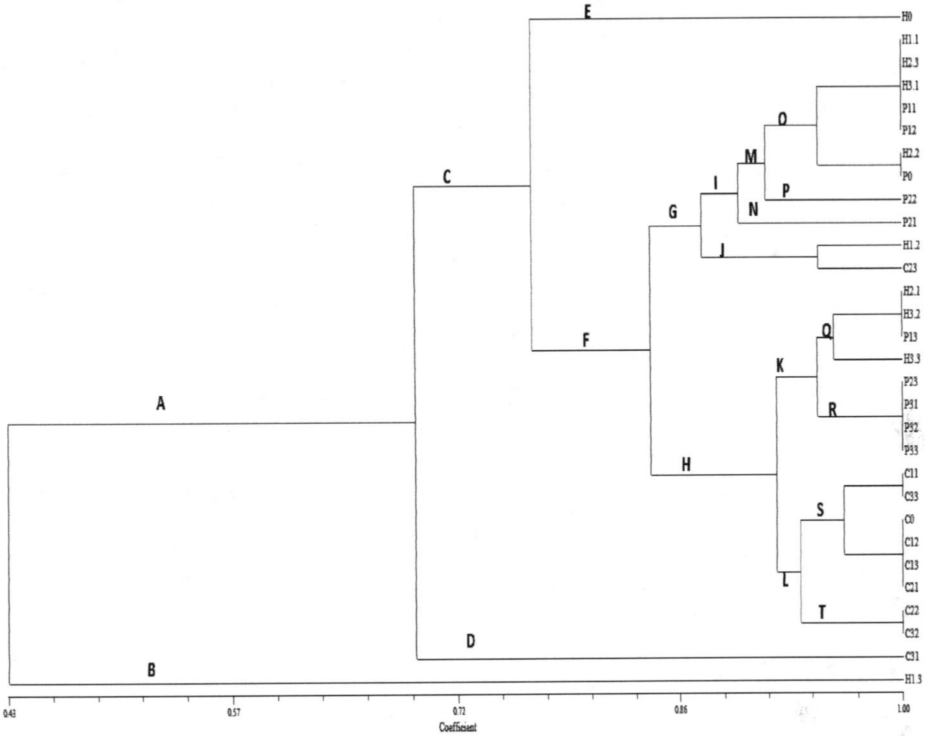

Figure 7. Dendrogram of 30 accession pigeon pea showing genetic similarity

Table 2. Random primer selected for RAPD analysis result of amplification

Primers	Average size of the fragments amplified	Total number of bands amplified	Total number of monomorphic bands	Total number of polymorphic bands	Percentage of polymorphic bands
OPA 5	300-1100	8	1	7	87.5
OPB17	325-1300	6	1	5	83.3

4. Conclusion

Genetic observations *C. cajan* by using two RAPD primer shows that 30 accessions that were exploited in the DNA fragment research have 14 bands with the size of 300bp until 1.3 kb. From 14 bands, 12 of them were polymorphic bands and 2 of them were monomorphic. The two primers produced 87.5% of polymorphism with OPA 5 and 83.3% of polymorphism with OPB 17. Cluster analysis of genetic similarity in 30 accessions of *c. cajan* displays separation of accession into the two main cluster, one of them shows clear genetical divergence from other clusters and some others split randomly. Based on the results of these studies can be concluded that to anticipate the climate changes which required plant varieties that are tolerant to these conditions. Pigeon pea has advantages such as drought resistant. Induction mutations with x-rays and gamma rays were used for the mutant varieties development in escalating the genetic diversity which employed as plant breeding agents and to produce more DNA markers that can be used to generate varieties that are tolerant to the climate change.

Acknowledgments
Thank you for Mrs. Dr. Endang Yuniastuti SP., M.Si has sponsored research fund.

References

[1] Cook B G, Pengelly B C, Brown S D, Donnelly J L, Eagles D A, Franco M A, Hanson J, Mullen B F, Partridge I J, Peters M, and Schultze-Kraft R 2005 *Tropical Forages: an interactive selection tool Cajanus cajun* (Brisbane: CSIRO, DPI and F(Qld), CIAT, and ILRI)
[2] Damor P R, Narwade A V, Jadav S K and Viradiya Y A 2016 Molecular Characterization of six pigeonpea varieties for drought tolerance by using RAPD markers *International Journal of Agriculture, Environment and Biotechnology Citation* **9(5)** 725-30
[3] Mahalaya S, Aris T S, Alberth S, Luther K dan Nakeus M 2015 *Cultivation techniques and livestock production* (Canberra: International Potato Center (CIP) and Australian Centre for International Agricultural Research (ACIAR))
[4] Williams J G K, Kubelik A R, Livak K J, Rafalski J A, and Tingey S V 1990 DNA polymorphisms amplified by arbitrary primers useful as genetic markers *Nucl. Acids Res* **18** 6531-35
[5] Waugh R 1997 *RAPD Analysis: Use for Genome Characterization, Tagging Traits and Mapping. Plant Molecular Biology a Laboratory Manual* (Berlin: Springer-Verlag) pp 154-75
[6] Welsh J and McClelland M 1990 Fingerprinting genomes using PCR with arbitrary primers *Nucleic Acids Research* **18** 7213-18
[7] Karp A, Edwards K, Bruford M, Funk S, Vosman B, Morgante M, Seberg O, Kremer A, Boursot P, Arctander P, Tautz D and Hewitt G 1997 Molecular technologies for biodiversity evaluation: opportunities and challenges *Nature Biotechnology* **15**625-28
[8] Yadaf K, Anurag Y, Veda P P, Upendra N D 2012 Genetic diversity of pigeon pea (*Cajanus cajan* (L.) Millsp.) cultivars and its wild relatives using randomly amplified polymorphic DNA (RAPD) markers *American Journal of Plant Sciences* **3** 322-30
[9] Souframanien J and Gopalakrishna T 2004 A comparative analysis of genetic diversity in blackgram genotypes using RAPD and ISSR markers *Theoretical and Applied Genetics* **109** 1687-93
[10] Beebe S, Skroch P W, Tohme J, Duque M C, Pedraza F and Nienhuis 2000 Structure of genetic diversity among common bean landraces of Middle American origin based on correspondence analysis of RAPD *Crop Science* **40** 264-73
[11] Jeffrey A, Thompson, Randall L and Nelson 1998 Identification of diverse soybean germplasm using RAPD markers *Crop Science* **38** 1348-55
[12] Cheghamirza K, Koveza O, Konovalov F and Gostimsky S 2002 Identification of RAPD markers and their use for molecular mapping in Pea (*Pisum sativum* L.) *Cellular and Molecular Biology Letters* **7** 649-55
[13] Kumar J, Choudhary A K, Solanki R K and Pratap A 2011 Towards Marker-Assisted Selection in Pulses: A Review *Plant Breeding* **130** 297-313
[14] Souframanien J, Manjaya J G, Krishna T G and Pawar E S 2003 Random amplified polymorphic DNA analysis of cytoplasmic male sterile and male fertile pigeon pea (*Cajanus cajan* (L.) Millsp.) *Euphytica* **129** 293-99
[15] Nadimpalli B G, Jarret R L, Phatak S C and Kochert G 1992 Phylogenetic relationships of the pigeon pea (*Cajanus cajan*) based on nuclear restriction fragment length polymorphism *Genome* **36** 216-23
[16] Yadav K, Singh B D, Srivastava C P, Chand R and Yadav A 2010 Analysis of genetic

divergence in pea (*Pisum sativum* L.) using quantitative traits and RAPD markers *Indian Journal of Genetics and Plant Breeding* **70** 363-69.

[17] Reddy L J, Upadhyaya H D, Gowda C L L and Singh S 2005 Development of a core collection in pigeon pea [*Cajanus cajan* (L.) Millspaugh] using geographic and qualitative morphological descriptors *Genetic Resource and Crop Evolution* **52** 1049-56

[18] Lakshmi M, Senthikumar P, Parani M, Narayana J and Parida A K 2000 PCR-RFLP analysis of chloroplast gene regions in *Cajanus* (Leguminosae) and allied genera *Euphytica* **116** 243-50

[19] Maluzynski M K, Nichterlein L, Van Zanten, and Ahloowalia B S 2000 Officially released mutant varieties the FAO/IAEA database *Mut. Breed. Rev.* **12** 1-84

[20] Ahloowalia B S, Maluzynski and Nichterlein 2004 Global impact of mutation-derived varieties *Euphytica* **135** 187-204

[21] Rohlf F J 1997 *NTSYS-pc Version 2.02i Numerical Taxono- my and Multivariate Analysis System* (New York: Applied Biostatistics Inc, Exeter Software)

[22] Saker M M, Youssef S S, Abdallah N A, Bashandy H S and Sharkawy M E 2005 Genetic analysis of some Egyptian rice genotypes using RAPD, SSR and AFLP *African Journal of Biotechnology* **4** 882-90.

[23] Tingey S V, Rafalski J A and Hanafey M K 1994 Genetic analysis with RAPD markers *Plant Molecular Biology* 491-98

[24] Sulistianingsih R, Purwantoro A, Mangoendidjojo W and Semiarti E 2012 Genetic variation of natural orchid *Phalaenopsis amabilis* (L.) Blume produce gamma-ray irradiation *A Scientific Journal for The Applications of Isotopes and Radiation* **8** 1-10

[25] Fox J E D, Brand J E, Barret D R and Effendi M 1994 Genetic variation in Santalum album in Timor. *Proceedings of a Regional Workshop for Pacific Island Countries* 93-110

IOP Conf. Series: Earth and Environmental Science **129** (2018) 012017 doi:10.1088/1755-1315/129/1/012017

Mapping the rainfall distribution for irrigation planning in dry season at pineapple plantation, Lampung Province, Indonesia (Study case at Great Giant Pineapple Co. Ltd.)

P Cahyono[1], N K Astuti[1], Purwito[1] and A Rahmat[2]

[1]Department of Research and Development, Great Giant Pineapple Company, Jl. Raya Arah Menggala Km 77, Terbanggi Besar, Lampung Tengah, Terbanggi Besar, Central Lampung Regency, Lampung 34163 Indonesia
[2]The United Graduate School of Agricultural Science, Gifu University, 1-1 Yanagido, Gifu, Gifu Prefecture 501-1193, Japan

Abstract. One of the problems caused by climate change is unpredictable of the dry season. Understanding when the dry season will start is very important to planning the irrigation schedule especially on large plantation. Data of rainfall for 30 years in Lampung, especially in Pineapple Plantation show that dry month occurs from June to October. If in two *decadals (ten days period)* rainfall less than 100 mm then it is predicted that next decadal will be dry season. Great Giant Pineapple Co. Ltd. has 32,000 hectares plantation area and located in three regencies at Lampung Province, Indonesia with varies rainfall between regions within a plantation. Therefore, monitoring the rainfall distribution by using ombrometer installed at 10 representative location points can be used to determine irrigation period at the beginning of dry season. Mapping method using the server program and data source is from 10 monitoring rainfall stations installed at the observed points. Preparation of rainfall distribution mapping is important to know the beginning of the dry season and thus planning the irrigation. The results show that 2nd decadal of April is indicated as the starting time of dry season, which is similar with Indonesian government for climate agency's result.

1. Introduction

Pineapple (*Ananas comosus* Merr.) is one of the commodity with large planting area and high production in Indonesia. This makes pineapple become the important economic plants in the tropic area [1,2]. Great Giant Pineapple Co. Ltd. (GGP) is the biggest pineapple exporter in Indonesia [3], with the total plantation area around 32,000 hectares with effective pineapple planting area approx. 20,000 hectares. GGP has a large area and belongs to the different regencies at Lampung Province. Climate change and global warming, in many ways, have adverse effects on the environment and humankind [4,5]. One of the indications of climate change is increases in the earth surface temperature in global scale [5] extreme weather condition and also changing in the precipitation pattern. Indonesia have dry season which is irrigation activities is needed to growing plant. Irrigation on pineapple plant is very important because it affects the growth and production [3].

The limitation of equipment in a large plantation area and unpredicted when the dry season will start caused by climate change, makes irrigation management and planning become very vital to support good production. Priority scale must be considered due to the large area with heterogeneous climatic condition. Moreover, the global climate change nowadays also influences the shifting of wet and dry season [6]. Mapping the distribution of rainfall will be important as base information to make

Content from this work may be used under the terms of the Creative Commons Attribution 3.0 licence. Any further distribution of this work must maintain attribution to the author(s) and the title of the work, journal citation and DOI.
Published under licence by IOP Publishing Ltd

a decision on irrigation scheduling and to understand when dry season starts especially under global climate change issue recently.

2. Methods

The experiment was conducted in Great Giant Pineapple plantation, Central Lampung, Indonesia. The coordinate of Great Giant Pineapple is 4°49'07"S latitude and 105°13'13"E longitude, with elevation 46 meters above sea level. Precipitation data were collected from 10 locations (Div II, Div III, Div IV, Div V, Div VI, PG2, Lab PG3, R&D, Kijung, Lakop). Observation time was April 2017 (*decadal* 1 (1-10 April), *decadal* 2 (11-20 April), and *decadal* 3 (21-30 April) and supported with 30 years climatological data from meteorology station belongs to the company. *Decadal* (Dasarian) is 10 days period used by Indonesian government for climate determination.

3. Results and discussions

Figure 1 shows the water of pineapple in a different age of plant. Water balance shows the availability of water for a year which calculate the factors affect water availability (input and output), for the months of water deficit and surplus. Figure 1 and Table 1 show that pineapple plantation had water surplus in January - May, and November - December. This means that the surplus months occur for 7 months, whereas water deficit occurs for 5 months duration, from June to October with the peak in August.

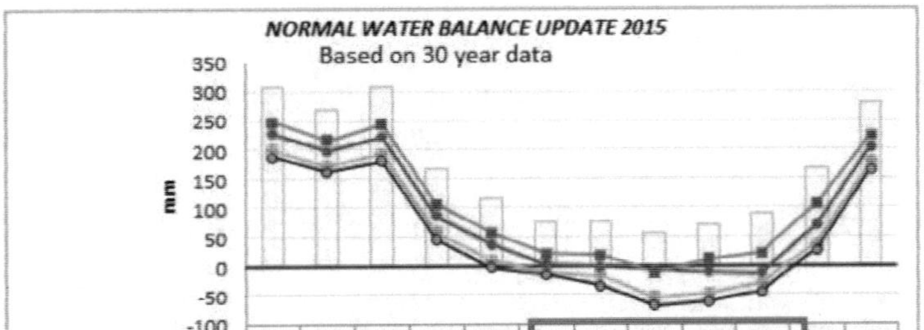

	jan	feb	mar	apr	may	jun	jul	aug	sep	oct	nov	dec
CH-P80%	307,4	269,6	306,0	166,2	116,5	74,3	74,3	56,8	70,3	87,9	65,6	279,3
< 3 MAT	198,9	171,8	191,6	57,7	9,7	11,8	-17,2	-54,9	-48,6	-32,2	7,9	176,4
4 - 10 MAT	226,6	196,8	220,8	85,4	37,0	3,3	-0,7	-6,2	-11,2	-13,2	58,8	202,7
> 10 MAT	247,1	215,3	242,5	105,9	57,2	21,3	18,4	-12,5	10,7	21,3	05,6	222,1
referen crop (Eto)	186,9	161,0	178,9	45,6	-2,2	14,8	-34,9	-68,3	-61,9	-45,4	25,7	164,9

Figure 1. Monthly water balance of pineapple in a different age of the plant.

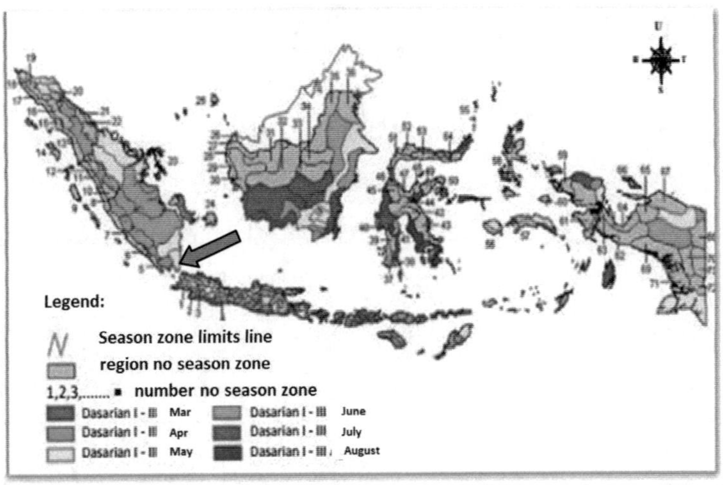

Figure 2. Map of the average starting time of dry season based on 30 years data (BMKG, 2011).

IOP Conf. Series: Earth and Environmental Science **129** (2018) 012017 doi:10.1088/1755-1315/129/1/012017

Table 1. Pineapple water requirement.

Month	Evaporation		Pineapple water requirement			
			Crop factor = 0,5 (small plant)		*Crop factor* = 0,3 (big plant)	
	mm/mth	mm/day	mm/mth	mm/day	mm/mth	mm/day
Jan	100.0	3.2	48.4	1.6	29.0	1.0
Feb	95.0	3.4	50.5	1.7	30.3	1.0
Mar	98.8	3.2	47.8	1.6	28.7	1.0
Apr	106.6	3.6	53.3	1.8	32.0	1.1
May	111.8	3.6	54.1	1.8	32.5	1.1
Jun	109.1	3.6	54.6	1.8	32.7	1.1
Jul	117.5	3.8	56.9	1.9	34.1	1.1
Aug	139.7	4.5	67.6	2.3	40.6	1.4
Sep	146.0	4.9	73.0	2.4	43.8	1.5
Oct	143.3	4.6	69.3	2.3	41.6	1.4
Nov	109.0	3.6	54.5	1.8	32.7	1.1
Dec	108.9	3.5	52.7	1.8	31.6	1.1

The movement of season starting time can easily indicated by the change of the precipitation pattern. Figure 2 shows that Province of Lampung has two different starting time of the dry season: around April and May, respectively. According to Figure 3,the starting time of the dry season moved forward 1-2 *decadal* in Lampung Province. The rainfall distribution maps during April in the plantation which refers to Figure 2 are presented in Figure 5.

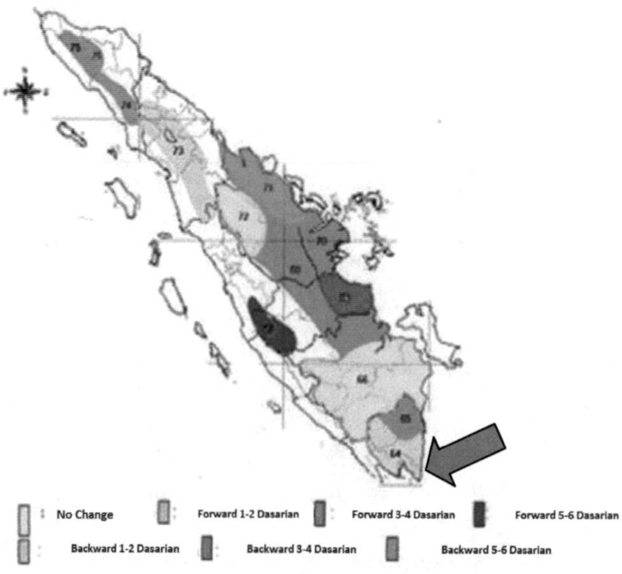

Figure 3. Movement of starting time of dry season based on 30 years data (BMKG 2011).

The provision of irrigation on a very large pineapple plantations is not the same between one location and another location. Therefore, on this 32,000 ha plantation, 10 rainfall monitoring stations were installed in representative points.

Figure 5 shows that mapping the rainfall distribution from *decadal* to *decadal* obtained there were still enough rain in the first *decadal* of April at all stations (more than 50 mm / *decadal*). Then the second *decadal* shows a drastic drop of rainfall, where all stations indicated the accumulated rainfall data were very low (less than 50 mm / *decadal*). *Therefore, decadal* 2 of April is indicated as the beginning of dry season.

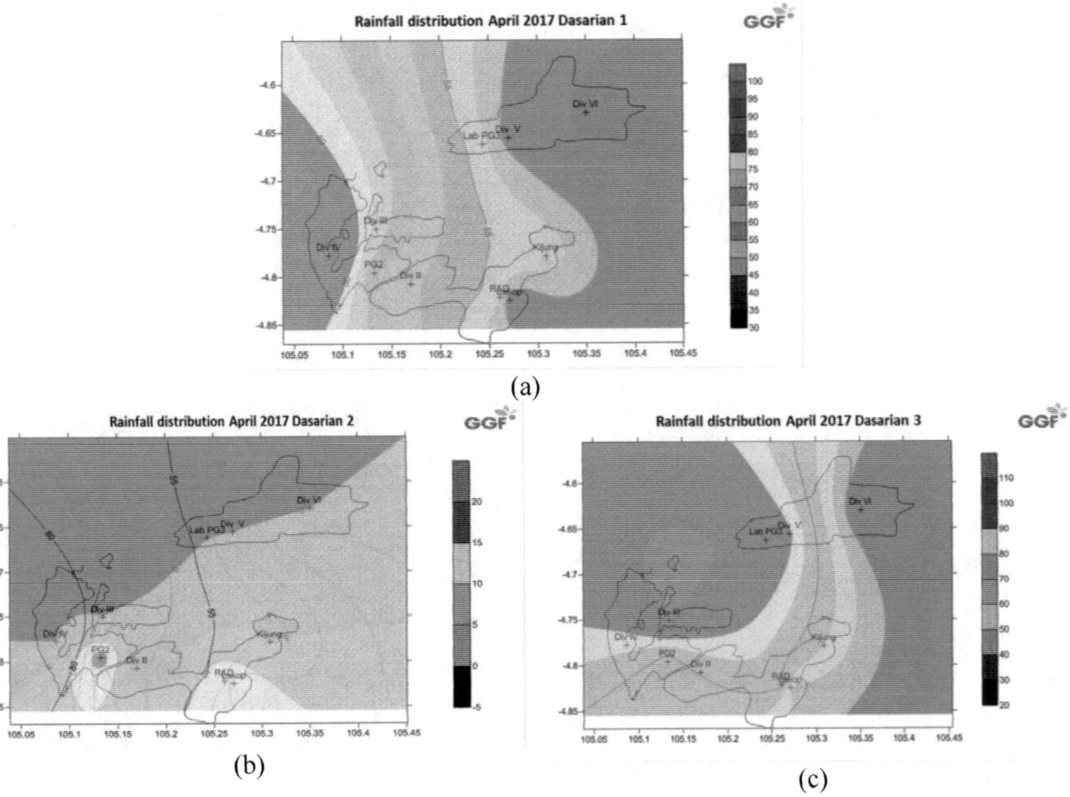

Figure 4. Mapping of rainfall distribution in pineapple plantation April (a) *Decadal* 1, (b) *Decadal* 2, and (c) *Decadal* 7.

4. Conclusion

In the pineapple plantation of Lampung Province, Indonesia, water surplus occurred for 7 months and water deficit occurs for 5 months (June to October) with the deficit peak is in August . 2nd *decadal* of April is indicated as the starting time of the dry season. By this results, the anticipation of climate change impact on water deficit on pineapple plantation can be determined

Acknowledgement

The authors thanks the Research Center and Development of Great Giant Food Company to support this research.

References

[1] Pornsuriya C, Wang H K, Lin F C and Soytong K 2008 First report of pineapple root tot caused by *Phythium graminicola* *J. Agri. Tech.* **4(1)** pp 139-50

[2] Martin D A N and Rahmat A 2017 Relationship of soil psycochemical properties and existence of Phytophtora sp in pineapple plantations *Indonesian J. Sci. Tech.* **2(1)** pp 81-6

[3] Rahmat A, Afandi, Manik T K and Cahyono P 2013 Effects of irrigation and cassava peel mulch on soil water content and pineapple growth *J. of Irigation* (in Indonesian) **8(2)** pp 99-114

IOP Conf. Series: Earth and Environmental Science **129** (2018) 012017 doi:10.1088/1755-1315/129/1/012017

[4] Ahmad N N N and Hossain D M 2015 Climate change and global warming discourses and disclosures in the corporate annual reports: A study on the Malaysian companies. *Procedia Soc. Behav. Sci* **172** pp 246-253.

[5] Rahmat A and Mutolib A 2016 Comparison air temperature under global climate change issue in Gifu city and Ogaki city, Japan. *Indonesian J. Sci. Tech.* **1(1)** pp 37-46

[6] Aldrian E, Karmini M and Budiman 2011 *Climate Change Adaptation and Mitigation in Indonesia* (In) (Jakarta: Indonesian Meteorology Climatology and Geophysical Agency (BMKG))

IOP Conf. Series: Earth and Environmental Science **129** (2018) 012018 doi:10.1088/1755-1315/129/1/012018

Evaluation of cropping pattern in rainfed areas based on studies of *pranata mangsa* and weather dynamics

M K Zaki[1,4], N T Furi[2], Jauhari Syamsiyah[3] and Sumani[3]

[1]Agronomy Dept., Graduate School Program, Universitas Sebelas Maret, Ir. Sutami Street Number 36A, Surakarta, Indonesia
[2]Agrotechnology Dept., Agriculture Faculty, Universitas Sebelas Maret, Ir. Sutami Street Number 36A, Surakarta, Indonesia
[3]Soil Science Dept., Agriculture Faculty, Universitas Sebelas Maret, Ir. Sutami Street Number 36A, Surakarta, Indonesia
[4]Corresponding author: Email: zakimuhamad30@gmail.com

Abstract. Weather dynamics such as the fifth time of the rainy season and drought are becoming more frequent. These conditions pose a significant impact on the strategies of cultivation such as cropping pattern and crop yields, especially in rainfed areas. One of the steps that can be taken is to return to local wisdom, such as *pranata mangsa*. This study aimed at analyzing the relationship of the variability of precipitation in rainfed areas with pranata mangsa and then to evaluate cropping patterns based on the result of the analysis. The study was conducted in rainfed areas of the District of Jumantono, Karanganyar Regency; and District of Teras and District of Ampel, Boyolali Regency in June until December 2014. The research method is a descriptive exploratory survey with purposive sampling based on moderate altitude (200-700 masl). The types of data that are used are primary and secondary. Data analysis was used correlation test. The results showed that precipitation in rainfed areas has a close relationship with paranata mangsa. These results explain that pranata mangsa still relevant to be used even though it has happened weather dynamics.

1. Introduction

Dynamics weather such as duration of rainy and dry seasons are more frequent. These conditions have a significant impact on cultivation strategies and agricultural production [1], especially rainfed agriculture that directly utilizes rainwater as an irrigation media. The agriculture anticipation and adaptation strategies are needed to keep conscious. Its climate change can be accomplisjed by shifting the planting period, changing the plants variation, switching the cropping pattern, and shuffling the planting place and location [2]. Cropping pattern judgement using rainfall data needs to be operated in order to adjustan appropiate condition. According to Gomez [3], the cropping pattern needs to suit to rainfall conditions to reduce air deficit in the dry season.

One way to react to the climate change is back to local wisdom in agricultural land managing included in cropping patterns decision. One of them that still operated is *pranata mangsa*. According to Trewartha and Horn [4], information about the *pranata mangsa* is quite necessary for farmers to avoid to choose the planting pattern in the rainfed land.

Cropping pattern that is usually exploited which based on *pranata mangsa* is not considered relevant anymore to the weather dynamics occurrence, therefore it is necessary to conduct a research which related to the cropping pattern evaluation based on *pranata mangsa* and weather dynamics. This

Content from this work may be used under the terms of the Creative Commons Attribution 3.0 licence. Any further distribution of this work must maintain attribution to the author(s) and the title of the work, journal citation and DOI.
Published under licence by IOP Publishing Ltd

research aims to analyze the relation of rainfall variability with the *pranata mangsa* application in rainfed agriculture and then evaluate cropping patterns based on the results analysis.

2. Materials and method

2.1. Materials

The research was operated from June to December of 2014 in rainfed agriculture area,medium plain (200-700 masl), Jumantono District, Karanganyar Regency; Teras and Ampel District, Boyolali Regency. Local weather dynamics analysis was supervised by looking at daily patterns of local rainfall, air temperature, and humidity based on descriptive statistical results. Rainfall variability was investigated by employing average data per decade rainfall over the last 30 years (1985- 2014). Analysis of local weather dynamics and rainfall variability were accomplished by utilizing Microsoft office excel. *Pranata mangsa* evaluation was done by practicing correlation test between rainfall area to *pranata mangsa* using SPSS 16.0 software with $\alpha = 0.05$. The timing and cropping patterns were determined employing the evaluation data of *pranata mangsa* institutions.

2.2. Method

The research was conducted by the descriptive exploratory method through survey and interview to public figures (PPL, Chairman of Gapoktan) and farmer in farmland. Local weather measurements (air temperature, humidity, and rainfall) were performed by observation method. The data type which employed was primary and secondary. The primary was obtained from local weather observations. Secondary data was earned from various institutions that provide daily rainfall data during the last 20-30 years (1985-2014) such as Jumantono Climatology Station, Territory of Bengawan Solo River Region, and Agricultural Extension Agency of Ampel District.

3. Results and discussions

3.1. Location characteristics

Current site was mostly used as rice field, tegalan, gardens, and settlements. Types of soil in most research sites were Vertisols and Alfisols. Based on Table 1 it can be seen that the average monthly rainfall in Kecamatan Jumantono, Karanganyar District was 187.87 mm/month. Whereas, average monthly rainfall in Kecamatan Teras, Kabupaten Boyolali was 184.07 mm/month. While, average monthly rainfall in Ampel District, Boyolali District was 204.87 mm/month.

Table 1. Rainfall Condition in Jumantono, Teras, and Ampel (1985-2014).

Lokasi	Jan	Feb	Mar	Apr	Mei	Jun	Jul	Agst	Sep	Oct	Nov	Des	Avarage
Jumantono	371.8	366.3	304.9	231.5	108.9	70.9	55.8	13.9	25.4	120.7	268.7	315.6	187.87
Teras	411.4	378.6	338.7	200.9	118.3	66.4	44.4	22.3	20.7	95.4	206.9	304.8	184.07
Ampel	500.3	409.6	344.0	234.9	120.9	126.8	48.1	21.6	28.8	106.6	202.3	314.5	204.87

Source: Data Analysis

Based on Oldeman climate classification, it can be seen that Jumantono Subdistrict, Karanganyar District, and Teras Subdistrict, Boyolali Regency was entered into type C3 climate while Ampel District, Boyolali Regency was classified into type C2 climate. Type C2 and C3 climates can be implemented on rice cultivation for one growing season and 2 crops for two cropping seasons, with second crop palawija not to fall in dry months [5].

3.2. Dynamics local weather and rainfall variability

Ampel sub-district, Boyolali regency has higher average temperature compared to Jumantono, Karanganyar and Teras sub-districts, Boyolali regency because it is located on higher area. Handoko [6] states that earth's surface temperatures are lower with increasing latitude as well as temperature decreased by altitude. The vertical spread of the earth surface temperature was an warming origin, therefore the higher the place will be the lower the temperature. District Jumantono, Karanganyar District has high temperatures with low air humidity, yet high rate of rainfall. This shows the dynamics weather due to the circumstances that occurred in District Jumantono contrary to normal situation. Under normal condition, it will be accompanied by high air humidity and low temperatures. Its position was depend on the temperature which determined the air capacity to accommodate moisture and the actual moisture content in place. In wet and hot evaporation areas will be resulting a high density of humidity, while mountainous areas in Indonesia are generally have high humidity because the site temperature is low, therefore the air capacity to hold water vapor is relatively small [6].

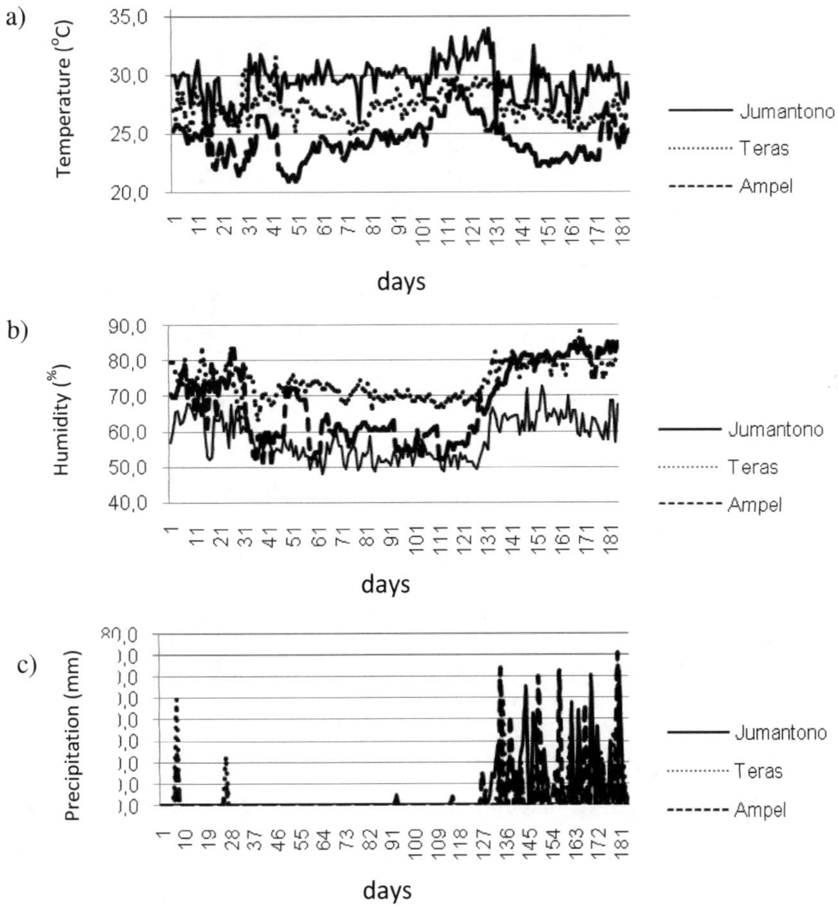

Figure 1. The local condition of a) Temperature, b) Humidity, c) Rainfall in July-December 2014.

3.3. Analysis of pranata mangsa between rainfall pattern

Pranata Mangsa is a local wisdom to determine the season calculation. Each listed season in the calendar is based on natural events. The *pranata mangsa* farmers utilization by them is now quite difficult to locate because most of them are considered that the *pranata mangsa* is not relevant to the shifting seasons occurrence. According to Simanjuntak *et al.* [7], the existing *pranata mangsa* system cannot be entirely guided for establishing the growing season initial period due to the climate change, yet the *pranata mangsa* can be evaluated because it is not a rigid pattern.

Table 2. Correlation between *pranata mangsa* and rainfall characteristic.

Location	*Pearson Correlation*	Sig.
Jumantono	0.874**	0.000
Teras	0.886**	0.000
Ampel	0.857**	0.000

Source: Data Analysis

Based on Table 2, it can be seen that the *pranata mangsa* by occurred rainfall in each study area has close relationship. The *pranata mangsa* assumption order was not relevant to the weather dynamics which was not in accordance with the analysis results. The analysis results of Table 2 explained that the *pranata mangsa* was still relevant and applicable despite weather dynamics and rainfall variability.

Pranata mangsa and rainfall region have a direct relationship. *Pranata Mangsa* rainfall pattern needs to be considered in cropping pattern application because it shifting can result in the groundwater availability of which greatly plant growth influences. Types selection and varieties of cultivated crops was also one of the factors that must be considered on *pranata mangsa* calendar.

3.4. Crop pattern evaluation

Planting time selection should be precisely performed due to cropping time and pattern were the agricultural production system spearhead. It were established by rainfall conditions. Water and rainfall systems greatly influenced the cropping pattern shape that can be developed [8] Evaluation of planting schedules and patterns at study sites was based on *pranata mangsa* and weather dynamics studies were presented: a) Jumantono District, Karanganyar Regency in the first season (Paddy - Horticulture - Bero), second season (Vegetables - Rice - Palawija - Bero), and third season (Rice - Palawija - Bero); b) Teras Sub-District, Boyolali Regency in the first season (Horticulture - Padi - Palawija - Bero), second season (Paddy - Horticulture - Bero), and third season (Rice - Palawija - Bero); c) Ampel District, Boyolali District in the first season (Vegetables - Rice - Palawija - Bero), second season (Rice - Horticulture - Bero), third season (Rice - Palawija - Bero).

4. Conclusions

High rainfall variations which were occurred in rain-fed areas such as District Jumantono, Karanganyar District; Teras and Teras sub-districts, Boyolali district has a close relationship with *pranata mangsa,* therefore *pranata mangsa* is still relevant to be applied despite the weather dynamics such as rainfall variability.

Acknowledgment

The authors gratefully acknowledged that the present research was supported by Higher Education Science and Technology Development Grant funded by Indonesia Ministry of Research and Higher Education.

References

[1] Irawan B 2006 *Climate Anomaly Phenomenon (El Nino and La Nina); Long-Term Trend and Its Influence on Food Production* (Bogor: Indonesian Center for Agriculture Socio-Economic and Policy Studies) pp 1-2

[2] Miranda T, Deny H, Herry Y, Gutomo B A and Ali Y A 2011 *Farmers Adaptation of Climate Change Conditions in Agricultural Management* (Jakarta: Lembaga Ilmu Pengetahuan

Indonesia (LIPI)) pp 3-5
[3] Gomez K A and Gomez A A 1983 *Procedure Statistic for Agriculture Research* (Philippines: University of Philippines) pp 23-7
[4] Trewartha G T and Horn L H 1995 *Introduction to Climatology* Fifth Edition (Yogyakarta: UGM Press) pp 17-9
[5] Dwiyono H 2009 *Meteorology and Climatology* (Malang: Universitas Negeri Malang) pp 31-3
[6] Handoko 1995 *Basic of Climatology* Second Edition (Jakarta: Dunia Pustaka Jaya) pp 2-6
[7] Simanjuntak B H, Sri Yulianto J P and Krsitoko D H 2010 *Preparation of New Model Pranata Mangsa Based on Agrometeorology using LVQ (Learning Vector Quantization) and MAP Alov for Effective Cropping Planning*. Completed Year 1 Competitive Grant Report (Salatiga: Universitas Satyawacana) pp 1-7
[8] Sutidjo D 1986 I*ntroduction to Agronomy Production System* (Bogor: Bogor Agricultural Institute) pp 3-8

IOP Conf. Series: Earth and Environmental Science **129** (2018) 012019 doi:10.1088/1755-1315/129/1/012019

Assessment of the Standardized Precipitation Index (SPI) in Tegal City, Central Java, Indonesia

Y Pramudya[1], T Onishi[2]

[1]The Graduate School of Natural Science and Technology, Gifu University 1-1 Yanagido, Gifu, Gifu Prefecture 501-1193, Japan
[2]Department of Applied Biological Science, Gifu University 1-1 Yanagido, Gifu, Gifu Prefecture 501-1193, Japan

Abstract. One of the adverse impacts of climate change is drought, which occurs more frequently in Tegal city, Indonesia. The application of drought index analysis is useful for drought assessment to consider adaptation and mitigation method in order to deal with climate change. By figuring out the level and duration of the drought. In order to analyze drought in the specific area, Standardized Precipitation Index (SPI) is an index to quantify the rainfall deficit for multiple timescales. In 2015, Indonesia experienced severe drought, which has not been analyzed, yet. Thus, it is important to assess a quantitative evaluation of the drought condition. The study shows that from all deficit periods, the most severe drought in duration and peak took place in 2015, with each drought index as follows: 1 month deficit or SPI-1 (-3.11) in 1985 (-2.51) in 2015, 3 month deficit or SPI-3 (-2.291) in 1995 (-1.82) in 2015, 6 month deficit or SPI-6 (-2.40) in 1997 and (-1.84) in 2015, 9 month deficit or SPI-9 (-1.12) in 2015, 12 month deficit or SPI-12 (-1.19) in 2015. The result underlines the potential that SPI exhibits in drought identification and the use of the rainfall strongly linked to drought relief policy and measure implementation in Tegal city.

1. Introduction

Indonesia, both the rainy and the dry seasons, become the causes of flood and drought in the country. Drought is a characteristic deficiency of the water availability or water supply which results in prolonged shortages in surface or groundwater. It sometimes declared after as few as fifteen days. El-Nino phenomenon affects the drought frequencies, but it is not always referred to a severe drought because there are other factors which generate drought in Indonesia. Such as land use land cover change and rainfall anomaly.

A research Pramudya et al [1] has reported about the conversion rate of wetland into settlements and other land use in Tegal city, Indonesia. It was reported that the recent wetland conversion rate-is rather high than in the last two decades. It was also reported that rainfall anomalies, tidal flooding occurrence before the 1990s, and water insufficiency for agriculture after 2000s are the factors of land use conversion in Tegal city which resulted by the major human impact on urban activities which had related to climate change. So that, the development and improvements method on meteorological and agricultural drought condition is still an interesting topic to discuss.

The Meteorology, Climatology, and Geophysics Agency in Indonesia or usually known as BMKG uses SPI (Standardized Precipitation Index) method in order to express the analyze the meteorological

Content from this work may be used under the terms of the Creative Commons Attribution 3.0 licence. Any further distribution of this work must maintain attribution to the author(s) and the title of the work, journal citation and DOI.
Published under licence by IOP Publishing Ltd

drought information. Some researchers already discussed SPI and it's calculating. Agnew [2] studied relationship between drought duration, McKee [3] carried out about SPI calculation method on drought frequency and drought timescale, then Tsakiris et al [4] estimated method of SPI over a geographical area and it is used for the estimation of drought, and WMO [5] an important work drought monitoring to facilitate effective decision-making. According to the World Meteorological Organization (WMO), SPI is a recommended tool to show the calculated index probability in order to record the amount of rainfall, drought negative index value, and wet positive index value. It can also be used to forecast the condition of the climate condition monitoring on a range of time by insisting on some modeling approaches.

Operationally, using an index for drought characterization serves the following purposes: drought detection and real-time monitoring [6]. Drought period estimation by the start and end of the time scale [7], drought assessment with allowing drought to determine drought levels and investigate drought responses measures [6] drought in a region represented by drought index characterization are acceptable [7]. Then, among variously interested drought index value correlating with quantitative drought impacts over geography variable scales and timescales that facilitating to communicate the information from the drought characterization.

In order to consider and evaluate the impact of the climate change using Standardized Precipitation Index (SPI) method as the evidence of climate anomaly, this study discusses the method improvements using SPI to describe meteorological and agricultural drought conditions. Also, it discusses anomaly phenomenon to predict and monitor dry season onset in deficit 1 (monthly), 3 (three monthly), 6-9 (seasonal) and 12 (annual).

2. Methodology

Rainfall observations from a specific station were used to establish SPI time series baseline and identify drought during 1983 to 2015. Future rainfall generated then examined for drought events using the SPI. Figure 1 presented annual rainfall time series in Tegal city station.

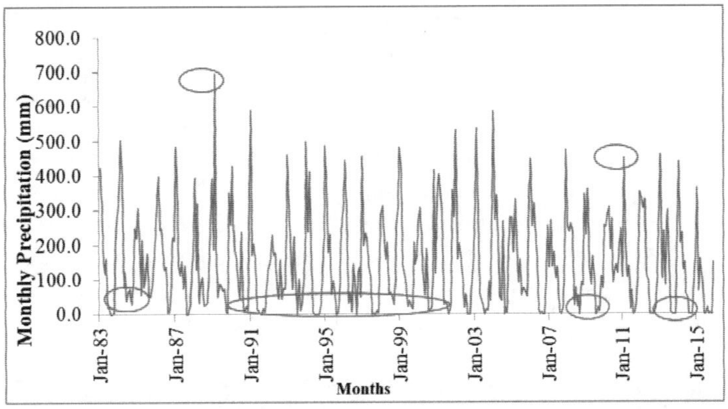

Figure 1. Time series of precipitation for the Tegal city meteorological station (1983-2015)

Table 1 shows different time scales of SPI with related effects [8,23].

Table 1. Phenomena reflected by specific-duration standardized precipitation indices (SPI) and their applications NDMC [8]

SPI Duration	Phenomena reflected	Application
1 month	Short-term conditions	Short-term soil moisture and crop stress (especially during the growing season)
3 month	Short and medium term moisture conditions	A seasonal estimation of precipitation
6 month	Medium-term trends in precipitation	Potential for effectively showing the precipitation over distinct seasons. e.g., for California, the 6-month SPI can effectively indicate the amount of precipitation from Oct. to Mar.
9 month	Precipitation patterns over a medium time scale	If SPI9 < –1.5 then it is a good indication that substantial impacts can occur in agriculture (and possibly other sectors)
12 month	Long-term precipitation patterns	Possibly tied to stream flows, reservoir levels, and also groundwater levels

2.1. Location of study and rainfall data
Tegal city, Indonesia has been decided as a research site. It has a total catchment area of about 39,467 km^2, formed by 4 sub-districts (6° 54' S and 109° 08' E). E. Yin [9] suggested that in order to get well-defined research objective, qualitative analysis and data collection analysis are important to be conducted. In the previous study, Pramudya Y et al. [1] used description and linear regression to identify the frequency of dry spell and wet spell as a drought indicator but there is still limited information about drought index analysis. This study is in an agricultural area of Tegal city which analyzed 30 years of daily precipitation time series data to understand the climate change by only correlating with the frequency of dry spell and wet spell and some interviewing method to the farmers, no further evaluation [1]. In decades, this research area is growing as fast as in terms of industrialization, rapid urbanization, modern road network, and new build-up areas, which in contrary agricultural activity is decreasing meanwhile settlement expansion. Sustainable development to reduce this fast rapidly urban is challenges to avoid dire consequences of ecosystem deficiency in surrounding area, pollution, forest fragmentation, land use and land cover changes. Under such circumstances, the implementation of a best-suited drought index on recorded climate outlook was deemed necessary. The rainfall data from Tegal city station (6° 87' S and 109°12' E, 54.0 m above sea level), was used to represent Tegal city research area. One of the reasons for this selection was due to its close proximity to the land cover and land use change phenomenon in other hand climate anomaly. The 33 years (1983-2015) data available from BMKG [10] had been subjected to homogeneity tests before perusal.

2.2. The Standardized Precipitation Index (SPI) as a drought indicator
The SPI computation in this research was following the method which proposed by McKee [3]. For the prediction of future drought events, two parts of SPI computation were carried out. SPI was computed based on the observed rainfall from BMKG [10] in the year 1983-2015 (33 years of data). Rainfall condition in Tegal city was described by using gamma probability function. This is the similar method which proposed by McKee [3] and it has also been proven in Sharma and Singh research [11,12] which studied rainfall in monsoon seasons. Then, in order to obtain the SPI value, the function will be normalized and standardized. It could be said that z-score of the distribution function to represent the deviation event from the mean of rainfall data as the SPI value.

The Standardized Precipitation Index by McKee et al. [3] is a probability index which described the representation of abnormal wetness and dryness and compares precipitation with its multiyear average. Determining the probability density function which described long time series of precipitation is the first step of SPI method. The series could be modified for much different time duration but typically series used for total precipitation is 1, 2, 3, 6, 9, 12, 24 months. The cumulative probability of an observed precipitation amount could be computed after determined the function of probability density. Then, inverse normal of the Gaussian function applied to the cumulative probability (mean zero and variance one) [13].

Drought monitoring could be explained by the SPI time series. Positive SPI values indicate a wetter than the typical period (accumulated precipitation is greater than median), and negative SPI values are represented dryer period with less precipitation than normal. Accumulated precipitation was corresponded by a value of zero, then the magnitude value which started from zero could be used to explain the risk management as the wet or dry event [14].

In order to analyze the drought severity, the accumulated value of SPI could be used. SPI threshold values define drought beginning and ending [8]. When SPI value is -1 or less, it means that drought occurs. Different duration of SPI reflected different phenomena [15]. SPI-1 is described short-term conditions as short as the term of soil moisture and vegetation during growing seasons. SPI-3 is described short-medium conditions, roughly approximating seasons, while the six-month SPI (SPI6) shows the precipitation across distinct seasons. When the nine months SPI (SPI9) is less than -1.5 substantial impacts occur in agriculture and other sectors, one-year SPI (SPI12) is tied to the evolution of stream flows and reservoir and groundwater levels [16].

To explore correlations of drought with the areas in Tegal city we used monthly SPIs of different duration: SPI1, SPI3, SPI6, SPI9, and SPI12. These SPI values were calculated on the basis of long-term series of monthly precipitation within 40 years (1983-2015) in Tegal city stations to describe climate conditions according to SPIs.

2.3 Case study area and data processing

The flexibility of time scale choosing is an advantage of SPI calculation. This study focused on five kinds of time scales, which are 1, 3, 6, 9 and 12 months (SPI-1, SPI-3, SPI-6, SPI-9, SPI-12). Those times have been decided because it is suitable times to describe meteorological, agricultural and hydrological drought respectively as the study purpose. The calculation of SPI values here follows the method by Asadi Z M et al [17] and demonstrated in the following steps:

First, calculate the cumulative gamma distribution, noted that gamma distribution has been used to define the function of frequency or probability density function. The formula proposed by McKee [3] is as follows:

$$G(x_k) = \int_0^{x_k} g(x_k)dx_k = \frac{1}{\beta^\alpha \Gamma(\alpha)} \int_0^{x_k} t^{\alpha-1} e^{-x_k/\beta} dx_k \tag{1}$$

Symbol α and β are defined as shape parameter and as scale parameter respectively, and it is estimated by using approximation by Thorn [18]. The value for the both of them is should bigger than zero, while x is the amount of precipitation over the consecutive months k (selected time scale) in millimeter. The function $\Gamma(\alpha$ refers to gamma function. When $x_k = 0$, the cumulative gamma distribution is undefined and the in order to encounter this situation, the cumulative probability. SPI calculations also included matching the gamma probability density function into the frequency distribution of rainfall for each station. Equations proposed by McKee [3] to optimize the following estimates of α and β values:

$$\alpha = \frac{1}{4A}\left(1 + \left(1 + \frac{4A}{3}\right)^{1/2}\right) \tag{2}$$

$$\beta = \frac{x}{\alpha} \tag{3}$$

Where:

$$A = \ln(x) - \frac{\sum \ln(x)}{n} \tag{4}$$

Noted that n is the number of observation data for rainfall. In order to observed rainfall events for each month and the timescale for each station, the cumulative probability is found by using resulting parameters. It is explained by McKee [3]as follows:

$$G(x_k) = \int_0^{x_k} g(x_k)dx_k = \frac{1}{\beta^\alpha \Gamma(\alpha)} \int_0^{x_k} t^{\alpha-1} e^{-x_k/\beta} dx_k \tag{5}$$

Due to undefined gamma function for $x_k = 0$, then the value of G (x_k) becomes:

$$H(x_k) = q + (1 - q).G(x_k) \tag{6}$$

Noted that q is the amount of zero rainfall and $H(x_k)$which refers to cumulative probability is standardized wherewith to obtain the SPI value. Lastly, to finish the calculating of SPI value, approximate conversion by Thorn [18] is believed to used.

The number of zeros rainfall data is symbolized with m, then the function of q could be estimated with m/n. The cumulative probability is then transformed into a normal standard random variable Z, with an average value zero and a variation of 1, the value obtained Z is the SPI value.

Normal standard value of random variable Z or SPI is accessible by using the approximation according to Abramowitz and Stegun [19] with the equation as follows:

Calculation Z or SPI for 0 <H (x_k) ≤ 5

$$Z = SPI = -\left(t - \frac{c_0+c_1t+c_2t^2}{1+d_1+d_2t^2+d_3t^3}\right) \tag{7}$$

Where:

$$\left(t = \left(\ln\left(\frac{1}{(H(x_k))^2}\right)\right)^{1/2}\right) \tag{8}$$

Calculation Z or SPI for 0 <H (x_k) ≤ 1.0 proposed by McKee [3]

$$Z = SPI = -\left(t - \frac{c_0+c_1t+c_2t^2}{1+d_1+d_2t^2+d_3t^3}\right) \tag{9}$$

Where:

$$\left(t = \left(\ln\left(\frac{1}{(1-H(x_k))^2}\right)\right)^{1/2}\right) \tag{10}$$

Where, c_0 = *2.515517*, c_1 = *0.802853*, c_2 = *0.010328*, d_1 = *1.432788*, d_2 = *0.189269*, d_3 = *0.001308*.

When the values are negative, it means that rainfall is less than a medium of historical precipitation. SPI values could use to classify the drought by certain ranges. They are : (1) Mild drought is described when the rainfall values of SPI between 0 and -0.99, (2) Moderate drought has the values between -1.00 and -1.49, (3) Severe drought has defined if the values between -1.5 and -1.99 and, (4) Extreme drought is described by the values less than -2.00.

McKee et al. [3] said that SPI is a popular meteorological drought index which based on precipitation data. This assumption is also supported by the aims of an Inter-Regional Workshop on Indices and Early Warning Systems for Drought was held in December 2009, the goal is to help in determining the best "meteorological" index and then recommend that all national meteorological services use this index (Lincoln Declaration on Drought Indices, [5]). This would make comparisons in drought severity among countries in the same region, and also among regions possible. The SPI was chosen by participants as the one to use [21]. For SPI, 30 years record is required but 50 years has been recommended [13]. Currently, this index has been widely adopted for research and operational modes.

Then, the SPI value is set to zero and as such. Wet periods and dry periods are described by the values above and below zero, respectively. National Standardized Precipitation Index [22] expected for any given values of drought in SPI from the normalized average is described as standard deviations its cumulative rainfall deficit deviates. McKee [3] also explained that drought is determined to occur if the value less than zero and it is consistently observed and reaches the value of –1 or less. Besides many limitations of SPI method, the ability of SPI method in calculating the drought levels in different time scales is an important advantage. McKee's index could be computed for any time period and typically applied in 1, 3, 6, 9 and 12 month periods due to time rainfall deficit gradually and variably affects different water resources such as groundwater. The reflect changes in different water features (such as in Tegal city) could be described by the multitude of SPI durations.

3. Result and Discussion

3.1. Study site

Tegal city, Central Java Province Indonesia has been decided as the site for this research and it lays on 6° 54' S 109° 08' E as shown in Figure 2.

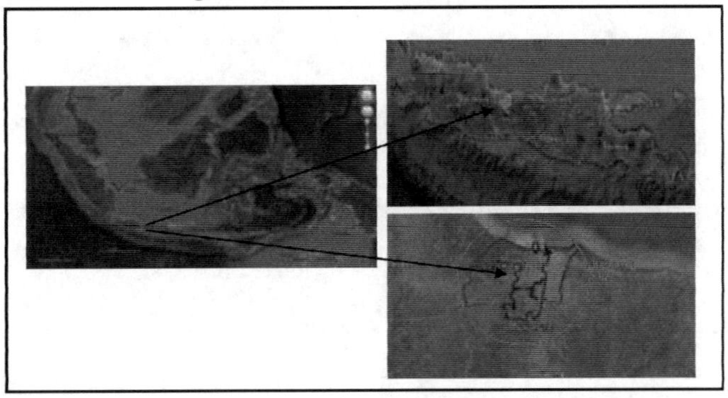

Figure 2. Research Site (Tegal City), Central Java Province, Indonesia.
Modified from Google Earth Pro 2016 [20]

3.2. Drought duration value of each deficit period (SPI 1, SPI 3, SPI 6, SPI 9, and SPI 12)

Based on the calculation of SPI drought index, it is found that the results study in all periods of severe drought deficit and in drought criteria has a tendency to drought deficit condition. Drought occurred in the year 2015 with the value of drought index respectively the period 1-month deficit or SPI-1 (-3.11) in 1985 and (-2.51) as shown in Figure 3 (a). The period of 3 month deficit or SPI-3 (-2.29) in 1997 and (-1.82) in 2015 Figure 3 (b), the period of 6 month deficit or SPI-6 (-2.40) in 1997 and (-1.84) in 2015 Figure 3 (c), the period 9 month deficit or SPI-9 (-1.12) in 2015 Figure 3 (d), the period 12 month deficit or SPI-12 (-1.19) in 2015 Figure 3 (e). So that, according to previously research by Pramudya Y et al. [1], it was proven that there are domination factors which could encourage land use conversion at Tegal city such erratic rainfall and weather anomalies, tidal flooding occurrence before the 1990s, and also water insufficiency for agriculture after 2000s, as the impact of climate change. Estimation of the SPI index described the appearance climate anomaly evidently at the research site within 33 years.

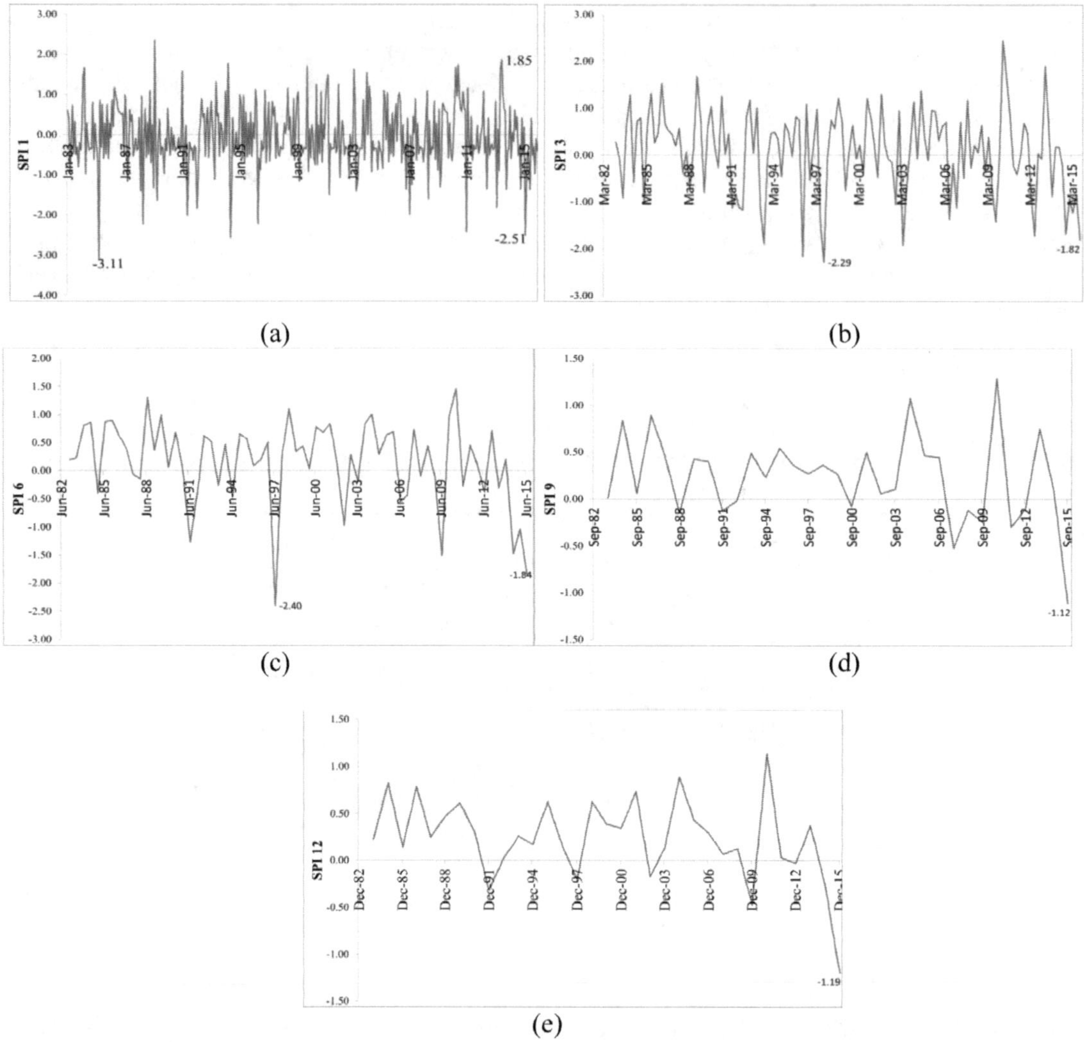

Figure 3. SPI values for Tegal city meteorological station with (a) 1-month time scale; (b) 3 month time scale; (c) 6-month time scale; (d) 9-month time scale; (e) 12-month time scale

4. Conclusion

The SPI value occurred in Tegal city has different drought characterization to each duration in all periods of severe drought deficit and in drought criteria has a tendency to drought deficit condition. Drought occurred in year 2015 with index value respectively, for SPI 1 are (-3.11) in 1985 and (-2.51) in 2015, for SPI 3 are (-2.29) in 1997 and (-1.82) in 2015, for SPI 6 are (-2.40) in 1997 and (-1.84) in 2015, for SPI 9 are (-1.12) in 2015, and for SPI 12 are (-1.19) in 2015.

By quantifying severity levels and declaring drought's start and end, SPI assessment currently aids in an improvement to describe erratic rainfall and weather anomalies condition in Tegal city. It needed knowledge of drought index and merge to the process of decision making for farming practices with regard to consider adaptation and mitigation of climate change in the urban city.

Adjusting planting time period and selecting proper crop varieties is one strategy in coping with climate variability in an agricultural area in Tegal city. The level of farmer knowledge mainly affects the use of drought index at the farm level in the region. Drought identification and the use of the rainfall strongly linked to drought relief policy and implementation. Better knowledge of drought index enabled farmers to interpret complex situations and adopt innovations hence minimizing the climate risk anomaly such as drought, flood and yield agricultural reduction followed by land use conversion at Tegal city.

References

[1] Pramudya Y, Komariah, Dewi WS, Sumani, Mujiyo, Sukoco T A and Rozaki Z 2016 *Remote Sensing for Estimating Agricultural Land Use Change as the Impact of Climate Change* (Proc. of SPIE) 9877 987720-1

[2] Agnew 2000 Using the SPI to identify drought (drought network news 1994-2001) *Paper I.* (Lincoln: University of Nebraska)

[3] McKee T B, Doesken N D and Kleist J 1993 *The Relationship of Drought Frequency and Duration of Time Scales* Conf. Eighth Conference on Applied Climatology, American Meteorological Society (Anaheim CA) pp 179-186

[4] Tsakiris G and Vangelis H 2004 Towards a drought watch system based on spatial SPI *Water Resources Management* **18** (1) 1–12.

[5] "Lincoln Declaration on Drought Indices World Meteorological Organization (WMO)", *www.wmo.int,* 2009. [Online]. Available: http://www.wmo.int/pages/prog/wcp/agm/meetings/wies09/documents/Lincoln_Declaration_Drought_Indices.pdf [Accessed : 22-5-2017]

[6] Niemeyer S 2008 *New Drought Indices* (Options Méditerranéennes, Série A: Séminaires Méditerranéens) 80 267–274

[7] Tsakiris G, Loukas A, Pangalou D, Vangelis H, Tigkas D, Rossi G and Cancelliere A 2007 *Drought Characterization Chapter 7* (Options Méditerranéennes) 58 85–102

[8] "Monitoring drought. The Standardized Precipitation Index. Interpretation of SPI Maps. National Climatic Data Center (NDMC)". *www.drought.unl.edu* 2006. [Online]. Available: http://www.drought.unl.edu/monitor/interp.html [Accessed: 22-5- 2017]

[9] Yin R K 2003 *Case Study Research: Design and Methods 3rd Edition* (Applied Social Research Methods) 5

[10] "Indonesia Agency of Meteorology, Climatology and Geophysics (BMKG) Data Online Database Center", www.bmkg.go.id, 2015, [Online]. Available: http://www.dataonline.bmkg.go.id. [Accessed: 8-04-2015]

[11] Sharma M A and Singh J B 2010 Use of probability distribution in rainfall analysis New York *Science Journal* **3**(9) 40-49

[12] Arora V K and Boer G J 2014 Terrestrial ecosystems response to future changes in climate and atmospheric CO2 concentration *Biogeosciences* **11** (15) 4157–4171

[13] Guttman N B 1999 Accepting the standardized precipitation index: a calculation algorithm *JAWRA Journal of the American Water Resources Association* **35** (2) 311–322

[14] Paulo A A, Rosa R D and Pereira L S 2012 Climate trends and behavior of drought indices based on precipitation and evapotranspiration in Portugal *Nat. Hazards Earth Syst., Sci.* **12** 1481-1491

[15] Zargar A, Sadiq R, and Naser B and Khan F 2011 A review of drought indices *Environ. Rev.* **19** 333–349

[16] Paulo A A and Pereira L S 2006 Drought concepts and characterization: comparing drought indices applied at local and regional scales *Water Int..* **31** 37-49

[17] Asadi Z M, Sivakumar B and Sharma A 2014 Assessment of global aridity change *Journal of Hydrology* **520** 300-313

[18] Thorn H C S 1958 A note on gamma distribution *Mon. Weather Rev.* **86** 117-122

[19] Abramowitz M and Stegun I 1964 *Handbook of Mathematical Functions with Formulas, Graphs, and Mathematical Tables* (Washington, U.S: Govt. Print. Off)

[20] "Tegal City Location Google Earth Pro", *www.earth.google.com* 2016 [Online]. Available: https://earth.google.com/web/@-6.8705707,109.1172396,0a,11540.0388811d,35y,0h,45t,0r/data=CkcaRRI_CiQweDJlNmZi

IOP Conf. Series: Earth and Environmental Science **129** (2018) 012019 doi:10.1088/1755-1315/129/1/012019

OWRmYmYzMjY0YzM6MHgzMDI3YTc2ZTM1MmJiZTAZ4qDOJtGEG8AhignFsQlIW0AqBVRlZ2FsGAIgASgC [Accessed: 21-6- 2016]

[21] Hayes M J, Svoboda M D, Wall N and Widhalm M 2011 The Lincoln declaration on drought indices: universal meteorological drought index recommended *Bull. Amer. Meteorol. Soc.* **92** (4)

[22] "National Standardized Precipitation Index Maps. Agriculture and Agri-Food Canada, Drought Watch" *www.agr.gc.ca* 2010 [Online]. Available: http://www.agr.gc.ca/pfra/drought/nlspidesc_e.htm [Accessed: 22-5- 2017]

[23] Moreira E E, Coelho C A, Paulo A A, Pereira L S and Mexia J T 2008 SPI-based drought category prediction using log-linear models *J. Hydro.* **354** 116-130

IOP Conf. Series: Earth and Environmental Science **129** (2018) 012020 doi:10.1088/1755-1315/129/1/012020

Effects of intermittent acid rain on proline and antioxidant content on medicinal plant "*Pereskia bleo*"

Sulandjari[1,2,4] and W S Dewi[3]

[1] Department of Agrotechnology, Universitas Sebelas Maret, Jl. Ir. Sutami 36A Kentingan Surakarta Central Java 57126, Indonesia
[2] Biotechnology and Biodiversity Research Center, Universitas Sebelas Maret, Jl. Ir. Sutami 36A Kentingan Surakarta Central Java 57126, Indonesia
[3] Department of Soil Science, Universitas Sebelas Maret, Jl. Ir. Sutami 36A Kentingan Surakarta Central Java 57126, Indonesia
[4] Correspondent author: sulandjari@staff.uns.ac.id

Abstract. Global warming due to CO_2 and other greenhouse gas emissions from human activities have led to climate change and environmental degradation. The acid rain, with the pH of rainwater below 5.6, is a serious environmental problem. Arising from air pollution and potentially harmful to health, it can damage old buildings and distract the growth and physiological metabolism of sensitive plants. How does the influence of climate change on medicinal plants such as *Pereskia bleo*? The leaf of *Pereskia bleo* (Kunth) DC. contains high antioxidants with benefits for anti-cancer, anti-tumor, anti-rheumatic, and anti-inflammatory. This research aims to investigate the influence of acid rain on the proline level and antioxidant content of *Pereskia bleo*. Having been carried out from June to August in Jogjakarta, this study was conducted through the use of artificial acid rain with pH 5.8, 4.9, 3.7 and 2.9, by adding sulfate acid (H_2SO_4) to rainwater. The interval of intermittent watering acid rain to the plants is once a day, twice a day, and once in three days with three replications for six weeks. The results showed that Acid rain with a pH less than 4.9 and the intermittent interval of acid rain twice a day and once in three days significantly suppresses growth and chlorophyll content. In contrast, it increases the proline and antioxidant levels as a tolerant action of the plant.

1. Introduction

Climate change is largely the result of human activity. Climate change is largely the result of human activity. Growth Industrialization has increased pollutants into the environment. Increased levels of global warming due to CO_2 and other greenhouse gas emissions from human activities have led to climate change and environmental degradation. Acid rain is the most serious environmental problem arising from air pollution. Natural rainwater is slightly acidic, with a pH value ranging from 5.6 to 7.0. As a result of the presence of CO_2 in the atmosphere, it dissolves in the rain when it falls. In addition, the advent of industrialization and transportation also plays a role in raising the acidity of rain. This increase in pH is not only caused by CO_2 emissions but also SO_2 and NOx as pollutant gases. SO_2 and NOx that reacts with water molecules in the atmosphere produce acid as H_2SO_4 and HNO_3 that is quickly dissolved in clouds and water. This acid rain potentially damages old buildings, health, water and plant environments [1,2,3]. Acid rain has already occurred in several Indonesian cities [4]. The pH of rain during the period 2001-2013 ranged from 4.3 to 5.6 in Bandung, Serpong, Jakarta, Kototabang, and Maros [5].

Content from this work may be used under the terms of the Creative Commons Attribution 3.0 licence. Any further distribution of this work must maintain attribution to the author(s) and the title of the work, journal citation and DOI.
Published under licence by IOP Publishing Ltd

Acid rain is a serious problem for environmental balance in the world that interferes with the life and development of plants, aquatic life, socio-economic development, and human health. Acid rain has a direct or indirect impact on plants, among others chlorosis on leaves, inhibition of photosynthesis and disruption of nutrient and mineral absorption [6,7,8]. There are several indirect effects of acid rain on soil fertility, such as the increase of soil acidity and the release of toxic minerals that would endanger the development of plant roots [2] and result in the disruption of plant growth and its development. Acid rain has a significant effect on wheat growth. The increasing rain acidity, with pH $\leq$ 4.5, can decrease total free amino acids and protein content [9].

Plants that are exposed to water stress will reduce mass significantly [10]. Water is an essential production factor for plants not only the quantity but also the quality. It is also considered as the most limiting factor for growing most plants and disrupting plant metabolism. Water quality, among others, is determined by acidity (pH), salinity and mineral content. Acid rain, with a pH of less than 5.6 has the potential to suppress plant growth [11] and against its metabolism [12].

Pereskia bleo is a medicinal plant that contains high antioxidants. It belongs to primitive cactus (Cactaceae) plants, which are leafy [13]. This plant contains high antioxidants [14], which are useful as anti-cancer [15], antimicrobial [16] and anti-inflammatory [17]. Drought stress reduces the efficiency of photosynthesis and decreases root dry weight and potential matric on the root zone. However, since the leaf area and leaf number are not affected, it can be concluded that Pereskia is tolerant of drought stress [18].

Proline is an amino acid that can increase plant tolerance to abiotic stress by lowering osmotic potential and maintaining cell turgor [19,20]. High proline content may be a good source of components that will help increase the overall antioxidant capacity of an organism and protect it from lipid peroxidation [21]. Plants under pressure produce both enzymatic and nonenzymatic defense mechanisms in chloroplasts and mitochondria [22]. Hernandez *et al.* [23] stated that the antioxidant is a component tolerance mechanisms in peas. Although there has been a lot of research on acid rain effect on plant growth, research of acid rain to *Pereskia bleo* has not been conducted yet. Therefore, this research aims to study the impact of acid rain on proline and antioxidant dynamics of *Pereskia bleo*.

2. Materials and methods

The research had been carried out from June to August, in Sleman, Jogjakarta. The age of *Pereskia bleo* plant material derived from shoots cuttings was eight months. This research was conducted through the use of artificial acid rain. The preparation of acid rain using rainwater collected in April 2017 at Sleman Yogyakarta was 7,720 S, 110.36 E. Elevation 242 m. The pH of rain obtained was 5.8 (Control). After the addition of 2 ml of 10% H_2SO_4 in 600 ml, 850 ml, and 1050 ml, the pH obtained were 2.91 ± 0.12; 3.7 ± 0.17 and 4.9 ± 0.11, respectively. The interval of intermittent watering acid rain to the plants is 1x1 day, 1x2 days, and 1x3 days, with three replications for six weeks. The comparison of soil medium and organic base fertilizer is 2:1, with the weight of 5 kg per polybag. In this research, media is placed in water saturated condition before treatment. Watering on intermittent is done using mini sprayer by adjusting nozzles and passing to leaves until soil media is saturated within 5 minutes with a volume of 1000 ml per plant. Parameter variables were used to plant height, a number of leaves, root volume; chlorophyll a, b and total, proline, and antioxidants. Chlorophyll was then analyzed by a spectrophotometer method whereas proline was analyzed by a ninhydrin method [25]. Meanwhile, antioxidants were analyzed by the DPPH method. The experimental design for both experiments was factorial randomized complete block design. The data were subjected to analysis of variance, and treatment means were compared using the least significant difference test (p = 0.05).

2.1. Extraction and estimation of chlorophyll

One gram of finely cut leaf sample was weighed and grounded with acetone. It was then centrifuged. The supernatant is transferred. The absorbance of the solution was measured at 645 nm against the solvent (80% acetone) as blank. The concentrations of chlorophyll a, chlorophyll b and total chlorophyll were then calculated using the following equation:

Chlorophyll a (mg/g) = 12.72(OD 663)-2.59(OD 645) (1)

Chlorophyll b (mg/g) = 22.9(OD 645)-(OD 663) (2)

Total Chlorophyll (mg/g) = 20.31(OD 645) + 8.05(OD 663) (3)

2.2. Proline analysis

Proline analysis was performed according to Bates, 1973 [24]. Diethyl ether extract was exempted from this test since proline is practically insoluble in diethyl ether. Briefly, 50 mg of extract were homogenized in 10 ml sulphosalicylic acid (3% w/v) and filtrated through filter paper. Two milliliters of the filtrate was mixed with 2 ml of acid ninhydrin solution (1.25 g ninhydrin + 30 ml glacial acetic acid + 20 ml 6 M H_3PO_4) and 2 ml of glacial acetic acid and kept at $\pm 100^0$ C for one hour. Then, the reaction was stopped by transferring the mixture to an ice bath. Four milliliters of toluene was added to the mixture and vortexed for 15–20 s. The toluene phase was aspirated, and the absorbance at 520 nm was measured using pure toluene as a reference. A calibration curve was prepared with pure proline. Results were expressed as μg proline/ gram extract.

2.3. DPPH radical scavenging assay

Radical scavenging activity of plant extracts against 2.2-diphenyl-1-picrylhydrazyl (DPPH) radical was determined spectrophotometrically [25]. The principle of the assay is based on the color change of the DPPH solution from purple to yellow as the radical is quenched by the antioxidant. Briefly, 1 ml of 0.1 mM DPPH in ethanol was mixed with 3 ml of extract solution with differing concentrations (5–250 lg/ml). Then, the mixture was vortexed. The samples were kept in the dark for 30 minutes at room temperature. Afterwards, the decrease in absorbance at 517 nm was measured. The absorbance of DPPH solution in the absence of plant extract was measured as the control. Ascorbic acid, BHT, and rutin were used as positive controls. DPPH radical scavenging activity was expressed using the formula: % DPPH radical scavenging activity = [(A0 _ A1)/A0] _ 100 where A0 was the absorbance of the control and A1 was the absorbance of the sample.

3. Result and discussion

3.1. Growth component

The influence of acid rain and intermittent on plant growth and development showed that leaf color change started at week 4, especially at the treatment of pH 3.7-2.9 with intermittent 1x3 days. Direct contact of low pH acid rain contains H_2SO_4 to the leaves, causing chlorosis spotting. The impact of treatment on intermittent 1x3 days also causes the plant to be not fresh. Analysis of growth component yielded at 6th week shows that there is an interaction between acid rain and intermittent that tends to suppress plant height at pH below 5.8 (control), (Figure 1). The number of leaves and root volume of the acid rain and intermittent treatment of each one is significant (Table 1 and Table 2). Acid rain has the potential to suppress plant height and damage the leaf surface. Acid rain also causes necrosis in G. Americana [14] and seriously damages the leaf surface [7], but this depends on the sensitivity of leaf surfaces of various plants [26]. The net photosynthetic rate of *P. massoniana* exposed to acid rain with pH 4 and 2.5 having decreases of 20% and 34% has shown that *P. massoniana* is susceptible to exposure to acid rain [27].

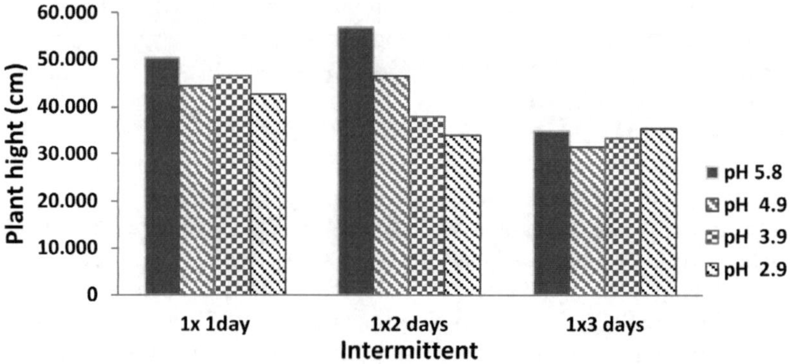

Figure 1. Interaction of acid rain and intermittent on plant hight.

Table 1. The effect of acid rain and intermittent on the volume of the root.

	pH acid rain				
Intermittent	5.8 (control)	4.9	3.7	2.9	Average
1x 1 day	29.33	26.33	26.00	19.00	25.17 b
1x 2 days	29.6	29.67	35.33	26.00	30.17 a
1 x 3 days	35.67	25.33	21.33	18.33	25.17 b
average	31.56 A	27.11 B	27.56 B	21.11 C	

Note: The same letter in the column or row is not significantly different at p = 0.05.

Table 2. The effect of acid rain and intermittent on the number of leaves.

	pH acid rain				
Intermittent	5.8 (control)	4.9	3.7	2.9	Average
1x 1 day	32.00	27.00	20.33	15.33	23.67 b
1x 2 days	33.00	30.33	26.00	22.00	27.83 a
1 x 3 days	33.00	28.67	19.00	17.00	24.42 b
average	32.67 A	28.67 AB	21.78 C	18.11 C	

Note: The same letter in the column or row is not significantly different at p = 0.05.

3.2. Chlorophyll content

There is a decrease of chlorophyll content (a, b and total) at pH≤ 4.9, except acid rain pH 2.5 with intermittent 1x3 days showed higher chlorophyll content than acid rain to pH 4.9. (Figure 2) The decrease in chlorophyll content is due to the leaf surface damage caused by chlorosis since its formation also requires water [28]. Acid rain with pH 4 and pH 3 decreases chlorophyll a and b significantly, with increasing acidity level (pH ≤ 5.0) [29]. Previous studies showed that acid rain decreases leaf chlorophyll content to 67% per pH of plant units in mainland China [30,31]. It suggests that simulated acid rain at pH 4.5, 3.5 and 2.5 significantly decreased the chlorophyll content than with control.

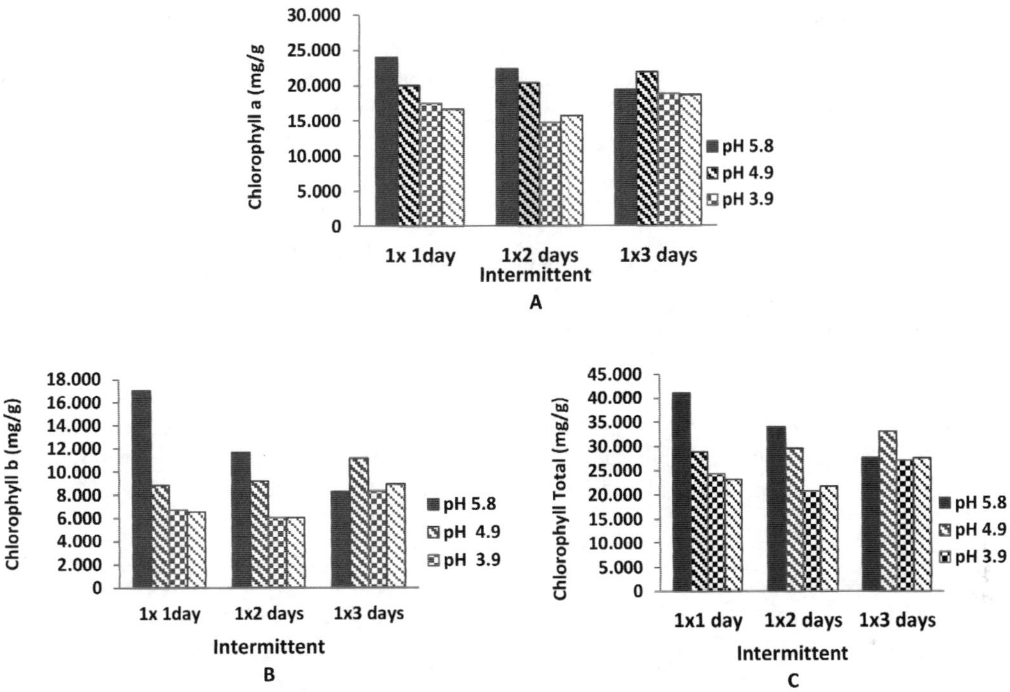

Figure 2. Interaction of acid rain and intermittent on Chlorophyll an (A), b(B) and total (C).

3.3. Proline and antioxidant content

In this research, acid rain and intermittent treatment had a significant effect on proline level and there was an interaction (Figure 3a). Acid rain with pH ≤ 4.9 -2.9 increases proline levels at all intermittent levels. Proline metabolism is coordinated in chloroplasts, cytoplasm, and mitochondria. Increased metabolism in stress-conditioned plants has been noted by some researchers [31]. It has been shown to be a molecular mechanism of defense against stress with accumulated proline [32,33,34]. A significant increase in proline content under stress is associated with increased rates of synthesis, reduced catabolism levels and loss of enzyme inhibitors on proline biosynthesis P5CS [35]. Abiotic and biotic stresses in plants are responded by increasing levels of proline. An increased level of proline due to stress is an indication of plant resistance to stress [36].

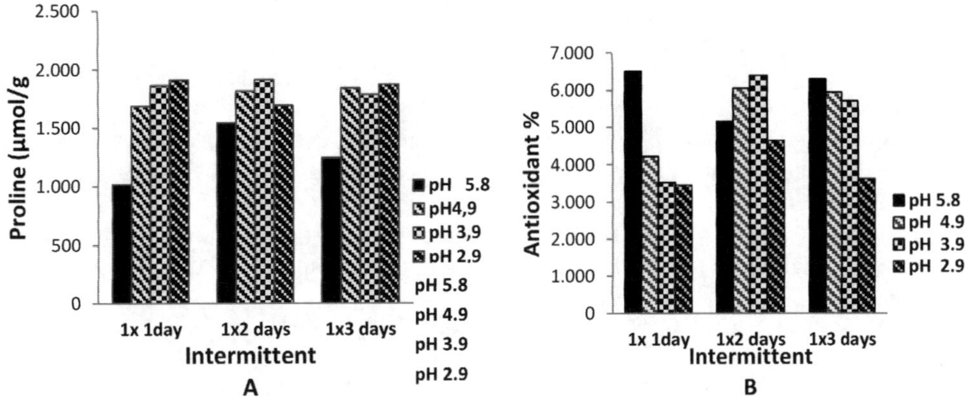

Figure 3. Interaction of acid rain and intermittent on proline content (A) and Antioxidant (B)

The increase of proline level is followed by an antioxidant percentage increase, at pH 3.7 - 2.9 with intermittent 1x2 days and 1x3 days (Figure 3b). Jahantigh *et al.* [37] stated that *Hyssopos officinalis* plant tolerance to salinity stress is characterized by the increase of proline and antioxidants. The antioxidant activity of catalase enzyme, peroxidase, and superoxide dismutase in drought-tolerant varieties increased significantly, observed during the different accumulation of proline on upland rice varieties. It can be concluded that antioxidant enzymes and accumulation of proline associated plant tolerance to drought stress [38].

4. Conclusion

Climate change has affected the activity of living organisms, such as the occurrence of acid rain on medicinal plants. The importance of medicinal plants is the content of secondary metabolites or their active ingredients. The content of active ingredients from medicinal plants is highly dependent on environmental factors. Environmental stresses can have a negative or positive impact on this active ingredient and depend on the activity of tolerance of each type of medicinal plant.

The results of this study indicate that acid rain with pH less than 4.9 and the intermittent interval of acid rain of one day, and once in three days significantly suppresses growth and chlorophyll content. In contrast, it increases the proline and antioxidant levels as the act of tolerance of the plant.

References

[1] Okochi H, Kameda H, Hasegawa S, Saito N, Kubota K and Igawa M 2000 Deterioration of concrete structures by acid deposition-an assessment of the role of rainwater on deterioration by laboratory and field exposure experiments using mortar specimens *Atmospheric Environment* **34(18)** pp 2937-45

[2] Savita D 2013 Acid rain-the major cause of pollution: its causes, effects, and solution *International Journal of Scientific Engineering and Technology* **2(8)** pp 772-5

[3] Punam K and Tomar Y S 2009 Effect of simulated acid rain on chlorophyll and ascorbic acid contents of *Mentha piperita L.* (Peppermint) *Agric. Sci. Dig.* **29** pp 1-6

[4] Dewi A S and Susandi A 2005 Projection of SO_2 in Indonesia as Implication of Global climate change: Impact and Health Cost (In) *Proceeding of the National Seminar on Environmental Research on University* (Bandung, Indonesia, 1-2 August 2005)

[5] Kurniawan A 2009 Effect of CO2 and SO2 gases, against rainwater pH at Bukit Kototabang *Megasains Bulletin* **2** pp 56-66

[6] Liu H, Yi L T, Yuf S Q and Yin X M 2015 Chlorophyll fluorescence characteristics and the growth response of Elaeocarpus glabripetalus to simulated acid rain *Photosynthetica* **53(1)** pp 23-8

[7] Sant'Anna-Santos B F, da Silva L C, Azevedo A A and Aguiar R 2006 Effects of simulated acid rain on leaf anatomy and micromorphology of *Genipa americana* L. (Rubiaceae) *Braz. Arch. Biol. Technol.* **49(2)** pp 313-21

[8] Zabawi A G M, Esa S M and Leong C P 2008 Effects of simulated acid rain on germination and growth of rice plant *J. Trop. Agric. and Fd. Sc.* **36(2)** pp 281-6

[9] Mai B R, Zheng Y F, Liang J, Liu X, Li L and Zhong Y C 2008 Effects of simulated acid rain on leaf photosynthate, growth, and yield of wheat *Chinese Journal of Applied Ecology* **19** pp 2227- 33

[10] da Silva E C, Silva M F A, Nogueira R J M C and Albuquerque M B 2010 Growth evaluation and water relations of *Erythrina velutina* seedlings in response to drought stress. *Braz. J. Plant Physiol* **22(4)** pp 225- 33

[11] Bhoomika S and Agrawal M 2004 Impact of simulated acid rain on growth and yield of two cultivars of wheat *Water, Air, and Soil Pollution* **152** pp 71-80

[12] Badridze G, Kacharava N, Chkhubianishvili E, Rapava L, Kikvidze M, Chanishvili Sh and Chigladze L 2015 Influence of ultraviolet irradiation and acid precipitations on the content of antioxidants in wheat leaves *Applied Ecology and Environmental Research* **13(4)** pp 993-

1013
[13] Edwards E J and Donoghue M J 2006 *Pereskia* and the origin of the cactus life-form *The American Naturalist* **167(6)** pp 777-93
[14] Hassanbaglou1 B, Hamid A A, Roheeyati A M, Saleh N M, Abdulamir A S, Khatib A and Sabu M C 2012 Antioxidant activity of different extracts from leaves of Pereskia bleo (*Cactaceae*) *Journal of Medicina Plants Research* **6(15)** pp 2932-7
[15] Malek S N A *et al.* 2008 Cytotoxic activity of Pereskia bleo (*Cactaceae*) against selected human cell lines *International Journal of Cancer Research* **4(1)** pp 20-7
[16] Karim A K and Sismindari 2011 *Mutiara Medika* **11(3)** pp 195-200
[17] Wahab S I A, Abdul A B, Mohan S M, AlZubairi A S, Elhassan M M and Ibrahim M Y 2009 Biological activities of Pereskia bleo extracts *International Journal of Pharmacology* **5(1)** pp 71-5
[18] dos Anjos Queiroz C R A, de Andrade R R, de Morais S A L and Pavani L C 2015 Growing Pereskia aculeata under intermittent irrigation according to levels of matric potential reduction *Pesq. Agropec. Trop.* **45(1)** pp 1-8
[19] Fangjun F, Xu X, Du X, Tong H, Luo L and Mei H 2012 Assessment of drought resistance among wild rice accessions using a protocol based on single-tiller propagation and PVC-tube cultivation *AJCS* **6(7)** pp 1204-11
[20] Paul E and Sandeep S 2010 Proline metabolism and its implications for plant-environment interaction *The Arabidopsis Book* **8** e0140
[21] Arife A K, Erdag B, Emek Y C and Uygun D A 2008 Antioxidant activity and proline content of leaf extracts from Dorystoechas hastata *Food Chemistry* **111** pp 400–7
[22] Mittler R 2002 Oxidative stress, antioxidants and stress tolerance *Trends Plant. Sci.* **7(9)** pp 405-10
[23] Hernandez J A, Jimenez A, Mullineaux P and Sevilia F 2000 Tolerance of pea (*Pisum sativum* L.) to long-term salt stress is associated with induction of antioxidant defenses *Plant Cell Environ.* **23** pp 853-62
[24] Bates L S, Waldren R P and Teare ID 1973 Rapid determination of free proline for water stress studies *Plant Soil* **39** pp 205-7
[25] Molyneux P 2004 The use of the stable free radical diphenylpicrylhydrazyl (DPPH) for estimating antioxidant activity *J. Sci. Technol* **26(2)** 211-9
[26] Nand L 2016 Effects of acid rain on plant growth and development *e-Journal of Science & Technology (e-JST)* **11(5)** pp 85-108
[27] Senmiao T and Zhang L 2014 Differential sensitivity of growth and net photosynthetic rates in five tree species seedlings under simulated acid rain stress *Pol. J. Environ. Stud.* **23(6)** pp 2259-64
[28] Vicas S I, Grosu E and Laslo V 2009 The effects of simulated acid rain on growth and biochemistry process in grass (Lolium perenne) *seria Agronomie* **52(1)** pp 277-83
[29] Bridget O and Adekunle O E 2015 The effect of simulated acid rain on growth component of cowpea (*Vigna unguiculata* L. Walps.) *Jordan Journal of Biological Sciences* **8(1)** pp 51-4
[30] Du E, Dong D, Zeng X, Sun Z, Jiang X and Vries W 2017 Direct effect of acid rain on leaf chlorophyll content of terrestrial plants in China *Science of The Total Environment* **605-606** pp 764-9
[31] Parida A K, Dagaonkar V S, Phalak M S and Aurangabadkar L P 2008 Differential responses of the enzymes involved in proline biosynthesis and degradation in drought tolerant and sensitive cotton genotypes during drought stress and recovery *Acta Physiologiae Plantarum* **30** pp 619-27
[32] Sharma S and Verslues P E 2010 Mechanisms independent of ABA or proline feedback have a predominant role in transcriptional regulation of proline metabolism during low water potential and stress recovery *Plant Cell and Environment* **33** pp 1838-51
[33] Szekely G *et al.* 2008 Duplicated P5CS genes of Arabidopsis play distinct roles in stress

IOP Conf. Series: Earth and Environmental Science **129** (2018) 012020 doi:10.1088/1755-1315/129/1/012020

regulation and developmental control of proline biosynthesis *Plant Journal* **53** pp 11-28

[34] Ben Hassine A, Ghanem M E, Bouzid S, and Lutts S 2008 An inland and a coastal population of the Mediterranean xero-halophyte species *Atriplex halimus* L. differ in their ability to accumulate proline and glycine betaine in response to salinity and water stress *Journal of Experimental Botany* **59** pp 1315-26

[35] Dalvi U S, Chavan U D, Kachare D P and Naik R M 2007 Short communication proline metabolism in sorghum and chickpea cultivars during water stress *Indian J. Plant Physiol* **12(3)** pp 287-9

[36] Szabados L and Savoure A 2010 Proline: a multifunctional amino acid *Trends in Plant Science* **15** pp 89-97

[37] Jahantigh O, F Najafi, H N Badi, Khavari-Nejad R A and Sanjarian F 2016 Changes in antioxidant enzymes activities and proline, total phenol and anthocyanine contents in Hyssopus officinalis L. plants under salt stress *Acta Biol Hung.* 67(2) pp195-204

[38] Lum M S, Hanafi M M, Rafii Y M and Akmar A S N 2014 Effect of drought stress on growth, proline and antioxidant enzyme activities of upland rice *J. Anim. Plant Sci.* **24(5)** pp 1487-93

Organic carbon sequestration under selected land use in Padang city, West Sumatra, Indonesia

Yulnafatmawita[1,3] and S Yasin[2]

[1] Soil Physics Laboratory, Agriculture Faculty Universitas Andalas, Kampus Unand Limau Manis Padang, 25163 Indonesia
[2] Soil Chemistry Laboratory, Agriculture Faculty Universitas Andalas, Kampus Unand Limau Manis Padang, 25163 Indonesia
[3] Corresponding author: yulna_fatmawita@yahoo.com

Abstract. Organic carbon is a potential element to build biomass as well as emitting CO_2 to the atmosphere and promotes global warming. This research was aimed to calculate the sequestered Carbon (C) within a 1-m soil depth under selected land use from 6 different sites in Padang city, Indonesia. Disturbed and undisturbed soil samples were taken from several horizons until 100 cm depth at each location. Soil parameters observed were organic carbon (OC), bulk density (BD), and soil texture. The result showed that soil OC content tended to decrease by the depth at all land use types, except under rice field in Kurao-Nanggalo which extremely increased at >65 cm soil depth with the highest carbon stock. The soil organic carbon sequestration from the highest to the lowest according to land use and the location is in the following order mix garden- Kayu Aro > mix garden- Aie Pacah > Rangeland- Parak Laweh >seasonal farming- Teluk Sirih > rice field- Kampuang Jua.

1. Introduction

Carbon (C) is considered to be one of the elements contained in the greenhouse gases (CO_2, CH_4, CFCs, HCFCs and HFCs). Therefore, an effort to sequester C becomes one among the answers for climate change control. As stated by Lal [1] that C-sequestration was the net rate of atmospheric CO_2 reduction by transferring into other long-lived global pools involving oceanic, pedologic, biotic, and geological strata. The C-sequestration is a method that might help to limit emissions and have potential to alleviate the risks of climate change [1,2].

A carbon sink is defined as storing CO_2 removed from the atmosphere or captured from emissions in other forms. Carbon is stored within terrestrial ecosystems for about 2110 Pg, almost three times the amount in the atmospheric CO_2, with 74% is stored by soils [1]. Soil organic carbon is estimated to be 684–724 Pg of C in the upper 30 cm, 1462–1548 Pg of C in the upper 100 cm soil profile [3].

The quantity of carbon sink in soil is determined by the types of land use and the management applied in each location. As found by Liu *et al.* [4] that the amount of biomass input into the soil was strongly affected by land-use. Plants capture CO_2 from and release O_2 to the atmosphere during photosynthesis process. Therefore, the more vegetation growing on a land more carbon is sequestered or removed from the atmosphere. Carbon stocks in soil decreased by 0.39 Mg/ha/y due to land conversion [5]. Moreno *et al* [6] found that the highest SOC stocks within a 4-m depth in tropical Colombia were found under primary forest (227.9 Mg ha^{-1}) > secondary forest (192.5 Mg ha^{-1}) > pasture (171.2 Mg ha^{-1}). Soil OC and easily decomposed OC in aggregates under grassland and forestland were higher than those under farmland.

Content from this work may be used under the terms of the Creative Commons Attribution 3.0 licence. Any further distribution of this work must maintain attribution to the author(s) and the title of the work, journal citation and DOI.
Published under licence by IOP Publishing Ltd

Johnson *et al* [7] stated that agriculture can be a source of greenhouse gases and it can also be a sink for CO_2 through C sequestration into biomass products and soil organic matter (SOM). SOM can be found intra or inter soil aggregates protects from microbial attacks and known as physically protected OM. On the other hand, it is exposed and easily attacked by degrading microorganisms. Liu [4] found that soil OC and easily decomposed OC under farmland in the Loess Plateau of China concentrated in < 1 mm macro-aggregates, in others in > 1 mm. Soil OC and easily decomposed OC within aggregates in grassland and forestland were higher than in farmland.

Land use change is considered as the primary sector that contributes to greenhouse gas emissions. The CO_2-eq emissions increased about six times in 2005-2009 compared to that in 2000-2005 (681 006.94 tons of CO_2-eq per year) as agricultural land conversed to settlement for approximately 11.12% in Bogor, Indonesia from 2000-2009 [8]. Novara *et al.* [9] in Mediterranean areas found that land use conversion, vegetation type, and management practices, which control biogeochemical and physical soil properties, affected CO_2 emissions and SOC sequestration. Chen *et al* ([10] found that suburban and rural areas in Southern China had higher soil C content than that in an urban area.

The number of OC sequestered in soil is also affected by the soil properties, besides the types and density of plants. Yulnafatmawita [11] and Yulnafatmawita *et al* [12] found that types of soil texture affected SOM content. Clay content was found to be positive-linearly correlated with SOM content. The higher the clay particle percentage in a soil the more the SOM content. It was found that there was 18.8 g SOM/kg soil having 82% clay [13] and 12.3 g SOM/kg soil having 67% clay [12] on the surface of 20 cm soil depth in Ultisols Limau Manis Padang.

Furthermore, the amount of SOM is also determined by the management applied to the land. Cultivated soil used to contain 50-75% of the original SOC pool [1]. Under clay textured soil being intensively cultivated, for example, the percentage of SOM could be lower than that under the loamy textured soil in a forest. A forest with plenty of woods is potential to catch carbon (in form of CO_2) from the atmosphere, as well as at land which is never cultivated accumulates carbon (in form of SOM) in the soil.

Land use in a city is dominated by building either for private or public facilities. In Padang City, a capital of West Sumatra Province, some land has been uncovering, especially the area close to the coastline, and some is still being covered by primary forest (±51%) based on statistical data [14]. The rest was for agriculture (rice field, mix garden, bush, seasonal farming), rangeland, animal husbandry and fishery, then for housing and other buildings, infrastructure, and other facilities. How much carbon was sequestered in different types of land use in an urban area under wet tropical climate is interesting to study. This research was aimed to identify the carbon sequestration under different types of land use in Padang city, Indonesia.

2. Material and methods

2.1. Research site

Our study was focused in Padang, a mid-sized city in Indonesia, covering an area approximately 694.96 km^2, having population 902,413 with the density was 1,299 people/km^2 [14]. It is located on the West Coast of Sumatra, Indonesia (0°44′-1°08’ S, 100°05’-100°34’E) (Fig. 1) with the altitude 0-1853 m *asl*. The region has a wet-humid tropical climate, receiving 3,552 mm of precipitation in 2016 and average annual daily minimum and maximum temperatures of 26.10°C and 31.30 °C, as well as 73% and 95% relative humidity, respectively. The climate of the city belongs to wet tropics, classified as *Af* (very wet) class according to Smith and Ferguson [15].

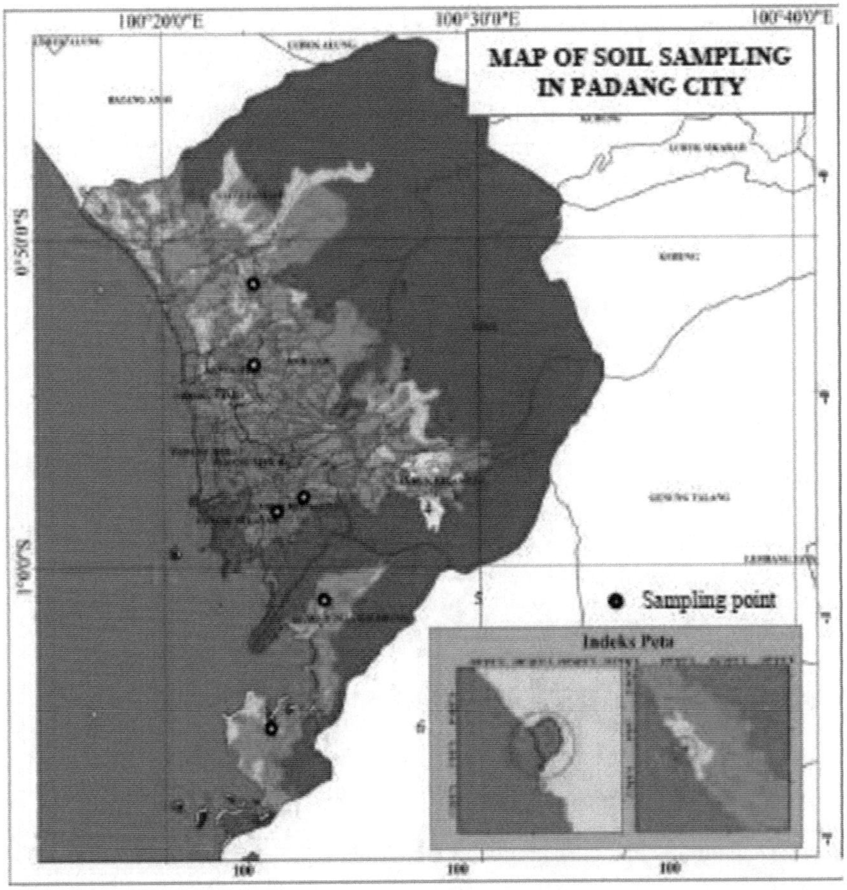

Figure 1. Map of research site and soil sampling points

2.2. Soil sampling

Soil sampling was taken from 4 different types of land use, rice field (Kurao Nanggalo and Kampuang Jua), mix garden (in Kayu Aro and Aie Pacah), seasonal farming (in Teluk Sirih), and rangeland (in Parak Laweh) in June 2016.

Disturbed (using mineral soil bore) and undisturbed (using 4 cm x 7 cm core) soil samples were taken from each sampling location from the topsoil until 100 cm soil depth. There were 6 sites being sampled, those were listed in Table 1. Soil bulk density (ρ_b) was measured by using the gravimetric method, and the data was calculated using the following formula:

$$\rho_b\,(Kg\ m^{-3}) = \frac{Dry\ Weight}{Total\ Volume} \quad (1)$$

Table 1. Types of land use selected in Padang City.

No	Land Use	Location	Code	Geographical Position
1	Mix Garden-AP	Aie Pacah	AP	S: 00°51'25.6" E: 100°22'54.4"
2	Rice field-KN	Kurao Nanggalo	KN	S: 00°53'55.2" E: 100°22'56.7"
3	Rice field-KJ	Kampuang Jua	KJ	S: 00°57'53.5" E: 100°24'30.6"
4	Rangeland-PL	Parak Laweh	PL	S: 00°58'16.2" E: 100°23'30.3"
6	Mix Garden-KA	Kayu Aro	KA	S: 01°01'1.5" E: 100°25'2.0"
5	Seasonal Farming-TS	Teluk Sirih	TL	S: 01°05'1.2" E: 100°23'22.3"

2.3. Soil preparation and analysis
Core soil samples were dried at 105 °C for 48 h and subsequently weighed for soil bulk density determination. The bulk soil sample was air dried under laboratory, then homogenized into a fine powder in an agate ball-mill. The ball-milled sample was then passed through a 2-mm sieve [16]. This was used for particle size distribution analyses. Then the 2 mm soil samples were further milled, sieved using 0.5 mm sieve for soil OC analyses. Soil OC content of the samples was determined using wet oxidation method [17]. Amount of OC sequestered in soil was employed the following formula:

$$Organic - C\ \left(\frac{kg}{ha}\right) = (\%\ Org - C)x\ BD\ x\ soil\ volume\ in\ 1\ ha \qquad (2)$$

3. Results and discussion

3.1. Research sites

3.1.1. Aie Pacah. Land use in Aie Pacah was in form of mix garden having some trees, such as jackfruit, bananas, petai, durian, mango, coconut trees, and some other tropical trees with quite low density. The land was not tilled for crop cultivation, therefore, the soil surface was covered by grass. According to the land owner, the area was less productive.

3.1.2. Kurao Nanggalo. Sampling area in Kurao Nanggalo was under rice field. The area was continuously used for rice cultivation all year long or without rotation. The rice biomass used to be burnt by farmers after harvest time. They used to cultivate the land 3 times a year for rice cultivation. Rice crops were fertilized using synthetic fertilization as recommended.

3.1.3. Kampung Jua. Sampling area in Kampung Jua was under rice field. The rice field culture was also without rotationFarmers here also used to burn rice biomass after being harvested.

3.1.4. Parak Laweh. Sampling area in Parak Laweh was under rangeland or open land, without any cash crops. Types of vegetation growing on the land were grass, imperata, melastoma, mimosa, etc. The short cycle life of the plants contributes the residue to be SOM after being degraded. This cycle run fast without any disruption to the soil.

3.1.5. Kayu Aro. Sampling area in Kayu Aro was under mix garden. The area of the land was cultivated for some tree crops. The area was not intensively cultivated like for vegetable crop farming.

3.1.6. Teluk Sirih. Sampling area in Teluk Sirih was under seasonal (annual crop) farming. The area of the land was cultivated for annual crops, or it was a rotation among seasonal crops especially food crops, such as chili, corn, etc.

3.2. Soil organic carbon concentration
Soil OC concentration at 6 different locations in Padang city is presented in Figure 2 a-f. In general, SOC concentration in Padang city was affected by the types of land use as well as the management applied to the land. The cultivated area including seasonal farming and rice field had the lowest amount of OC concentration. Tillage in seasonal cop farming causes SOM oxidation and CO_2 release into the atmosphere. As reported by Yulnafatmawita [18] the CO_2 emissions from ORP land use under rainforest in a temperate climate in Australia increased by the increasing of energy applied in cultivating the soil. Rice field had low SOM content due to farmer habit in burning biomass, returning it to the soil. Farmers just applied synthetic fertilizer as the nutrient sources for the rice.

Figure 2 presents soil OC concentration at each location. It can be seen that the percentage of OC concentration in soil profile tended to decrease by deeper soil depth from the soil surface, except under rice field in Kurao Nanggalo (Figure 2-a). Higher OC content on soil surface was found to be true since the source of soil OC was mainly derived from the soil surface, plant litters, waste of agriculture product, and animal decay. Yulnafatmawita and Hermansah [19] reported that lowering soil depth from surface soil decreased the soil organic C concentration in Ultisol Limau Manis under wet tropical area.

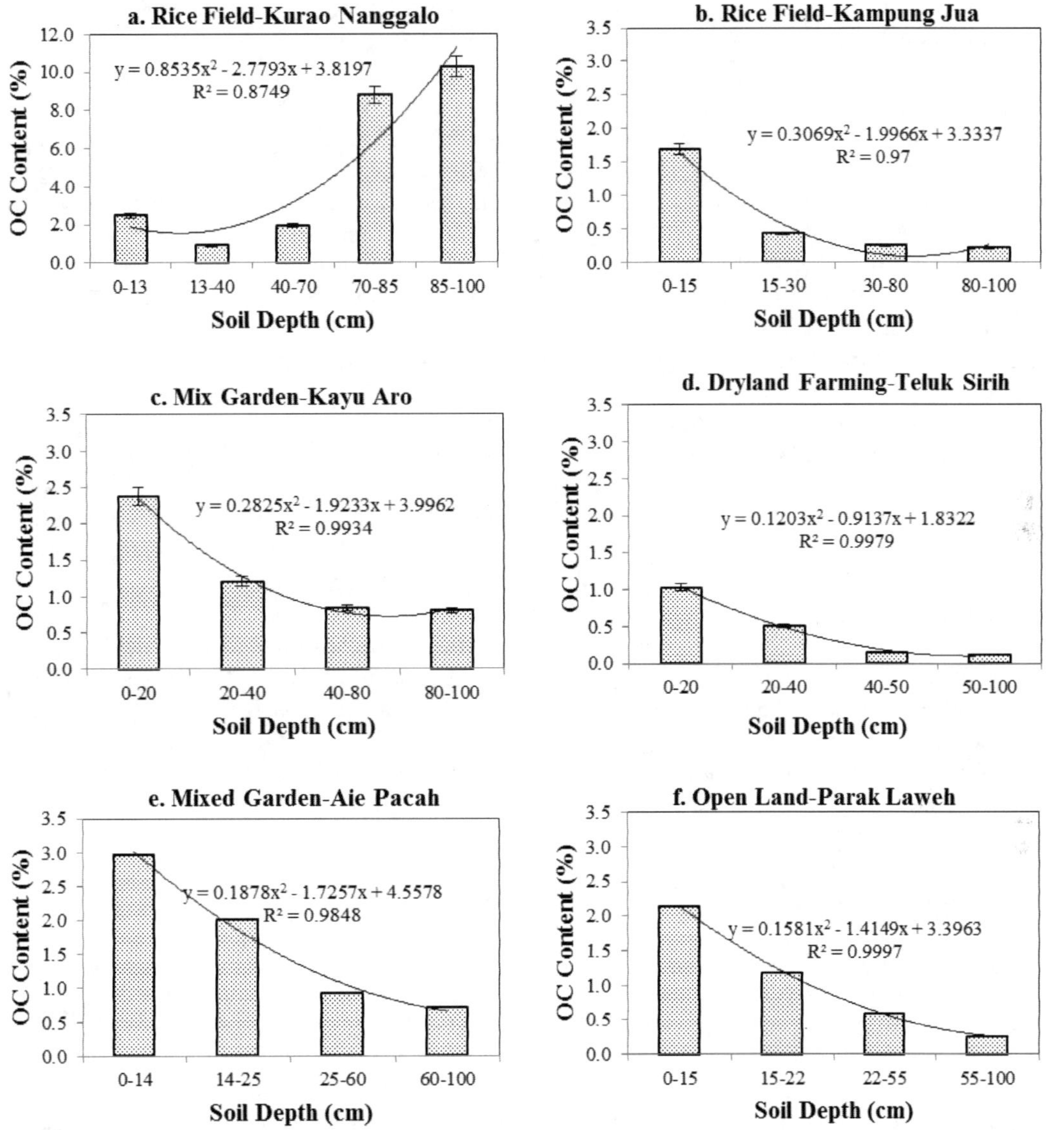

Figure 2. a-f. Soil Organic C percentage within a-100 cm soil profile under several types of land use in Padang City.

The OC found in lower depth within soil profile could be derived from soil microorganisms, root exudates and root senescent, and some are OM leached from the soil surface, even though there was only a little amount of SOM leached from the topsoil. The movement of the SOM was affected by the soil texture type, besides the amount of rainfall received in the area. Based on Yulnafatmawita *et al* [20] vertical movement of OC at coarser soil texture was higher than that at finer soil texture under super wet (annual rainfall reached 6,500 mm) tropical area in Bukit Pinang-Pinang West Sumatra, Indonesia.

Unlike common trend, organic C percentage under rice field in Kurao Nanggalo (Figure 2-a) extremely increased at depth >70 cm. The percentage of SOM at soil depth 85-100 was about 4.12 times higher than that at the top 13 cm soil surface. The OC form in the deeper soil profile was almost in pure OM. It seems that the area was previously a depression land and then filled with mineral matter due to flood routinely happened in that location. Since the rice field was not cultivated until 70 cm depth, the OC on that depth was sequestered. Nahlik and Fennessy [21] found that in the conterminous US wetland could store 11.52 PgC, which was mostly found in > 30 cm soil depth.

At the soil surface, OM content on the top 13 cm soil depth under rice field in Kurao Nanggalo was 2.49%. This is higher than the OC content under the same land use in Kampung Jua (1.69%) (Figure 2-b), which almost reached 1.5 times. The same tendency was also found under the second horizon of rice field, the OC content at a rice field in Kurao Nanggalo was approximately 2.07 times higher than that in Kampung Jua. This could be due to the higher sand percentage in Kampung Jua (20.1%) than that in Kurao Nanggalo (7.9%). The higher the sand particle percentage in a soil texture is the less the OM possibly bound by the mineral matters and the easier the OM degraded during good soil aeration, such as after rice harvesting. Besides that, finer particle soils have more micropores in which the OM contained in it was hard to be accessed by degrading organisms. Yulnafatmawita [18] reported that SOM content positively correlated to the finer soil particles.

Among the types of land use being study, mix garden-AP (Figure 2-e) contributed more percentage of OC (2.97%) at the top (0-14 cm) soil profile in Aie Pacah. This was 1.2 -2.9 times higher than OC contents at the topsoil profile in Kurao Nanggalo (Figure 2-a), Teluk Sirih (Figure 2-d), Parak Laweh (Figure 2-f), Kampung Jua (Figure 2-b), and Kayu Aro (Figure 2-c). Higher OC percentage under mix garden (Figure 2-c dan 2-e) was due to soil surface was covered by grass. The residue of grass contributed to high SOC by time. Since grass has a short life cycle, it could produce more biomass per year compared to perennial plants. Faster decomposition rate under seasonal and full sunlight received by the soil causes the grass residue decomposed and transform into SOC in relatively short time. As reported by Yulnafatmawita and Hermansah [19] that the more the grass covering soil surface was the higher the soil OC content under wet tropical area.

Organic C percentage under rangeland in Parak Laweh (Figure 2-f) was still higher than that in a rice field in Kampung Jua, seasonal farming in Teluk Sirih, and mix garden in Kayu Aro. Since the land was abandoned for the last few years, and it was just grown by bushes like *imperata, mimosa, melastoma*, as well as without any cultivation or soil disruption, the percentage of SOM in the rangeland increased by time.

3.3. Soil organic carbon sequestration

Soil OC sequestration at 6 different locations in Padang city is presented in Figure 3. Among the locations, OC sequestered under rice field in Kurao Naggalo was found to be very high. It was much higher than that under rice field in Kampung Jua, due to extremely high OC at depth ≥70 cm. This high OC content at lower soil depth was probably due to OM accumulation long time ago, much far beyond the rice field created. Therefore, the OC sequestration in Kurao Nanggalo is incomparable to other types of land use in Padang city to the depth of 100 cm soil profile.

Mix garden such as in Kayu Aro and Aie Pacah could sequester more OC. It was much higher than that under rice field in Kampung Jua, under dryland (seasonal crop) farming in Teluk Sirih, and under rangeland (open land) in Parak Laweh. Higher SOM sequestration in Aie Pacah was due to the short cycle of grass as well as management was given to the land. The area was not tilled because the crops being grown was perennial trees, as well as bananas which do not need soil tillage. Since the plants were not closely planted each other, the open land becomes covered by grass (*Axonopus campestris*) permanently causing high OM contribution to the soil. As reported by Deng et al [5] that change of land use types from farmland and forest into grassland increased soil C stocks 0.30 Mg/ha/y and 0.68 Mg/ha/y, respectively. Zhang *et al* [22] found that SOC stocks increase as cultivated land is conversed to grassland and therefore improve the potential for SOC sequestration in the surface soil over a decade period of time.

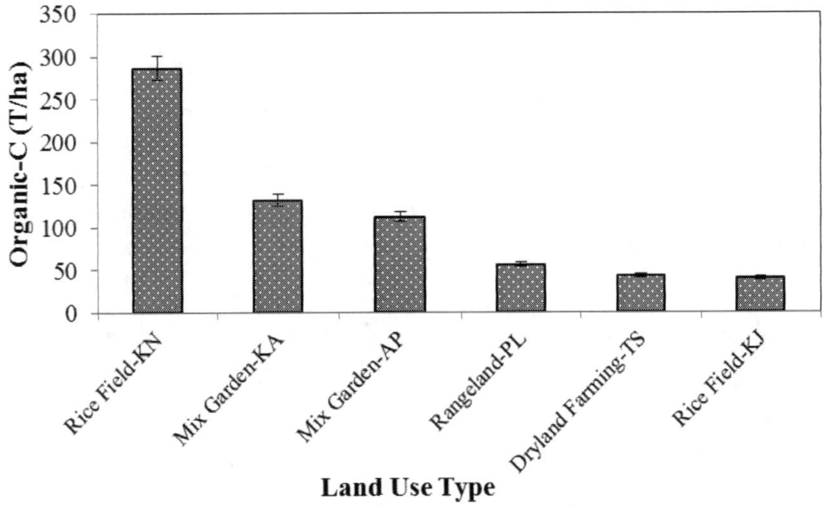

Figure 3. Organic-C sequestered within a-100 cm soil profile under several types of land use in Padang City.

Seasonal farming in Teluk Sirih sequestered low OC due to the types of crops being planted as well as the management given to the soils. Even though the seasonal crops can yield more SOM/year, less crop residue was left on the soil surface, and then the residue was degraded fast since the soil was tilled seasonally. Deng *et al* [5] found that soil C stocks significantly decreased (0.68 Mg/ha/y and 0.89 Mg/ha/y), respectively as grassland and forest conversed to farmland. Additionally, compared to Kayu Aro, Teluk Sirih had coarser soil texture on the top 20 cm soil. Coarse-textured soil means the soil has a large proportion of macropores causing good soil aeration. This condition increases decomposing organism activity in degrading and even oxidizing SOM, consequently, the percentage of SOM decreases.

3.4. Soil properties

Soil properties especially soil OC content, particle size distribution, bulk density, and texture at 6 different sampling locations were presented in Table 2. On the top 20 cm soil profile, soil bulk density was found to be in the medium criteria (> 600-<1140 Kg m^{-3}), it is not too soft and too hard to till or cultivate. It means that the soil could be used or continued for crop production. The value of soil bulk density determines the amount of OC sequestered within the soil profile. Ahukaemere [23] found that the average amount of C and N sequestered in the soil varied (from 3142.60-6139,74 g.m^{-2} to 101.33-388.6 g Nm^{-2}), it was affected by the soil bulk density besides the thickness of the soil horizon.

Table 2. Some soil properties at the top 20 cm soil profile.

Code	Land use	Location	BD	OC	Sand	Silt	Clay	Texture Class
			Kg m^{-3}	%				
AP	Mix Garden-AP	Aie Pacah	910	2.68	3.80	41.60	54.60	Silt Clay
KN	Rice Field-KN	Kurao Nanggalo	720	1.93	7.90	71.10	21.00	Silt Loam
KJ	Rice Field-KJ	Kampuang Jua	820	1.38	20.10	39.00	40.90	Clay
PL	Rangeland-PL	Parak Laweh	770	1.90	9.30	70.80	19.90	Silt Loam
TS	Mix Garden-KA	Kayu Aro	1090	2.38	9.60	39.10	51.30	Clay
KA	Seasonal farming-TS	Teluk Sirih	1130	1.03	16.60	71.60	11.80	Silt Loam

Soil OC also correlates with soil texture. The relationship between sand as well as clay and OC content is presented in Figure 4. Figure 4 showed that OC content was a negatively linear correlation with sand (R^2=0.7582) (Figure 4-a), and positively linear correlation with clay (R^2=0.5171) (Figure 4-

b) content of the soil tested. Lower OC content under highly sand containing soil (coarse texture) was due to good aeration in the soil, enough O_2 available, causing degrading microorganism activities become intensive. Therefore, plant residue cannot be accumulated in the coarse soil.

Higher OC content at clay dominated textured soil was due to the high specific surface area of the clay particles causing a large amount of OC could be adsorbed and protected from the degrading microorganisms. As found by Yulnafatmawita [11] that there was a positive linear relationship (R^2=0.8633) between clay and SOM content of a soil. Clay materials adsorb and occlude carbon in soil pores, and are not accessible to the main decomposing organisms, leading to stable complexes between Al-sites and organic compounds [24].

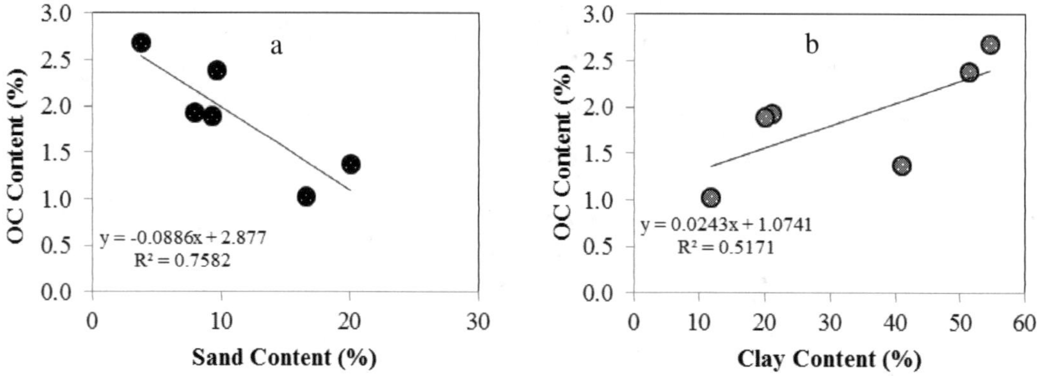

Figure 4. The relationship between soil particles ((a) sand and (b) clay) and OC content of the soil.

4. Conclusion

Different types of land use lead to different OC sequestered, which is also affected by was a land cover, land management, and the soil characteristics. Other than special case under rice field in Kurao Nanggalo, amount of OC sequestered was the highest under mix garden in Kayu Aro (132.11 t/ha/m depth) and Aie Pacah (112.54 t/ha/m depth). Then, it was followed by rangeland in Parak Laweh (55.30 t/ha/m depth). While the lowest OC sequestration was found under seasonal farming in Teluk Sirih (42.78 t/ha/m depth) and rice field in Kampung Jua (39.74 t/ha/m depth). In order to mitigate CO_2 emission to the atmosphere as a control for climate change, mix garden with grass covering the soil surface could be suggested to practice in a city. Then, OC sequestration in farmland could be increased if the farmers would like to improve their style in managing the soil.

Acknowledgements

Authors wishing to acknowledge Seprianto, Youngky Kurniawan, Farras Fadilla, Zaki Jumaika, Ultari Wahyuni for field soil sampling and to Desni Asrita as well as Erlinda for soil analyses in the laboratory.

References

[1] Lal R 2008 *Philos Trans R Soc Lond B Biol Sci.* **363**(**1492**) pp 815–30
[2] Ming T, Richter R de, Shen S and Caillol S *Environ. Sci Pollut. Res. Int.* **23**(**7**) pp 6119-38
[3] Batjes N H 2014 *Europ. J. Soil Sci.* 65(1) pp 10-21
[4] Liu M Y, Chang Q R, Qi Y B, Liu J and Chen T 2014 *CATENA* **115** pp 19-28
[5] Deng, Zhu Gy, Tang Zs and Shangguan Zp 2016 *Global Ecol. & Conserv.* **5** pp 127-38
[6] Moreno F, Oberbauer S F and Lara 2017 *Managing Forest Ecosystems: The Challenge of Climate Change* ed F. Bravo, V. LeMay, and R. Jandl 2nd Edition (Switzerland: Springer International Publishing) pp 367-83

IOP Conf. Series: Earth and Environmental Science **129** (2018) 012021 doi:10.1088/1755-1315/129/1/012021

[7] Johnson J M F, Franzluebbers A J, Weyers S L and Reicosky D C 2007 *Environ. Pollut.* **150** pp 107-24

[8] Setiawan G, Syaufina L and Puspaningsih N 2015 *J. Nat. Resourc. and Environ. Manag.* **5(2)** p 141

[9] Novara A, Gristina L and Mantia T L 2010 *Geophysical Res. Abstracts* **12** EGU2010-8468-3

[10] Chen H, Zhang W, Gilliam F, Liu I, Huang J, Zhang T, Wang W and Mo J 2013 *Biogeosciences* **10** pp 6609-16

[11] Yulnafatmawita 2005 Fractionation of Soils based on Bonding Energy and Aggregate Size: A Method for Studying the Effect of Structural Hierarchy on Degradation Process *Dissertation* Univ. of Queensland, Australia 219 (unpublished)

[12] Yulnafatmawita and Adrinal 2014 *Agrivita Agric. Sci. J.* **36(1)** pp 57-64

[13] Yulnafatmawita, Adrinal and Anggriani F 2013 *J. Tropical Soils* **18**(1) pp 33-44

[14] "Padang Dalam Angka" *https://padangkota.bps.go.id.* 2016 [Online] Available: https://padangkota.bps.go.id/index.php/publikasi/86 [Accessed 20-04-2017]

[15] Schmidt F H and Ferguson J H A "Rainfall Types Based on Wet and Dry Period Ratios for Indonesia with Western New Guinea [Monograph] (Verhandelingen No. 42, Kementerian Perhubungan, Djawatan meteorologi dan geofisika, Djakarta, 1951)" *http://library.wur.nl* 1951 [Online] Available: http://library.wur.nl/WebQuery/clc/436396 [Accessed 12-04-2017]

[16] Gee G W and Bauder J W 1986 *Methods of Soil Analysis Part 1* ed A. Klute ASA-SSSA 2nd Edition (Madison, Wisconsin: Soil Science Society of America, American Society of Agronomy) pp 383-411

[17] Nelson D W and Somner L E 1996 *Methods of Soil Analysis Part 3* ed D L Sparks ASA-SSSA 3rd Edition (Madison, Wisconsin: Soil Science Society of America, American Society of Agronomy) pp 961-1010

[18] Yulnafatmawita, So H B, Menzies N W and Dalal R C 2003 *Proceeding of the16th Triennial Int. Soil Tillage Res. Org.* M. Bell (Brisbane, Australia: International Union of Soil Sciences) pp 1396-402

[19] Yulnafatmawita and Hermansah 2016 *IJASEIT* **6(2)** pp 165-69

[20] Yulnafatmawita, Adrinal dan Hakim A F 2011 *J. Solum* **8(1)** pp 35-42

[21] Nahlik A M and Fennessy M S 2016 *Nat. Commun.* **7** 13835

[22] Zhang J H, Li F C, Wang Y and Xiong D H 2014 *Environ Manag.* **53(2)** pp 274-83

[23] Ahukaemere C M 2016 *Malaysian J. Soil Sci.* **20** pp 37-48

[24] Virakornphanich P, Wada S I and Wada K 1988 *J. Soil Sci.* **39(4)** pp 529-37

IOP Conf. Series: Earth and Environmental Science **129** (2018) 012022 doi:10.1088/1755-1315/129/1/012022

The ecological impacts of primary education facilities based on a child-friendly neighborhood unit criteria in Surakarta

E F Rini[1,2,3], R A Putri[1,2], Mulyanto[2] and N Handayani[1]

[1]Urban and Regional Planning Study Program, Universitas Sebelas Maret, Jl. Ir. Sutami 36A,Kentingan, Surakarta, Central Java 57126 Indonesia
[2]Center of Information and Regional Development (PIPW), LPPM, Universitas Sebelas Maret, Jl. Ir. Sutami 36A,Kentingan, Surakarta, Central Java 57126 Indonesia
[3]Corresponding author: ermafitria.plano@gmail.com

Abstract. A city should accommodate the citizen needs, especially for children. The absence of elementary school in a neighborhood unit (NU) will increase the use of transportation by children in the NU, every day at the same time. This activity will produce large quantities of - carbon dioxide (CO_2) that can trigger climate change. This article aims at discovering the ecological impacts of CO_2 emitted from the transportation used by children when commuting to their school, based on the conformity of each NU to the criteria of the a child-friendly city. Quantitative and spatial analysis techniques were employed in these four stages: (1) dividing the NU; (2) constructing the NU's typology based on a child-friendly criteria; (3) identifying the characteristic of children movements in each NU when accessing their elementary school; and (4) analyzing the ecological impacts (in CO_2 form). The result shows that 88.14% of CO_2 emissions in Surakarta can be reduced by interventions through the fulfillment of all NU's child-friendly criterias.

1. Introduction

The idea of child friendly city has become a concern in Indonesia since the enactment of the Minister of Women Empowerment and Child Protection Regulation Number 11 and 12 of 2011 on Cities and Children, including the availability of social facilities that accommodate the rights of children. Elementary School is one of the children's social facilities that should be available in a neighborhood unit (NU), considering the distance of children's walking capabilities [1,2,3,4,5,6]. The availability of elementary school in the NU can give the opportunity for children to access primary education services by walking. On the other hand, its unavailability will result in the increasing number of transportation, as the children use them to commute to their school. However, the simultaneous use of transportation (especially motorcycle) in a certain period of time will give ecological impacts on the region; for instance, the air pollution caused by the residual burning of motorcycle energy. Therefore, it is hyphotesized that the exhaust gases originating from transportation activities are the major contributors to greenhouse gas emissions, including CO_2. Since the characteristics of CO_2 gas that can directly react in the air and increase the temperature of Earth's surface, there is a need to reduce the source of CO_2 gas emissions.

Surakarta as a child-friendly city is the awardee of Nindya category in 2011 from the Ministry of Women's Empowerment and Child Protection [7]. For the purpose of discovering the concept of a child-friendly city and its relation to the existing ecological impacts, this city became an interesting object to study. Based on the assessment of NU in Surakarta (2016) and the number of population

Content from this work may be used under the terms of the Creative Commons Attribution 3.0 licence. Any further distribution of this work must maintain attribution to the author(s) and the title of the work, journal citation and DOI.
Published under licence by IOP Publishing Ltd

IOP Conf. Series: Earth and Environmental Science **129** (2018) 012022 doi:10.1088/1755-1315/129/1/012022

served, it is known that there is only a 36.27% of the 246 elementary schools that meet the standard capacity of elementary school services [8]. This indicates that the characteristics of the Surakarta NUs still has not fulfilled the requirements of the child's decent NU. Based on this background, this study aims at determining the ecological impacts of daily population movement activities in accessing basic education service facilities (elementary schools) in terms of applying the criteria of child-friendly NUs.

2. Methodology

2.1. Neighborhood Unit Distribution in Surakarta

For the process of identification and intervention, in this study, Surakarta is divided into several NUs. This division process is conducted through the use of ArcGIS mapping software. There are three aspects that serve as the main input: (1) a social scale facility around the elementary school distance (equal to 800 m), (2) the environmental administration boundaries (RT and RW), and (3) the physical boundaries.

2.2. NU Typology Formulation Based on child-friendly criteria

The criteria of NU child-friendly city used to assess the characteristics of NU in Surakarta are:

Table 1. The criteria and indicator used to identify suitability for each NU in Surakarta.

	Criteria	Indicator	
		NU zone characteristic	
		Suitable	Not suitable
1	Service capacity to residents	≤1600 residents/elementary school (ratio ≤0.000625)	>1600 residents/elementary school (ratio > 0.000625)
2	Children's safety and comfort in accessing social infrastructure	All criteria are suitable	There is a criterion which is not suitable
	a. Walking distance on foot to reach the school by children	≤800m to reach on foot	> 800m to reach on foot
	b. Free from unwanted traffic across the roads	All of the internal roads (which connected to the local or collector roads) meet indicators: • 3.5-6 meters wide of road • has a difference of width or pavement between the different road function to reduce the vehicle speed when entering the NU area	Not meet the two indicators
	c. Separator between pedestrian and main street	Has separator between pedestrian and main street on local roads, collectors, and arteries	No separator between pedestrian and main street
	d. Children's safety in crossing the street	- Utilize with zebra cross or pedestrian bridge - Utilize with school zone mark(ZoSS)	Not meet the two indicators
3	Access of elementary school utilization by all children including disabled	Inclusion school	Not accepting children with special needs

From the criteria above, the NUs can be formulated into eight typologies of NUs that are derived from a combination of the results of the conformity of those three criteria.

2.3. The Identification of Child Movement Characteristic in Accessing Elementary School
Movement characteristics in this study are the distance between school and the transportation. In this step, it can be known the distance traveled by each student to their school. From those data, we can analyze the average distance traveled by each respondents, either who choose to go to school inside or outside the NU. For the characteristics of transportation preferences, the respondent data will be classified based on the NU typology. The data are then analyzed to know the tendency of the transportation preferences of the respondents who attend school in and outside the NU.

2.4. Ecology Impact Analysis Based on Movement Characteristics

2.4.1. CO_2 Emission Based on Movement Characteristics. The energy used by the students to go to school produces a combustion gas that impacts on air condition around the student movement areas. To understand the impacts of the combustion gas, it is necessary to identify the sources of air pollutants and the number of emissions released.

Table 2. Vehicle Emission Factors [9].

Category	CO_2 (g/km)
Motorcycle	0.398
Passenger car	0.663
Urban Buses Standard	2.65

Meanwhile, the following formula can be used to calculate the amount of emissions in a region based on the vehicle types and its distance [10]:

$$TE = \sum_j \left(N_j \times M_j \times EF_{i,j}\right) \quad (1)$$

TE = Total emissions of gas in a region (g)
Nj = Number of vehicles type j used in a region
Mj = average distance per vehicle type j (km)
Efi, j = emission factor for pollutant i and vehicle type j (g/ km)

2.4.2. The Average of Existing CO_2 Gas Emissions in Each Typology. Based on the formula above, the process of calculating CO_2 emissions of existing gas begins by calculating the number of CO_2 emissions in each vehicle type in each NU. After knowing the amount of emissions of existing CO_2 gas in each type of vehicle in each NU, the total amount of CO_2 emissions in each typology is calculated to know the total number of CO_2 emissions in the typology.

2.4.3. The Average of CO_2 Gas Emissions After Intervention on Each Typology. Interventions are made if the three criteria of child-friendly NUs are fulfilled in all NU. The intervention is formulated by multiplying the number of students on each NU intervened by contants. The constants are obtained from the average of each student's emissions on the ideal NU.

3. Result

3.1. The Distribution of NU in Surakarta
Surakarta has totally 263 elementary schools (or those that are equivalent with public elementary schools), private elementary schools, Madrasah Ibtidaiyah (MI) and schools for children with special needs (SLB). Based on the coverage analysis of elementary schools (800 m); overall, the existence of elementary schools in Surakarta has been able to serve almost all areas of Surakarta. The area that is not served by the existence of elementary schools/equivalent is situated in the West Surakarta,

precisely in the westernmost Karangasem Village. However, the area is still served by the existence of other elementary schools that are administratively located outside Surakarta.

In deciding the limit of the NU, administrative boundaries are used with local service scales, such as neighborhood units (NU) and community units (CU). Some limitations in determining NU are necessary because each NU can consist of 30-50 family with four persons in each family. Thus, the provisions in a NU is related to the standard of elementary school services serving (<1600 people) which can be seen from the administrative boundaries of NU and CU. Ideally, the NU has alleys inside and surrounded by road functions higher than alley. But it is not used as the consideration in determining the boundary of NU in Surakarta.

The results of the analysis of those three aspects revealed that there are 108 NUs that are close to elementary schools inside or outside the NU itself, and one NU that does not have any school facility.

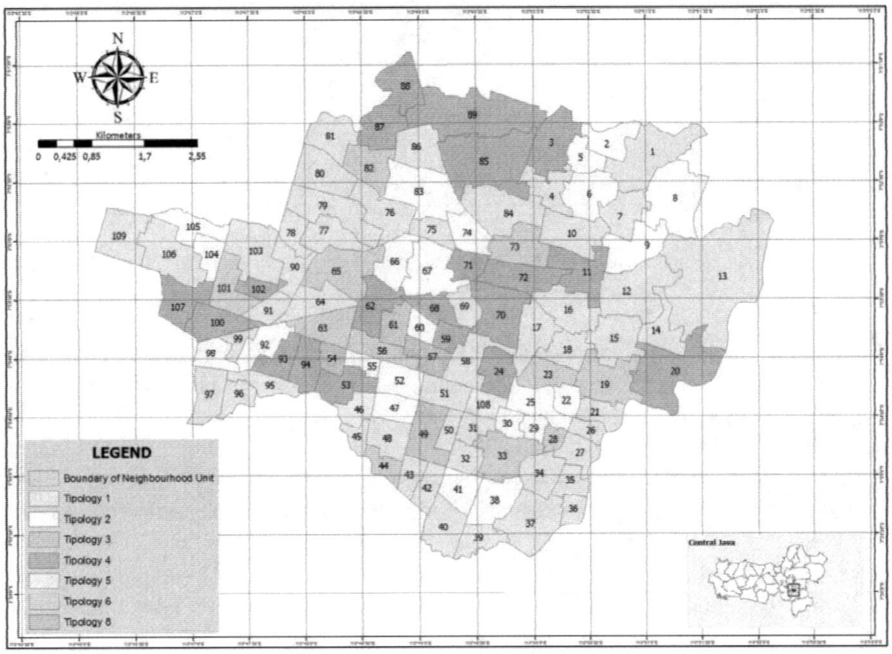

Figure 1. Neighborhood unit distribution in Surakarta.

3.2. Neighborhod Unit Typology in Surakarta based on Child-friendly city criteria
The method of grouping the NU in Surakarta to each typology starts with the identification of each NU based on the three criteria of child-friendly city.

3.2.1. Elementary School Services Capacity Towards the Total Population. The elementary school service capacity criteria are used to identify the existence of elementary schools in a NU as well as its capacity to serve a certain population. A NU can meet the criteria of the capacity of elementary school services to the population if its entire population is able to be served by the elementary schools within the same NU. In calculating the number of residents, it is assumed that every house is inhabited by five family members. Meanwhile, one elementary school unit is supposed to serve 1600 residents. The result indicated that there are 38 NUs that meet the criteria of elementary school service capacity towards the total population. On the other hand, there are totally 71 NUs that do not meet that criteria.

3.2.2. The Children's Safety and Comfort in Accessing Social Infrastructure. A NU can be said to meet the criteria of children's safety and comfort in accessing social infrastructure if it satisfies the following four sub-criteria:

IOP Conf. Series: Earth and Environmental Science **129** (2018) 012022 doi:10.1088/1755-1315/129/1/012022

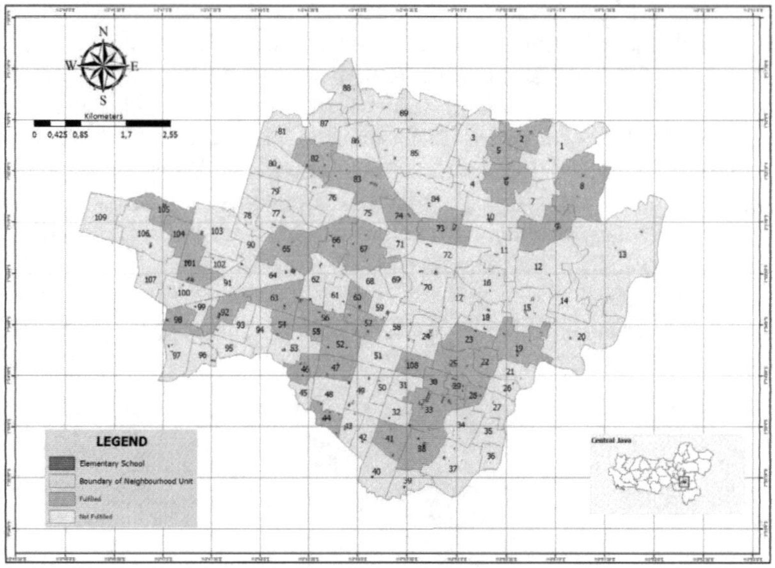

Figure 2. The suitability of NUs in Surakarta with the criteria of elementary school services capacity towards the total population.

3.2.2.1. The distance to reach elementary school. Ideally, the maximum distance for students to walk to their school is 800 m, which is the distance between the school and NU. From the result of analysis, there are totally 108 NUs that meet the criteria whereas one NU do not satisfy the criteria.

3.2.2.2. Free from unwanted traffic across the road. There is one NU that is traversed by the alley and limited by the road function above the neighborhood. The difference of road function as the boundary of the NU must also be accompanied by traffic calming utilities to enter the NU area. Those traffic calming utilities are the difference of width and pavement between the roads and portal at the entrance of the NU. From the result of identification, it is known that there are 30 NUs that meet the criteria of the free traffic jam, while 79 NUs do not satisfy the criteria.

3.2.2.3. The existence of a separator between pedestrian and main street. It can be road markings or the clarity of the boundary between the main road and the pedestrian, either with a barrier or high street difference. The analysis showed that 76 NUs meer the sub-criteria of separation between pedestrian and the main road, while 33 other NUs do not.

3.2.2.4. Children's safety in crossing the street. The fulfillment of criteria for children's safety in crossing the street can be identified from the existence of child safety support facilities in the traffic. The facilities can be in the form of zebra cross, pedestrian bridges, and school zone marks (ZOSS). Based on the result of identification, there are 33 NUs that were appropriate whereas 76 other NUs were considered to be inappropriate.

Based on the four criteria above, the result of the assessment showed that six neighborhoods meet all those four sub-criteria while the other 103 NUs only meet several sub-criteria. The following is a map of the NUs suitability in Surakarta with the criteria of free traffic.

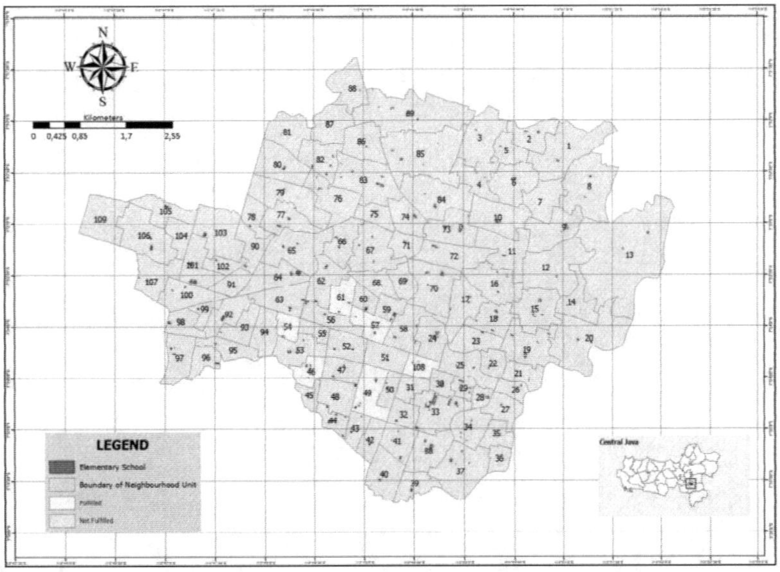

Figure 3. The suitability of NUs in Surakarta with the criteria of children's safety and comfort in accessing social infrastructure.

3.2.3. The Access to Elementary School Utilization. The identification of elementary schools existence that can be accessed by all children, including the disabled, is conducted through listing the existing inclusion schools in Surakarta. The inclusion school is a public elementary school (not elementary schools for disabled) appointed by the government of Surakarta to educate children with special needs.

The result of analysis revealed that there are 26 NUs that meet these criteria, while the other 83 NUs do not. The following is a map of the conformity of the NU in Surakarta with the criteria of the access of elementary schools utilization by all children.

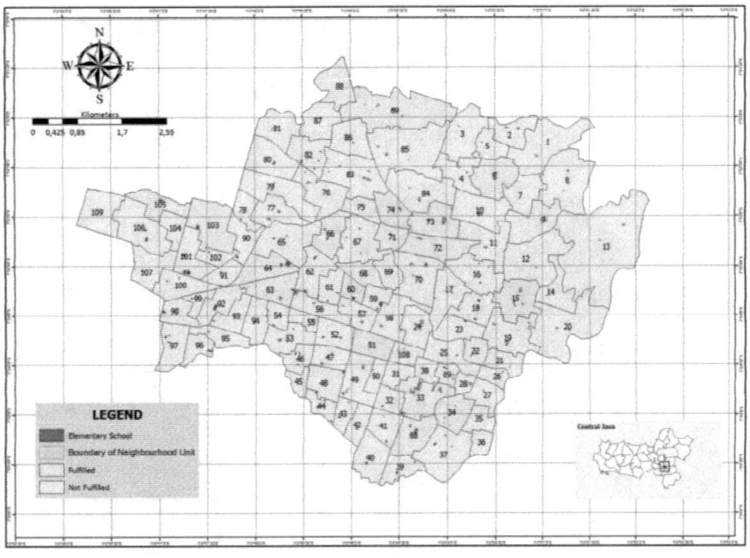

Figure 4. The conformity of NU with the criteria of elementary school utilization by all children including disabled.

After identifying the NU conformity in Surakarta based on three criteria of the child-friendly neighborhood, the NU is then classified into a typology that is in accordance with the results of the criteria assessment. The following are the distribution of NU based on the typology that has been formulated before.

Table 3. Neighborhood unit grouping based on typology.

Typology	Conformity criteria (1-2-3)	Neighbourhood unit	Total	Percentage
1	NS-NS-NS	1-4-7-10-12-13-14-15-16-17-18-21-26-27-31-32-34-35-36-37-39-40-42-43-45-48-50-51-58-64-69-75-76-77-78-79-80-81-84-86-90-91-95-96-97-99-103-106-109	49	44.95%
2	S-NS-NS	2-5-6-8-9-22-25-29-30-38-41-47-52-55-60-66-67-74-83-92-98-104-105	23	21.10%
3	NS-S-NS	49-61	2	1.83%
4	NS-NS-S	3-11-20-24-53-59-62-68-70-71-72-85-87-88-89-93-94-100-102-107	20	18.35%
5	S-S-NS	46-108	2	1.83%
6	S-NS-S	19-23-28-33-44-56-63-65-73-82-101	11	10.09%
7	NS-S-S	-	0	0%
8	S-S-S	54-57	2	1.83%

S = Suitable NS = Not suitable

3.3. The Identification of Child Movement Characteristic in Accessing ElementarySchool in Surakarta

3.3.1. Distance Characteristics. Identification of distance characteristics can be analyzed by using the distance data of respondent of elementary school students/equal from their house to school. The distance characteristic is divided into 2 groups, first is a respondent who goes to school inside residence and second is he/she who goes to school outside his/her NU residence.

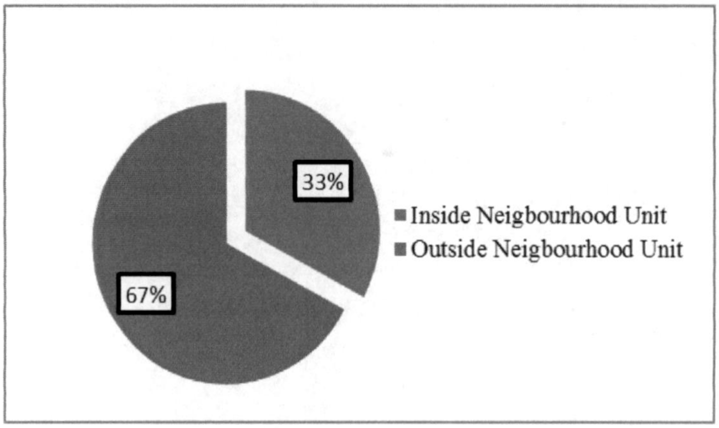

Figure 5. Students' movement diagram.

Table 4. Students' average distance on each typology.

Typology	Distance average (Meter)	
	inside *NU*	outside *NU*
1	232	1,696
2	264	1,688
3	297	1,307
4	302	1,917
5	301	1,437
6	292	1,672
7	-	-
8	291	1,390

The distance characteristic shows that the respondents likely tend to choose schools that are outside the NU residence. This is the character of most students in Indonesia who prefer to study in favorite schools. The number of movement outside the NU is greater than the movement inside the NU, with the average distance of respondents to the school reached more than 1 km or beyond the reach of elementary school services (800 meters). This distance can also indicate the use of vehicle by respondents. The farther the distance traveled, the greater the chance of using the motor vehicle. Therefore, it increases the number of emissions released by the vehicle.

3.3.2. Transportation Preference Characteristics. In this discussion, it will be identified the tendency of transportation preference used by respondents to go to school.

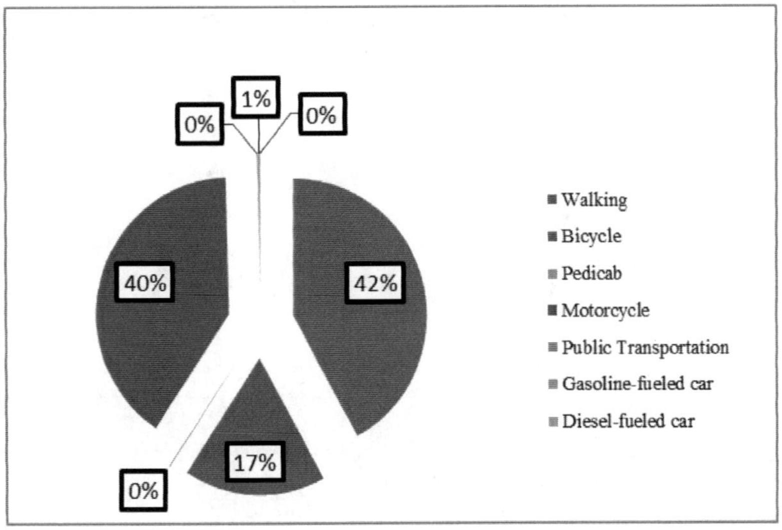

Figure 6. Respondents' preference of Transportation Mode Diagram inside the NU Zone.

The most preferred modes of transportation by the respondents who attend the same neighborhood school unit are walking (42%), motorcycle (40%), and bicycle (1%). The findings indicated that, although the distance between home and school in a NU is just less than 1000 meters, the number of motor vehicle usage is quite high.

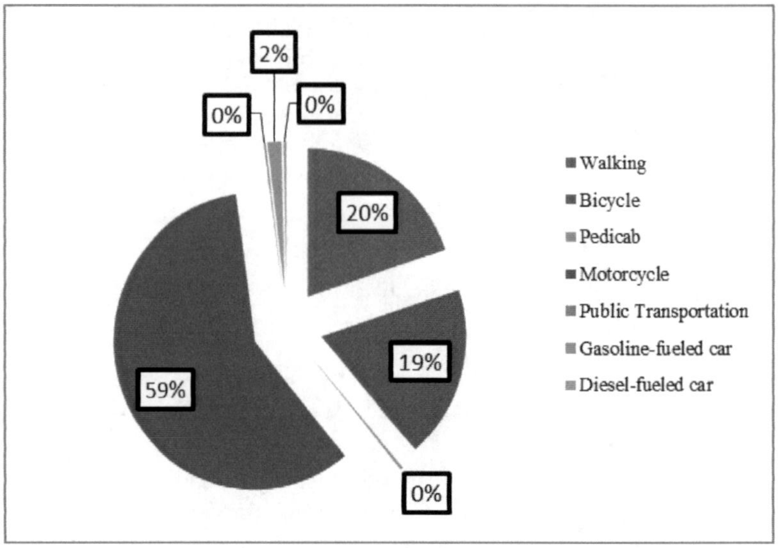

Figure 7. Respondents' transportation preference diagram outside the NU zone.

In the group of respondents who choose to go to school outside the NUs, the use of motorcycles shows a fairly high number, because it has a percentage of more than 50%. Based on the data above, it can be seen that the respondents who go to school outside the NU tend to choose motorized vehicles to go to their school.

Table 5. The preference type of transportation used by students.

Typology	Types of Transportation (%)						
	Walking	Bicycle	Pedicab	Motorcycle	Public transportation	Gasoline-fueled car	Diesel-fueled car
1	26	19	1	49	1	2	2
2	28	17	2	51	-	2	-
3	25	17	-	56	-	2	-
4	28	19	-	51	-	2	-
5	37	20	-	43	-	-	-
6	33	16	1	50	-	-	-
7	-	-	-	-	-	-	-
8	24	13	-	63	-	-	-

From Tabel 5, it can be seen that the use of motor vehicles, in general, is still the main choice of transportation used to go to school. Even in typology **8**, where there is a NU that meet all criteria of child-friendly, the use of motor vehicles reaches more than 50%.

3.4. Ecological Impact Analysis based on Movement Characteristics

3.4.1. Existing CO_2 Gas Emissions. Transportation is one of the activities that produce air pollutants in the form of CO_2, contributing to a global warming index [11]. The movement of children to elementary school also has great potential in increasing the amount of air pollutants, especially when the children's companions use motor vehicles. The following is the calculation of the average amount of existing CO_2 gas emissions in each typology based on the use of motor vehicles.

Table 6. Average Existing CO_2 Gas Emissions.

Typology	The average of CO_2(g/day)
1	2,959,728.04
2	203,007.63
3	112,615.69
4	274,571.70
5	250,074.146
6	316,205.91
7	-
8	49,578.86
Surakarta City	1,463,211.80

3.4.2. CO_2 Gas Emissions After Intervention. Interventions are conducted in a condition where the entire NU in each typology has met all criteria of a child-friendly NU. Then, the number of CO_2 in every NU in each typology is calculated with the following results.

Table 7. The Average of CO_2 Gas Emissions After Intervention

Typology	The average of CO_2(g/day)	The average changes of CO_2(%)
1	174,730.02	-94.10
2	187,023.61	-7.87
3	111,550.48	-0.95
4	166,934.31	-39.20
5	141,123.32	-43.57
6	192,183.90	-39.22
7	-	-
8	49,578.86	0
Surakarta City	173,582.83	-88.14

From the results of calculation, it can be seen that all typologies have decreased the number of CO_2 after all criteria of child-friendly NUs are met in every NU in each typology.

Typology 1 shows the decreasing number of CO_2 emissions after the intervention. The characteristics of NUs in typology 1 do not meet all criteria of child-friendly. However, the data shows that the intervention process in this typology plays a role in decreasing the amount of CO_2 emissions.

Other typologies also show that the number of CO_2 emissions are decreased after the intervention. On the other hand, the NU group on typology 8 is not intervened because it has already fulfilled all criteria of child-friendly NU.

The data also reveals that the number of CO_2 emissions released by motor vehicles from the school children's shuttle activity in Surakarta is decreased. That can be done by fulfilling all three criteria of child-friendly in each NU.

4. Conclusion

The findings showed that Surakarta residences do not consider the child-friendly NU criteria in choosing an elementary school. This can be seen from the majority of student who prefer to study in the schools outside their NU. It certainly leads to greater emissions due to the movement of students when commuting to and from their schools every day. Simulated interventions in each typology showed that fulfillment of all three child-friendly NU criteria can reduce overall CO_2 emissions in Surakarta by 88.14%.

Acknowledgments
Authors would like to thank all the elementary school students who fulfilled the questionaire as a part of the data in this research.

IOP Conf. Series: Earth and Environmental Science **129** (2018) 012022 doi:10.1088/1755-1315/129/1/012022

References

[1] Chiara D J, Panero Jand Zelnik M 1995 *Time Saver Standards for Housing and Residential Development* 2nd Edition (United States of America : Mc Graw-Hill Inc.) p 207
[2] Gallion A Band Eisner S 1986 *The Urban Pattern* (New York, USA : Van Nostrand Reinhold Company Inc.)p 298
[3] Mc Millan T E 2005Urban Form and a Child's Trip to School : The Current Literature and Framework for Future Research *Journal of Planning Literature 05* **19(4)** pp 440-56
[4] Perry C 1929 *Neighborhood and Community Planning, Chapter II The Neighborhood Unit* Ed: Regional Survey of New York and its Environs 1974 Volume VII, Monograph One, p 34 21-140. (New York: Arno Press)
[5] Porteous J D 1977 *Environment and Behaviour: Planning and Everyday Urban Life* (Massachusetts: Addison-Wesley Publishing Company) p 72
[6] Reiner T A 1968 *The Place of The Ideal Community in Urban Planning* (Philadelphia: University of Pennsylvia Press) p 60-62
[7] "Solo achieve the best city award program implementation towards child-friendly cities"*www.solopos.com* 2011 [Online] Available: http://www.solopos.com/2011/07/21/solo-raih-penghargaan-kota-terbaik-penyelenggaran-program-menuju-kla-108036 [Accessed: 29-08-2017]
[8] Putri R A, Subulussalam M D, Rahayu M J and Ramadhani A K 2016 A Spatial analysis for assessing the suitability of elementary school as social infrastructure at the neighbourhood unit scale in supporting child-friendly Surakarta *Geoplanning: Journal of Geomatics and Planning* **3(1)** pp33-52
[9] European Environment Agency 2009EMEP/EEA emission inventory guide book 2009 *Technical Guidance to Prepare National Emission Inventories* (Luxembourg: Publications Office of the European Union) Part 1.A.3.b.i-iv p 25 [Online] Available: https://www.eea.europa.eu/publications/emep-eea-emission-inventory-guidebook-2009 [Accessed: 20-10-2017]
[10] European Environment Agency 2013 EMEP/EEA air pollutant emission inventory guide book 2013 *Technical Guidance to Prepare National Emission Inventories* (Luxembourg: Publications Office of the European Union) Part 1.A.3.b.vii-vii p 13
[11] Samiaji T 2009 Efforts to Reduce CO_2 in the Atmosphere *Semi Popular Science Magazine: Aerospace News* **10(3)** pp 92-95

IOP Conf. Series: Earth and Environmental Science **129** (2018) 012023 doi:10.1088/1755-1315/129/1/012023

The rapid measurement of soil carbon stock using near-infrared technology

B H Kusumo[1,2], Sukartono[1] and Bustan[1]

[1] Department of Soil Science, Faculty of Agriculture, University of Mataram, JL. Majapahit No. 62 Mataram, Lombok, Indonesia.
[2] Corresponding author: bambanghk@gmail.com

Abstract. As a soil pool stores carbon (C) three times higher than an atmospheric pool, the depletion of C stock in the soil will significantly increase the concentration of CO_2 in the atmosphere, causing global warming. However, the monitoring or measurement of soil C stock using conventional procedures is time-consuming and expensive. So it requires a rapid and non-destructive technique that is simple and does not need chemical substances. This research is aimed at testing whether near-infrared (NIR) technology is able to rapidly measure C stock in the soil. Soil samples were collected from an agricultural land at the sub-district of Kayangan, North Lombok, Indonesia. The coordinates of the samples were recorded. Parts of the samples were analyzed using conventional procedure (Walkley and Black) and some other parts were scanned using near-infrared spectroscopy (NIRS) for soil spectral collection. Partial Least Square Regression (PLSR) was used to develop models from soil C data measured by conventional analysis and from spectral data scanned by NIRS. The best model was moderately successful to measure soil C stock in the study area in North Lombok. This indicates that the NIR technology can be further used to monitor the change of soil C stock in the soil.

1. Introduction

A global soil pool stores about 1500 Pg carbon (C) that is approximately three times higher than that in an atmospheric pool [1]. The disturbance of this pool through leaching, soil erosion and respiration of soil organisms may cause soil C loss that may turn into CO_2 to the atmosphere, causing global warming [1]. C loss through respiration is predicted to be the largest loss [2]. Monitoring the soil C stock will be very useful in observing CO_2 flux to the atmosphere.

However, monitoring soil C stock using conventional procedures is time-consuming and expensive. Large numbers of soil samples have to be collected, transported to the laboratory, dried, sieved and analyzed with appropriate and systematic procedures. Tedious procedures of laboratory measurements take a long time in handling a large number of samples. So it requires a rapid technique that is simple and does not need chemical substances.

Near-infrared (NIR) technology has become an important analytical technique in recent years [3]. It works based on the vibration of covalent bonds of small atoms such as C-H, O-H, and N-H. Soil reflectance of near infrared is a complex spectrum and contains rich information of chemical and physical composition of materials [4,5]. This technique has been successfully able to measure or map soil C and N [6,7,8,9]. It is also able to measure soil moisture content, organic matter, CEC, total C, ammonium N, nitrate N, total N, available P, and P absorptive coefficient [10], root density [11,12,7], and biochar properties [13].

Content from this work may be used under the terms of the Creative Commons Attribution 3.0 licence. Any further distribution of this work must maintain attribution to the author(s) and the title of the work, journal citation and DOI.
Published under licence by IOP Publishing Ltd

IOP Conf. Series: Earth and Environmental Science **129** (2018) 012023 doi:10.1088/1755-1315/129/1/012023

However, no information has been found related to the ability of this technique in measuring and mapping C stock in Indonesian soils. This study presents the ability of this technique to rapidly measure and map soil C stock in Kayangan sub-district, North Lombok, Indonesia.

2. Materials and methods

2.1. Soil sample collection, preparation and analysis

Topsoil (0-10 cm depth) from 305 locations were collected using soil corer (2.54 cm diameter) in Kayangan agricultural area, North Lombok, Indonesia (Figure 1). The coordinate of each sample was recorded, and then soil samples were transported to the laboratory to be dried (air dry), crushed and sieved to pass 0.2 mm diameter sieve. Part of each sample was analyzed for total soil C (Walkley and Black).

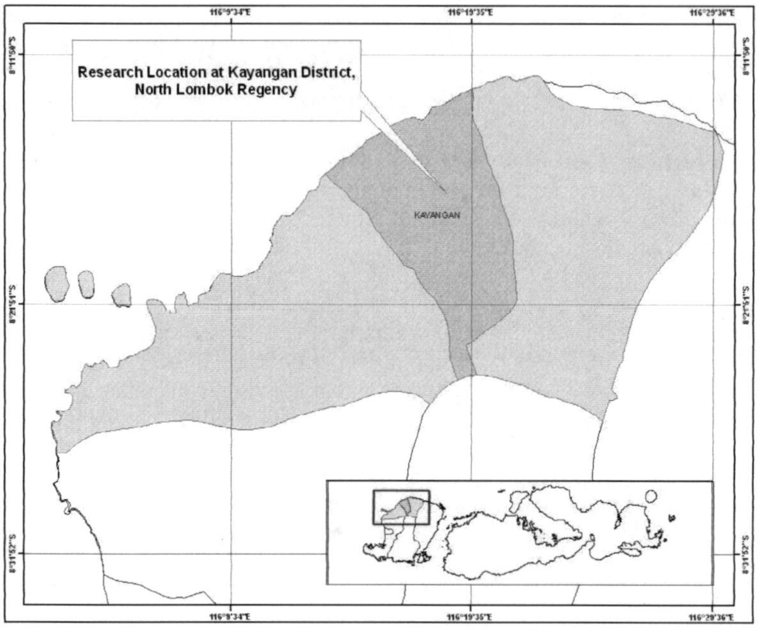

Figure 1. Location of soil sample collection.

2.2. Spectral acquisition and spectral pre-processing

Spectral reflectance (UV, visible and NIR range; 350 – 2500 nm) of another part of each sample were acquired using ASD FieldSpec 3 V-NIR Spectrometer (Analytical Spectral Device, Boulder, CO, USA). The spectral data were then imported into a software [ParLeS; 14] for spectral pre-processing namely: transformation to log (1/R) – R, wavelet detrending, and smoothing using a Savitzky-Golay filter. The smoothed data were thereafter processed into the first derivative, and then finally treated using mean centering [13].

2.3. Developing calibration models and parameters indicating accuracy of the model

Partial Least Square Regression (PLSR) was used to develop calibration models between the pre-processed spectral data and the reference analytical data of soil C. To avoid overfitting, the number of factors (principal components) that produce low root mean square error (RMSE) and low Akaike Information Criterion (AIC) were used to develop the models [14]. Then the models were tested using one-leave-out cross-validation. Afterwards, the ability of the PLSR models to predict soil C was assessed using the following statistics [13]: (i) RMSE (root mean square error) of measured and predicted soil C, (ii) coefficient determination (R^2), and (iii) RPD (ratio of prediction to deviation);

RPD is calculated as standard deviation of the reference data divided by root mean square error (SD/RMSE). The best prediction model is showed by the largest RPD and R^2, and the smallest RMSE.

3. Results and discussion

3.1. Summary of Soil Carbon

The summary of soil property (soil C) collected from Kayangan agricultural area, North Lombok, is shown in Table 1. The range (min-max) of soil C is narrow, from very low to medium, with no samples containing high C. Low content of C in this soil is probably related to the low return of organic matter and also due to coarse soil texture with low clay content. Soil with less clay and coarse texture has less ability to protect organic matter from biological degradation.

Table 1. Soil property.

Soil property	Range		Median	Mean	Variance	Standard deviation	Coefficient of variation (%)
	Min.	Max.					
Total C (%)	0.39	2.28	1.01	1.03	0.092	0.304	29.45

3.2. Soil Spectral Shape

As the C content is one of the most influential factors on the spectral shape, the soil spectral shape is shown based on the variation of C content (very low < 1% C; low 1-2% C; and medium 2-3% C) (Figure 2). Soil with very low C has a higher reflectance at around 750 nm compared to soils with low and medium C, showing a brighter color. Soil with higher C shows lower reflectance (darker color) at around visible band, indicating a higher amount of organic matter. Soil with very low organic matter content has a sharper angle at around 750 nm, compared to low and medium organic matter content; this was also found by Dematte et al. [15] and Kusumo et al. [6]. Strong absorption at around 1400 and 1900 nm (Figure 2) is the first overtones of the O-H bond of water and the combination of the H-O-H bend and O-H stretching, respectively [17]. Strong absorption at around 2200 nm is a combination of metal O-H bend plus O-H stretch [17].

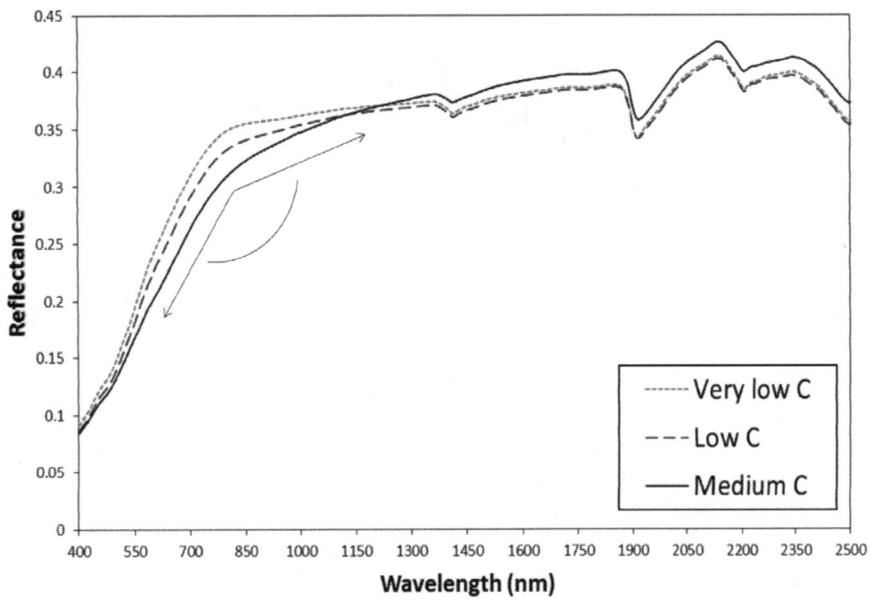

Figure 2. The spectral shape of soils with very low, low and medium amount of carbon.

3.3. Accuracy Measurement of Soil C using NIRS

Prediction values of soil C using leave-one-out cross-validation are shown in Table 2. NIRS technique was able to moderately predict soil C (RPD around 2.00). Both methods produce the average of soil C content 1.03%. According to Chang et al. [17], the prediction values of soil properties with R^2 0.5-0.8 and RPD 1.4-2.0 are considered moderately successful. While Malley et al. [18] considered moderate accuracy if the R^2 0.7 – 0.8 and RPD 1.75 – 2.25. Some factors may influence the accuracy of near-infrared prediction models, such as less accuracy of laboratory analysis as the reference data, spectral outliers or both. Other chromophores (such as water, decomposition level of organic matter, clay and non-clay soil minerals, carbonates, iron oxides, particle size) [19] may also influence the accuracy.

Table 2. Prediction values of soil C using leave-one-out cross-validation.

Properties	Prediction values (leave-one-out cross-validation)		
	R^2_{CV}	$RMSE_{CV}$	RPD_{CV}
C Total	0.756	0.151	2.01

The relationship between laboratory measurement and NIRS prediction of soil C is shown in Figure 4. The moderate accuracy of NIRS measurement on soil C (R^2_{CV} 0.75-0.76 and RPD_{CV} 2.00-2.01) shows that this technique may be used to measure and map soil C in the Kayangan agricultural area North Lombok. Some previous researchers also found the moderate accuracy of C measurement using NIRS technique [6,17].

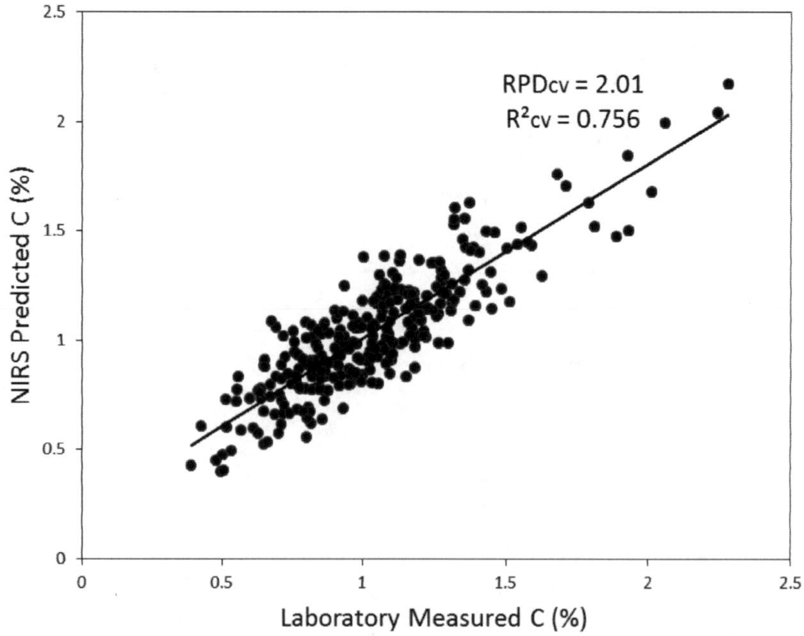

Figure 3. The relationship between laboratory measurement and NIRS prediction of soil C; from the C variation values, both methods produce the average of soil C content = 1.03%.

3.4. Successful Mapping of Soil Carbon

The map of soil C content produced from laboratory analysis and NIRS analysis is shown in Figure 5. It can be seen that soil C can be successfully mapped using NIRS technique. The map can show the degradation of the soil C content from lower (brighter blue color) to higher C content (darker blue color).

The successful mapping soil C using NIRS technique may benefit for the other purposes. The map can be used as a management tool for rapid monitoring the C stock in soil and for C sequestration strategy. It may also be used as a guidance in adding organic matter. It is suggested to add more organic matter in a location with a low soil organic matter content. It may be used as well as the information in mapping soil fertility status. The high fertile soil is shown by a high content of organic matter (soil organic C).

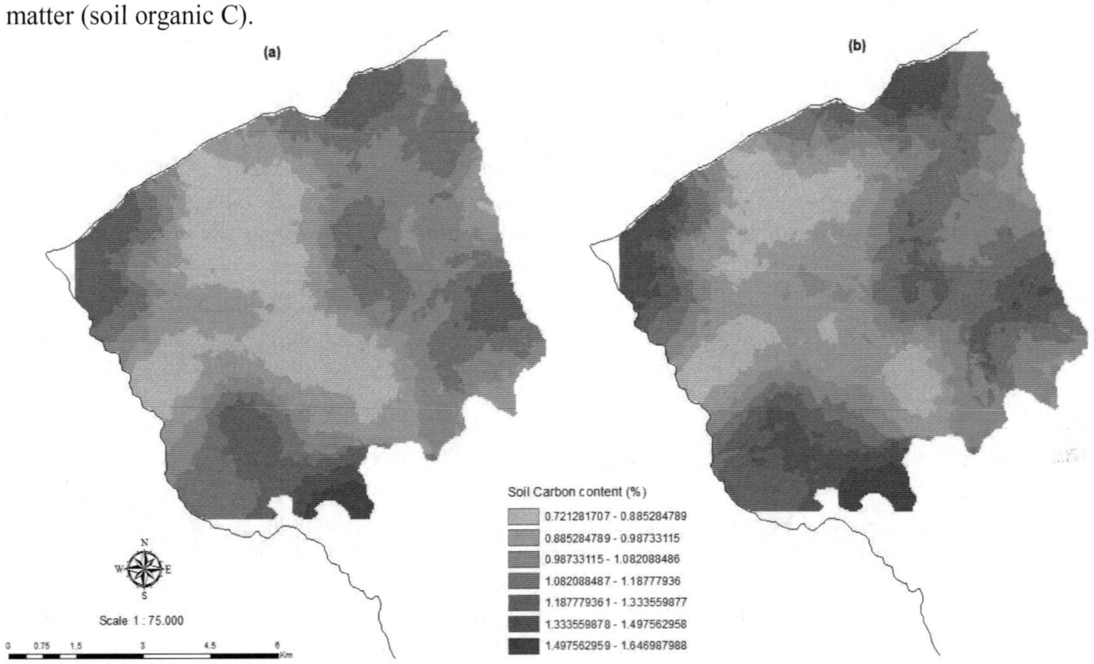

Figure 4. Distribution of soil C content in the study area: (a) measured by conventional analysis and (b) predicted by near-infrared technology.

With the average of soil bulk density 1.00 g cm^{-3} and the average of soil C content of laboratory analysis (Walkley and Black) and NIRS 1.03% C, thus the soil C stock per ha of 10 cm soil depth is 10,300 kg. As NIRS technique produced similar soil C stock (10,300 kg/ha/10 cm soil depth) to the laboratory analysis, so this technique may be used as a rapid measurement of soil C stock.

4. Conclusion

The near-infrared technology was able to be used for a rapid measurement and mapping of soil C stock in Kayangan agricultural area, North Lombok, Indonesia. The ability of this rapid technique to measure and map soil C content can give benefit for monitoring C stock in the soil, which in turn it may be used to monitor soil C (as CO_2) emitted to the atmosphere or sequestered to the soil. Then the information may be used to study the potential risk of global warming caused by depletion of soil C stock. This technique may also be employed to warn the critical level of C stock that may cause soil fertility and environmental problem.

References

[1] Lal R 2003 Offsetting global CO_2 emissions by restoration of degraded soils and intensification of world agriculture and forestry *Land Degradation and Development* **14** pp 309-22

[2] Rees R M, Bingham I J, Baddeley J A and Watson C A 2005 The role of plants and land management in sequestering soil carbon in temperate arable and grassland ecosystems *Geoderma* **128** pp 130-54

[3] Stenberg B, Viscarra Rossel R A, Mouazen A M and Wetterlind J 2010 Chapter Five-Visible and Near Infrared Spectroscopy in Soil Science *Advances in Agronomy* **107** pp 163-215

[4] Workman J J and Shenk J 2004 *Understanding and using the near-infrared spectrum as an analytical method. In Near-infrared spectroscopy in agriculture*. Agronomy Monograph no. 44. 'ed Al-Amoodi L, Roberts C A, Workman J J *et al.* (Madion, Wisconsin, USA: American Society of Agronomy, Inc., Crop Science Society of America, Inc., and Soil Science Society of America, Inc. 44 Agronomy series: 3-10, Madion, Wisconsin, USA)

[5] Ozaki Y, Morita S and Du Y 2007 *Spectral analysis. Near-infrared spectroscopy in food science and technology* 'ed Ozaki Y, McClure W F and Christy A A (New Jersey: Wiley-Interscience, John Wiley & Sons, Inc. Hoboken)

[6] Kusumo B H, Hedley C B, Hedley M J, Hueni A, Tuohy M P and Arnold G C 2008 The use of diffuse reflectance spectroscopy for in situ carbon and nitrogen analysis of pastoral soils *Australian Journal of Soil Research* **46** pp 623-35

[7] Kusumo B H, Hedley M J, Hedley C B and Tuohy M P 2011 Measuring carbon dynamics in field soils using soil spectral reflectance: prediction of maize root density, soil organic carbon and nitrogen content *Plant and Soil* **338**(**1-2**) pp 233-45

[8] Minasny B, McBratney A B, Malone B P and Wheeler I 2013 Digital mapping of soil carbon *Advance in Agronomy* **118** pp 1-47

[9] Vågen T G and Winowiecki L A 2013 Mapping of soil organic carbon stocks for spatially explicit assessments of climate change mitigation potential *Environmental Research Letters* **8** (**1**) 015011

[10] Kodaira M and Shibusawa S 2013 Using a mobile real-time soil visible-near infrared sensor for high resolution soil property mapping *Geoderma* **199** pp 64-79

[11] Kusumo B H, Hedley M J, Hedley C B, Hueni A, Arnold G C and Tuohy M P 2009 The use of Vis-NIR spectral reflectance for determining root density: evaluation of ryegrass roots in a glasshouse trial *European Journal of Soil Science* **60** pp 22-32

[12] Kusumo B H, Hedley M J, Hedley C B, Arnold G C and Tuohy M P 2010 Predicting pasture root density from soil spectral reflectance: field measurement *European Journal of Soil Science* **61**(**1**) pp 1-13

[13] Kusumo B H, Camps Arbestain M, Mahmud A, Hedley M J, Hedley C B, Calvelo Pereira R, Wang T and Singh B P 2014 Assessing biochar stability indices using near-infrared (NIR) reflectance spectroscopy *Journal of Near Infrared Spectroscopy* **22** pp 313-28

[14] Viscarra Rossel R A 2008 ParLeS: Software for chemometric analysis of spectroscopic data *Chemometrics and Intelligent Laboratory Systems* **90** pp 72-83

[15] Demattê, J A M, Campos R C, Alves M C, Fiorio P R and Nanni M R 2004 Visible-NIR reflectance: a new approach on soil evaluation *Geoderma* **121**(**1/2**) pp 95-112

[16] Clark R N 1999 *Spectroscopy of rocks and minerals, and principles of spectroscopy. In Remote sensing for the earth science: Manual of remote sensing* 'ed Rench A N (New York: John Wiley & Sons)

[17] Chang C W, Laird D A, Mausbach M J and Hurburg C R J 2001 Near-infrared reflectance spectroscopy – principal component regression analysis of soil properties *Soil Science Society of America Journal* **65** pp 480-90

[18] Malley D F, Martin P D and Ben-Dor E 2004 *Application in analysis of soils. Near-Infrared Spectroscopy in Agriculture*, Agronomy Monograph no. 44. 'ed Barbarick K A, Roberts C A, Dick W A *et al* (Madison, Wisconsin, USA: American Society of Agronomy, Inc., Crop Science Society of America, Inc., Soil Science Society of America, Inc.) pp 729-84

[19] Ben-Dor E, Irons J R, Epema G 1999 *Soil reflectance. In Remote sensing for the earth science: Manual of remote sensing* 'ed Rench A N (New York: John Wiley & Sons)

IOP Conf. Series: Earth and Environmental Science **129** (2018) 012024 doi:10.1088/1755-1315/129/1/012024

The diplomacy of scientific research in the South China Sea: the case of join to oceanographic marine scientific research expedition between Vietnam and the Philippines

I A Satyawan

Center for ASEAN Studies, Sebelas Maret University, Jl. Ir. Sutami 36A Kentingan, Surakarta, Central Java 57126 Indonesia

Corresponding author: agungsatyawan@staff.uns.ac.id

Abstract. The South China Sea is one of the hot-spot areas in the world. This area is claimed by China, Malaysia, Brunei, Taiwan, Vietnam and the Philippines. It also noted, the South China Sea is rich in biodiversity as well as oil and gas. On the other side, environmental degradation is still happening in the South China Sea due to the reluctance of surrounding states to conduct a preservation program and mitigating action on climate change effects. Joint Oceanographic Marine Scientific Research Expedition between Vietnam and the Philippines is a breakthrough to start collaboration actions as well as to conduct Science Diplomacy.

1. Introduction

The South China Sea (SCS) is defined by the International Hydrographic Bureau as the semi-enclosed body of water surrounded by China, Taiwan, the Philippines, Malaysia, Brunei, Indonesia, Vietnam, Singapore, Thailand, and Cambodia. It has an area of some 3,3 million km² connecting Pacific and the Indian Ocean. It is also recognized as a Large Marine Ecosystem (LME) having specific characteristics of oceanography, biography, and ecology [1]. There are three groups of islands in the SCS namely Pratas Islands, Paracel Islands, Spratly Islands and one submerged bank called Macclesfield Bank. All things considered, these are tiny islands and mostly uninhabited.

The SCS has at least three significances. The first is sea-lane connecting Southeast and Northeast Asia. More than 41,000 ships a year pass through the SCS. Over half of the world's annual merchant fleet tonnage passes through the region's waters. More than 80 % of the oil imported by Japan, South Korea, and Taiwan transits through the SCS [2].The second is known as the world's highest level of biodiversity. It contains 1,027 fish, 91 shrimp and 73 cephalopod species in the northern part of continental shelf; approximately 205 fish and 96 shrimp species in the continental slope, and more than 520 fish species around the islands and reefs of the southern part [2]. The third is the potential of hydrocarbon, oil, and gas. Based on a survey report conducted by the Chinese Ministry of Geology and Mineral Resources, oil deposit in the SCS has amounted about 130 billion barrels. This amount is comparable to 112 billion barrels for Iraq, which ranks the second of the world oil reserves after Saudi Arabia. China also estimated that SCS had more than 2,000 trillion cubic feet of natural gas reserves [3,4].

Given these significances, sovereignty and jurisdiction over all or part of the SCS are claimed by six states including China, Taiwan, Malaysia, Brunei, Vietnam and the Philippines. Of course, diplomatic debates and military skirmishes among the adversaries occasionally happen. Therefore, the SCS is known as one of the hot-spot places in the world.

Content from this work may be used under the terms of the Creative Commons Attribution 3.0 licence. Any further distribution of this work must maintain attribution to the author(s) and the title of the work, journal citation and DOI.
Published under licence by IOP Publishing Ltd

Although having rich in biodiversity and uniqueness, the SCS now faces a serious problem relating to the environmental degradation due to pollution and population pressures in surrounding states. This condition is exacerbated by the effects of climate change with rising sea temperatures resulting coral reef damage. Moreover, the SCS is located in conflict areas so that its surrounding states are reluctant to make efforts to preserve the environment and its impacts on climate change.

Vietnam and the Philippines have made a breakthrough in building trust by conducting marine scientific research in the SCS since 1996.This project was named Joint Oceanographic Marine Scientific Research Expedition (JOMSRE). It was hoped that JOMSRE could enrich marine scientific knowledge in the SCS as a foundation to establish cooperation for preserving the environment and mitigating from the damage of climate change. In the long run, collaboration among parties could reduce tension in the SCS.

This article analyzes the effort of Vietnam and the Philippines in managing their dispute through JOMSRE. It is divided into five sections. The first section is introduction then followed by theoretical framework and method as the second and the third section. The fourth section is finding and discussion and the last section is the conclusion.

2. Theoretical framework

2.1 Diplomacy

In general, diplomacy means the conduct of international relations by negotiation rather than by force or applying activities by other peaceful means [5]. If it applies scientific research as a mean, it is called Science Diplomacy. Langenhove [6] indicates three variants of Science Diplomacy, namely Diplomacy for Science, Science in Diplomacy, and Science for Diplomacy. Diplomacy for Science is talking about facilitation of diplomacy tools for scientific collaboration. Science in Diplomacy is concerning scientists' contribution to support foreign policy. The goal of such diplomacy is to conduct foreign policy through the use of scientific knowledge. Science for Diplomacy means using science or scientific activities to improve relations between states. It can be done when tensions between states are high and they are not able to find a solution. Scientific collaboration is an alternative tool to pave the way to solving the problem. This article is guided by the term of Science for Diplomacy.

2.2 Confidence Building Measure (CBM)

The phenomenon of CBM is not new in security and military studies. Up to now, the term of CBM is applied in non-security studies and even mushroomed in academic literature, it is no single definition of the CBM. Basically, CBM aims to reduce tensions and suspicions between and among adversaries. The United Nations states that the ultimate goal of CBM is "to strengthen international peace and security and to contribute to the development of confidence, better understanding and more stable relations between nations, thus creating and improving the conditions for fruitful international co-operation" [7(p.6)]. To enhance assurance and trustworthiness, it needs concrete measurements such as building communication channels and transparency [8].

3. Method

This article is based on descriptive qualitative method because its aim is to describe certain phenomena. Main data and information are collected through library research. Data are validated by using triangulation technique using sources' triangulation. The interactive model is applied to data analysis which conclusion is draw from collecting, reducing and displaying data [9].

4. Finding and discussion

JOMSRE was initiated by President of the Philippines Videl Ramos and President of Vietnam Le Dec Anh in 1994. At an international conference on JOMSRE 2008 results, President Ramos said that JOMSRE is an effort to transform SCS into a peaceful and prosperous sea. He further said: "The South

China Sea serves as the unifying element for the better understanding between the Philippines and Vietnam and all other countries in the neighborhood" [10].

The Embryo of JOMSRE is the annual workshop on Managing Potential Conflict in the South China Sea organized by Indonesia since 1990. The workshop is an informal meeting of experts, diplomats, and bureaucrats from the countries surrounding the SCS and attendees present on their personal capacity. The workshop discussed several issues such as marine scientific research and environmental protection. In 1991, the participants of the Workshop launched Bandung Joint Statement essentially recommended the government of each participant to promote the rational utilization of living resources, to protect and preserve the marine environment, and to conduct marine scientific research. The Joint Bandung Statement was later formalized as the code of conduct to reduce tension between Vietnam and the Philippines in 1995 [11].

Meanwhile, the United Nations Convention on Law of the Sea (UNCLOS) in 1994 came into force. In article 123 of the UNCLOS stipulates that: "States bordering an enclosed or semi-enclosed sea should co-operate with each other in the exercise of their rights and in the performance of their duties under this Convention". They should: To coordinate the management, conservation, exploration, and exploration of the living resources of the sea; To co-ordinate the implementation of their rights and duties with respect to the protection and preservation of the marine environment; To coordinate their scientific research policies and undertake where appropriate joint programs of scientific research in the area; To invite, as appropriate, other interested States or international organizations to cooperate with them in furtherance of the provisions of this article.

Based on the spirit of the Bandung Statement and the UNCLOS, in 1994 the Philippines President Fidel Ramos and Vietnam President Le Duc Anh agreed for both countries to cooperate in marine scientific research and environmental protection in the SCS. The implementation of the agreement was the setting-up of JOMSRE. Therefore, it can be said that JOMSRE is the beginning of formalized cooperation in the South China Sea and management mechanism for marine resources and it is hoped to spill-over to other fields as well as other countries in the SCS [12].

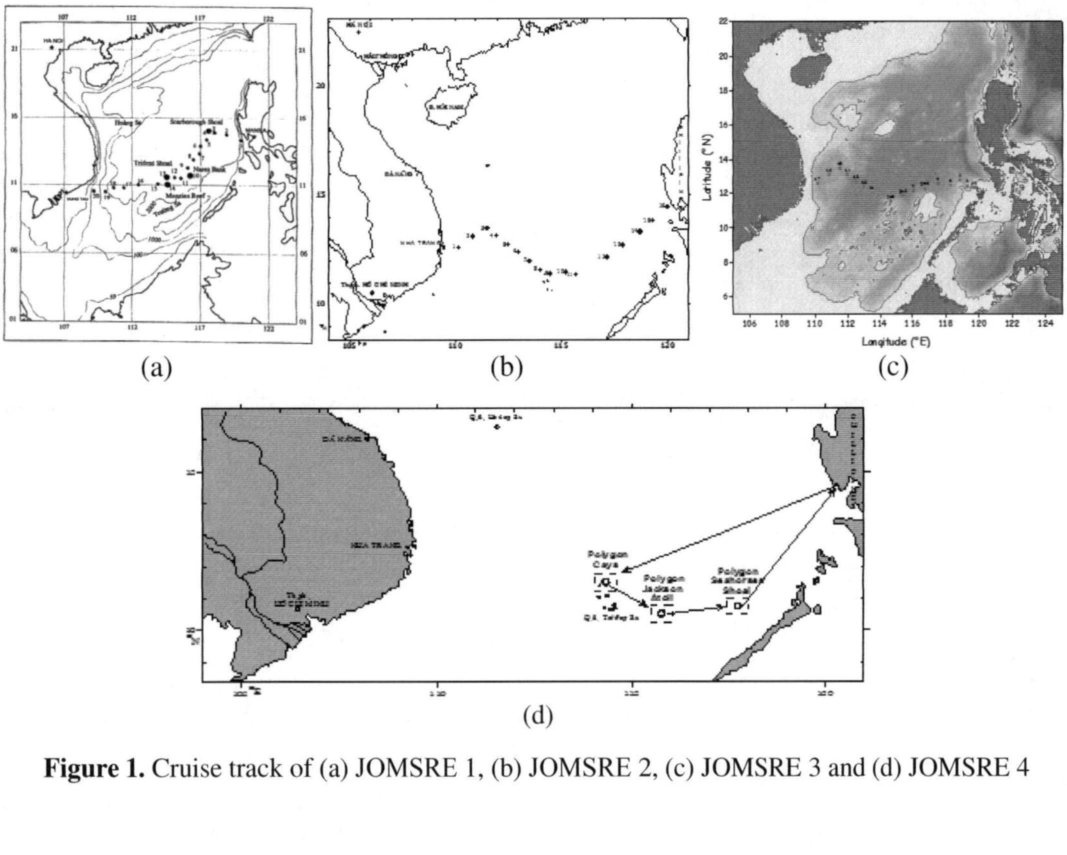

Figure 1. Cruise track of (a) JOMSRE 1, (b) JOMSRE 2, (c) JOMSRE 3 and (d) JOMSRE 4

JOMSRE was begun in 1996 to 2007 and divided into four phases. The first phase was 18 – 09 May 1996, the second phase was 27 May – 2 June 2000, the third phase was 6-19 April 2005, and the fourth phase was 7-12 April 2007. Marine scientists from the two countries were working together cruising various spots in the SCS, especially in the water of Spratly Islands. They studied marine biology, marine geology, marine chemistry, and oceanography. JOMSRE I cruised the South China Sea from Manila to Ho Chi Minh City with focusing on four coral reefs in the Spratly Islands namely Scarborough Shoal, Trident Shoal, Nares Bank, and Menzies Reef. JOMSRE II tracked from Nha Trang in Vietnam to Manila while JOMSRE III started from Subic Bay to Cham Rah Bay with stopping sites in Trident Shoal and North Danger Reef. JOMSRE IV investigated southern part of Spratly Island located in the north of Palawan Island. The following figures show the tracks of JOMSRE expeditions [13].

During cruising and analyzing the finding, JOMSRE concluded firstly that densities and stocks of the marine species associated with coral reefs have been drastically reduced. Secondly, some species of shark were difficult to be found. Thirdly, the biomass of fish species in 2007 has been reduced to one-third compared to 1990s. Fourthly, two-thirds of the fish stock was a decline in 11 years (1996-2007[14].

Moreover, in March 2006 the Philippines and Vietnam agreed to expand JOMSRE participation to include scientists from other interested States, particularly the ASEAN states surrounding the South China Sea, and China. At the fourth phase, a scientist from Laos participated in this activity.

Data collected and analyzed by JOMSRE, of course, contributed to further understanding of biodiversity, geology, and oceanography of the SCS. It is also important that the result can be used as the scientific basis for cooperative programs such as adaptation and mitigating the impacts of climate change [15], prevention of the impacts of both natural and man-made hazards, and formulation policies to enact sustainable ocean management in the SCS. Through the JOMSRE, cooperation among scientists from adversary states is able to reduce tensions in the SCS dispute.

5. Conclusion

The problem of the phenomenon of international relations is now becoming more complicated, so it is very easy to cause conflict. Therefore, diplomacy activities need to be expanded including conducting collaboration scientific activities. JOMSRE is an example of Science Diplomacy to initiate cooperation in the hot-spot area.

Through cooperation among scientists from various disputing countries, they can build trust and confidence and in the long run, it is essential to reduce suspicion. It is also possible to pave the way for the permanent solution of the dispute.

The most important is the results of JOMSRE that could be developed for further cooperation in preserving the environment of the SCS. It also can be used to plan for mitigating regional action of climate change effect.

Acknowledgments

I would like to thank Institute of Research and Community Services Sebelas Maret University for granting me to conduct this research.

References

[1] Morton B 2001 the South China Sea *Marine Pollution Bulletin* **42** 1236-63
[2] Khemakorn P 2006 "Sustainable Management of Pelagic Fisheries in the South China Sea Region" [online]. Available: http://www.un.org/depts/los/nippon/unnff_programme_home/fellows_pages/fellows_papers/khemakorn_0607_thailand.pdf. [Accessed: 12-Jul-2017]
[3] Salameh M 1995 China, oil and the risk of regional conflict *Survival* **37** 133-46
[4] Leifer M 1995 Chinese economic reform and security policy: The South China Sea connection *Survival* **37** 44-59

[5] Berridge G R 1995 *Diplomacy: Theory and Practice* (London: Prentice Hall)
[6] Langenhove L V 2016Science Diplomacy: New Global Challenges, New Trend [Online]. Available: https://www.rsis.edu.sg/wp-content/uploads/2016/04/CO16082.pdf. [Accessed: 20-Jul-2017]
[7] "United Nations Office for Disarmament Affairs", un.org, 1982 [Online]. Available: https://unoda-web.s3-accelerate.amazonaws.com/wp-content/uploads/assets/HomePage/ODAPublications/DisarmamentStudySeries/PDF/SS-7.pdf. [Accessed: 24-Jul-2017]
[8] Zulfqar S 2013 Efficacy of confidence-building measures (cbms) in India-Pakistan relations *IPRI J.***13** 106-16
[9] Miles M B and Huberman M 1994 *Qualitative Data Analysis*, 2nd *ed.* (London: Sage)
[10] Alcala A C 2008 *Proceedings of the conference on the results of the Philippines-Vietnam Joint Oceanographic and Marine Scientific Research Expedition in the South China Sea (JOMSRE-SCS I to IV)* (Dumaguete City: Ocean friends Foundation)
[11] Song Y H 2000 Codes of conduct in the South China Sea and Taiwan's stand *Marine Policy* **24** 449-59.
[12] Hoesslin K V 2005 A view of the South China Sea - from within report on the Joint Oceanographic Marine Scientific Research Expedition (III) in the South China Sea *Culture Mandala: The Bulletin of the Centre for East-West Cultural and Economic Studies***7** 1-5
[13] "Vietnam Academy of Science and Technology, Institute of Oceonragrphy", vnio.org.vn, 2012. [Online]. Available: http://www.vnio.org.vn/Home/tabid/107/ctl/Details/mid/579/ItemID/798/language/en-US/Default.aspx. [Accessed: 03-Aug-2017]
[14] Alcala A C 2008 *Proceedings of the Conference on the Results of the Philippines-Vietnam Joint Oceanographic and Marine Scientific Research Expedition in the South China Sea (JOMSRE-SCS I to IV)*(Dumaguete City: Ocean friends Foundation)
[15] Sembiring M 2017 Climate change adaptation: Case of South China Sea [Online]. Available: https://www.rsis.edu.sg/wp-content/uploads/2017/06/CO17116.pdf. [Accessed: 10-Aug-2017]

Climate change adaptation in Tanjung Mas – Semarang: a comparison between male- and female-headed households

W Handayani[1,4], M R Ananda[2], L Esariti[1] and M Anggraeni[3]

[1]Department of Urban and Regional Planning, Diponegoro University, Jl. Prof. Sudharto, SH Tembalang, Semarang, Central Java 50275 Indonesia
[2]Initiative for Regional Development and Environmental Management (IRDEM), Diponegoro University, Tembalang, Semarang, Central Java 50275 Indonesia
[3]Initiative for Urban Climate Change and Environment (IUCCE), Jl. Tirto Agung Barat V/21 Pedalangan, Banyumanik, Semarang, Central Java 50268 Indonesia
[4]Corresponding author: wiwandari.handayani@pwk.undip.ac.id

Abstract. Mainly due to its complexity, the effort to mainstream gender in addressing climate change issues has been far from the satisfying result. However, there is an urgent call to accommodate gender lens issues and to become more gender sensitive in an attempt to have an effective intervention in responding climate change impact. To enrich the reports on gender and climate change adaptation in city-based case, this paper aims to elaborate climate change adaptation in Tanjung Mas – Semarang city focusing on the gender perspective analysis in male- and female-headed households. The quantitative descriptive method is applied to carry out the analyses, including adaptive strategy and gender role analyses. The research result indicates there are not any significant differences in the climate change adaptation strategies applied in male- and female-headed households. This shows that women in the female-headed households, with their double burden, performed well in managing their roles. Therefore, in particular perspective, it may not be relevant to state that woman and female-headed households are likely to be more vulnerable compared with their counterparts.

1. Introduction

Climate change is a global phenomenon that is calling for further attention. The impact of climate change significantly influences human activities in various sectors mostly for those who live in a coastal area [1],[2][3]. Most rapid growing cities in developing countries are located close to the sea. These high-density cities are at risk due to different climate change hazards mostly flood and water surges [3]. Accordingly, there has been a growing concern on the policy-making process as well as academic research on climate change adaptation to minimize damage and loss as the impact of the climate change. However, mainly due to its complexity, the effort to mainstream gender in addressing climate change issues has been far from the satisfying result [5][6].

There are at least two themes to relate gender and climate change. Those are “vulnerable” and “virtue” [7]. Women are likely to be more vulnerable than men. Women mostly in the developing and less developed countries have relatively lower education, limited ability to find support and information, and most importantly has so little control to the family outcome compared to the men in the family. Even though it is not formally stated, there are more women categorized as poor compared to men as their counterpart. Despite women are commonly defined as vulnerable, women are, however, have more consciousness on environmental issues. They are more sensitive, closer to natural

Content from this work may be used under the terms of the Creative Commons Attribution 3.0 licence. Any further distribution of this work must maintain attribution to the author(s) and the title of the work, journal citation and DOI.
Published under licence by IOP Publishing Ltd

IOP Conf. Series: Earth and Environmental Science **129** (2018) 012025 doi:10.1088/1755-1315/129/1/012025

surroundings, better prepared for dealing with changes, and likely to be more supportive to deal with new/different policies. Therefore, there is an urgent call for the government at every level to accommodate gender lens issues and to become more gender sensitive to create more robust policy in responding climate change phenomena [8].

An important first step to accommodate gender lens in climate change analyses is based on a comparison between male- and female-headed households [9][10],[11]. There are growing numbers of female-headed households across the globe for various reasons [12],[13],[14]. Based on several studies, it is generally indicated that female headed-households are likely to be more vulnerable than the male headed-households as they are poorer because of the lower capability of the women as the breadwinner to earn incomes [14],[15]. However, the current research to connect climate change adaptation and types of households are mostly focusing on rural area and agricultural based activities [10],[11][13]. Even though both women who live in the rural and urban are classified as the most vulnerable, however, there are some differences in the strategy options as well as norms/values applied. To enrich the discourse on gender and climate change adaptation in city-based case, this paper aims to elaborate climate change adaptation in Tanjung Mas – Semarang city focusing on the gender perspective analysis in male- and female-headed households. The climate change adaptation analyses are focused on the way of the people at the household level respond to the flood as most obvious climate change impact emerge in the City.

Semarang City has quite a significant climate change impact. The city has been experiencing coastal inundation for 30-70 cm on average, and reaching 100 cm in certain area, as well as land subsidence for about 13.5 cm annually [16]. Tanjung Mas is one of the most vulnerable prone areas because of flood due to the coastal inundation and land subsidence. It is the densest urban village and becomes one of the pockets of poverty in the city. Therefore, to have more detail findings, the study area is only focusing on the neighborhood of Tanjung Mas.

2. Study Area and Methods

Tanjung Mas is an urban village located within the administrative boundary of Semarang city, the capital of Central Java Province. Fig. 1 illustrates the study area. It consists of six *Rukun Warga (RW)*[1] with 9,604 inhabitants and 1700 households [17]. The sex ratio of the population is 0.88, indicated there are more female compared with male living in the area. Most of them work as blue collar workers in the neighboring industries and construction laborers in the Tanjung Mas Harbour and therefore, categorized as poor. Flood happens almost every day before developing Banger polder and seawall. It is only since 2014 the flood decreases significantly up to 2-3 times a week in the tied season with 3-20 cm high of the flood. The condition in RW 1 is slightly better than in RW 12-16 as its location is not directly alongside to the sea.

Analyses used the quantitative descriptive method, with 98 households as the respondents. Primary data obtained from questionnaires in combination with interviews with particular nominated stakeholders. The number of respondents set by using Slovin formula [18]:

$$n = \frac{\mathbf{N}}{\mathbf{Ne^2 + 1}} \tag{1}$$

n: number of respondents (i.e. households)
N: number of population
e : error tolerance (i.e. 10%)

[1] Rukun Warga in an adminitrative boundary in neighborhood level. It consists of several sub-neighbourhood called as Rukun Tetangga.

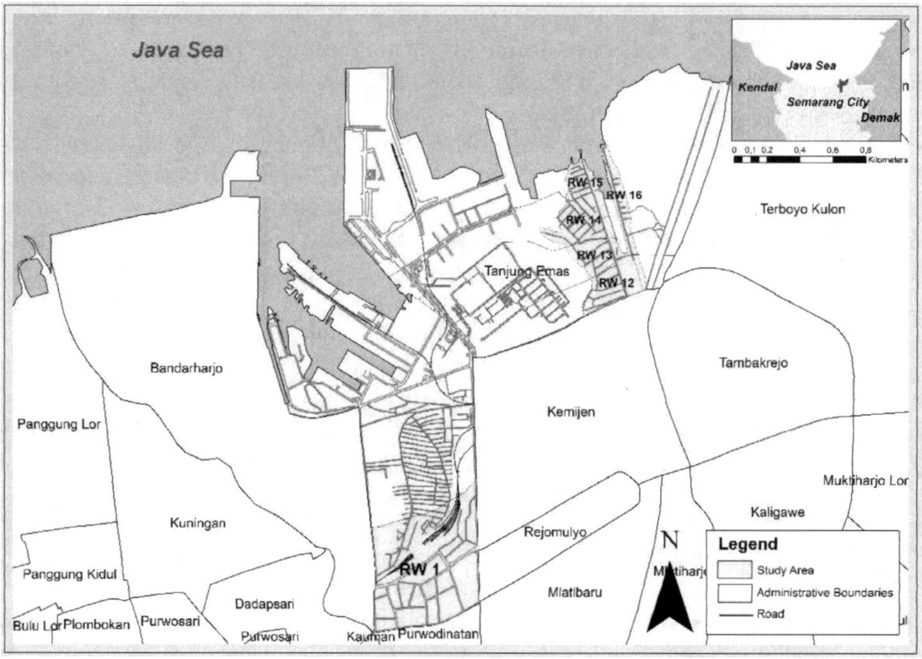

Figure 1. Study Area

The sampling method was proportional random sampling which means the number of respondents is proportionally based on the number of households in each selected RW. The respondents are divided into two main categories; 53 respondents are classified as male-headed households and 45 respondents are defined as female-headed families. The questionnaires are then compiled and interpreted in two main analyses, i.e. adaptive strategy and gender role. Types of adaptation and role of gender are summarized based on the results of the literature review.

3. Result and Discussion

3.1. Adaptation Strategy in the Selected Areas in Tanjung Mas

Adaptation strategies are those actions initiated by each household to cope with the flood in the six selected RWs in Tanjung Mas. There are three types of adaptation, categories as physical, social, and economic (see Table 1). As illustrated in the Table, there are not any significant differences on the strategies applied in male- and female-headed family. In physical aspect, elevating the house is the most common adaptation strategy. They are likely to elevate the house every 5-10 years. The cost ranging from US$ 350 – US$ 6,000 to elevate at least 3-4 meters high. Among the 98 respondents, about 50% both from male- and female-headed households borrow the money from the bank, cooperatives, and even money lenders to ensure that they are able to renovate their house. Both types of the family prefer to stay rather than move to better (no-flooded) area. There is only a slightly higher percentage of female-headed households those are move but only temporarily (during the disaster) compared to the male-headed households. The main reason why they like to stay is mainly that the current house is close enough to their workplace. Besides, they do not have enough money to move as the money has been utilized to elevate the house. Most of them are also born in the area so they have strong bonding with the place and may find difficulties to adapt to the new place.

Table 1. Adaptation Strategies in the Selected Areas in Tanjung Mas

Adaptation Form	Types of Family	YES	NO	Total
Physical				
Elevating the house	Male-headed households	85%	15%	100%
	Female-headed households	80%	20%	100%
Evacuate to temporarily place	Male-headed households	6%	94%	100%
	Female-headed households	13%	87%	100%
Social				
Being a member of social organization related to disaster	Male-headed households	19%	81%	100%
	Female-headed households	9%	91%	100%
Improving knowledge/information on the disaster (i.e. flood)	Male-headed households	74%	26%	100%
	Female-headed households	64%	36%	100%
Borrow money from neighbors	Male-headed households	51%	49%	100%
	Female-headed households	56%	44%	100%
Economic				
Alternative job during disaster	Male-headed households	25%	75%	100%
	Female-headed households	24%	76%	100%
Working family members	Male-headed households	57%	43%	100%
	Female-headed households	38%	62%	100%

Similar to the physical aspect, both types of households also have similar adaptation strategy in the social aspect. Only a few of them are engaged in the social organization. Male-headed households are more likely to be a member of social organization compared to the female-headed households. There are least four main local organizations established in the area; women organization so-called *PKK (Pendidikan dan Kesejahteraan Keluarga)*, youth organization named *Karang Taruna*, neighborhood watch known as *Siskamling*, and Save and Rescue or *SARDa*. Other adaptation strategies in the social category are improving knowledge/information and borrowing money from neighbors. All the respondents considered that knowledge is important. There are not any differences between male- and female-households in gaining and implementing their knowledge. They mostly learn and get the knowledge related to disaster from their experience. All the respondents do not receive information regularly from an appropriate official capacity building activity. Some of the interesting and practical knowledge is how they put electronics and other valuable assets on the 2nd floor and building a "small barriers" at the door to slow down water intrusion to the house. The last adaptation strategy in social aspect is borrowing money. An interesting finding that is closely related with common custom in the neighborhood is that more than half of the respondents in both types of households consider borrowing money from neighbors as a choice to deal with the disaster. There are higher percentages of female-headed households having this strategy compared with the male-headed households. On average, the money is more or less 30% of their monthly income (US$ 50), and. use to fulfill their basic daily needs.

There are also comparable adaptation strategies for both types of households in economic aspect. Most of them do not have an alternative job during the disaster and therefore, most of them are very vulnerable as they do not have an alternative to fulfill their daily needs during the disaster event. However, there are more family members work in the male-headed households compared to the female-headed households. Accordingly, mainly in economic, the female-headed households are more vulnerable compared with the male-headed households. Despite the similarity, the financial capacity is important in the adaptation strategies.

3.2. Gender Role in the Selected Households in Tanjung Mas

Following an explanation of the adaptation strategies, Table 2 explains the gender role in the selected households in Tanjung Mas. As found in the previous research, there are three categories in gender role, namely productive, reproductive, and community responsibility which is focusing on disaster preparedness and response [19][20]; with two types of families, i.e. in male- and female-headed households. There are at least two obvious findings. First, the female has a significant role as the bread-winner and the decision maker in female-headed households only. Second, the female has a dominant role in domestic works for both types of households. There is only slightly higher role of man in taking care of the children in comparison two other types of domestic responsibilities such as washing dishes and cleaning the house. Those two profiles suggest the households applied traditional family value where the man has more concern on the economic issues while the woman more into the responsibility for managing the household works.

Regarding the gender role related to the disaster (i.e. flood), Table 2 shows the role in three phases; pre-disaster, during disaster, and post-disaster. It can be seen in Table 2 that obviously in t of male-headed households, men have a very significant in all stages (pre, during, and pot disaster), and the women have a role mostly only in the female-headed family. However, in particular role, namely accessing knowledge on the disaster, rescuing family members and assets and almost all post-disaster activities, there is slightly an equal role between men and women mostly in the male-headed households.

3.3. Brief Discussion

This study has shown adaptation strategies at the household-level in one most vulnerable neighborhood in Semarang coastal area to deal with the flood, regarded as a regular or common disaster. Following the findings on the adaptation option, there is an analysis of the gender role to further elaborate implementing the strategy.

Female-headed households in the study area are likely to be more vulnerable at least for two reasons. The first is because the women in the family who performs as the bread-winner is low educated, so the second is women are likely to have lower income compared with men in the male-headed family. Fig. 2 and Fig. 3 present the comparison of income and education between male and female-headed households.

Despite female-headed households are more vulnerable for several reasons, however, both types of households have similar adaptation strategies. Both households are likely to elevate their house as the priority to live in the area. The main differences on the strategies are only in regard to the social form of adaptation such as taking part in the social organization and improving knowledge those are preferable only for the male-headed households and decision to stay in the temporary houses during a flood is preferable only for the female-headed households.

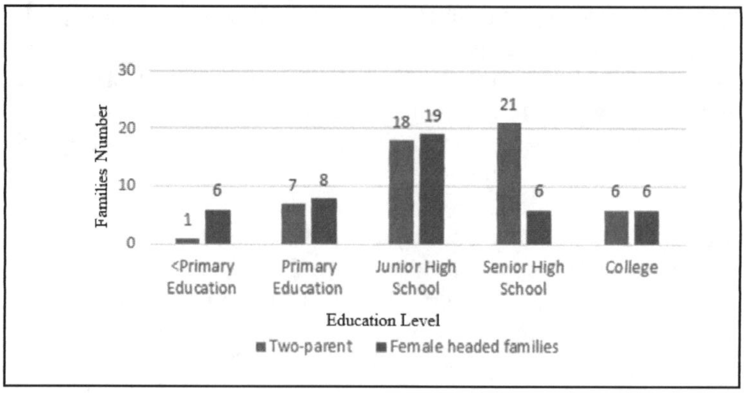

Figure 2. Education Level in Tanjung Mas, 2015

IOP Conf. Series: Earth and Environmental Science **129** (2018) 012025 doi:10.1088/1755-1315/129/1/012025

Table 2. Gender Role in Selected Households in Tanjung Mas

		Male	Female	Male Dominance	Female Dominance	Equal	Total
Productive Profile							
Breadwinner	Male-headed households	28.3%	0.0%	37.7%	3.8%	30.2%	100%
	Female-headed households	6.7%	75.6%	0.0%	11.1%	6.7%	100%
Family decision maker	Male-headed households	37.7%	0.0%	58.5%	1.9%	1.9%	100%
	Female-headed households	4.4%	75.6%	6.7%	13.3%	0.0%	100%
Domestic Responsibilities							
Taking care of the children	Male-headed households	1.9%	47.2%	11.3%	34.0%	5.7%	100%
	Female-headed households	0.0%	88.9%	0.0%	11.1%	0.0%	100%
Preparing food	Male-headed households	1.9%	77.4%	0.0%	18.9%	1.9%	100%
	Female-headed households	0.0%	88.9%	0.0%	11.1%	0.0%	100%
Washing dishes	Male-headed households	1.9%	64.2%	3.8%	28.3%	1.9%	100%
	Female-headed households	0.0%	84.4%	0.0%	15.6%	0.0%	100%
Washing clothes	Male-headed households	1.9%	71.7%	0.0%	24.5%	1.9%	100%
	Female-headed households	0.0%	84.4%	0.0%	15.6%	0.0%	100%
Cleaning house	Male-headed households	1.9%	52.8%	9.4%	34.0%	1.9%	100%
	Female-headed households	0.0%	80.0%	0.0%	20.0%	0.0%	100%
Cooking	Male-headed households	1.9%	79.2%	0.0%	17.0%	1.9%	100%
	Female-headed households	0.0%	88.9%	0.0%	11.1%	0.0%	100%
Disaster preparedness & response							
Pra Disaster							
Building protection	Male-headed households	20.8%	0.0%	64.2%	0.0%	15.1%	100%
	Female-headed households	2.3%	69.8%	9.3%	4.7%	14.0%	100%
Knowledge on disaster	Male-headed households	17.0%	0.0%	30.2%	0.0%	52.8%	100%

171

IOP Conf. Series: Earth and Environmental Science **129** (2018) 012025 doi:10.1088/1755-1315/129/1/012025

		Male	Female	Male Dominance	Female Dominance	Equal	Total
	Female-headed households	17.8%	44.4%	0.0%	0.0%	37.8%	100%
Being a member of social organization	Male-headed households	7.5%	0.0%	3.8%	0.0%	1.9%	13.2%
	Female-headed households	2.2%	11.1%	0.0%	0.0%	6.7%	20.0%
During Disaster							
Rescuing family members	Male-headed households	1.9%	0.0%	67.9%	0.0%	30.2%	100%
	Female-headed households	0.0%	40.0%	15.6%	15.6%	28.9%	100%
Rescuing assets	Male-headed households	3.8%	0.0%	52.8%	0.0%	43.4%	100%
	Female-headed households	0.0%	46.7%	11.1%	24.4%	17.8%	100%
Basic needs fulfilment	Male-headed households	28.3%	3.8%	30.2%	1.9%	35.8%	100%
	Female-headed households	4.4%	77.8%	2.2%	6.7%	8.9%	100%
Post Disaster							
Cleaning the house	Male-headed households	0.0%	0.0%	54.7%	0.0%	45.3%	100%
	Female-headed households	0.0%	60.0%	6.7%	4.4%	28.9%	100%
House renovation	Male-headed households	24.5%	0.0%	47.2%	0.0%	28.3%	100%
	Female-headed households	4.4%	42.2%	6.7%	11.1%	35.6%	100%
Living cost	Male-headed households	34.0%	0.0%	26.4%	3.8%	35.8%	100%
	Female-headed households	0.0%	75.6%	13.3%	11.1%	0.0%	100%

172

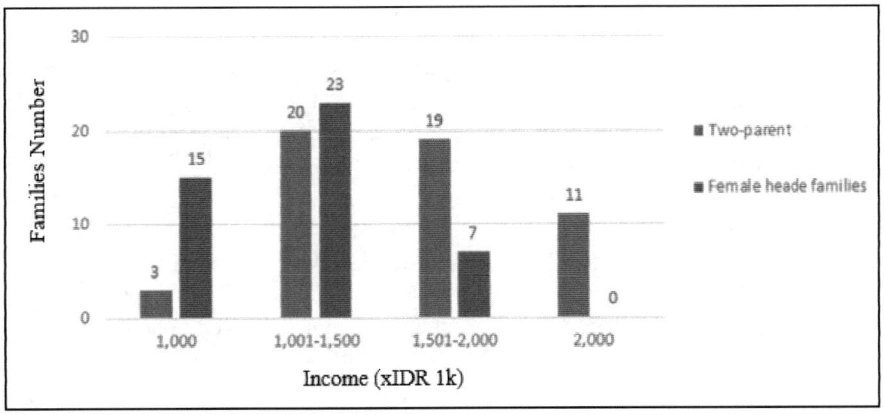

Figure 3. Level of Income in Tanjung Mas, 2015

The situation is an imbalance in the female-headed family where the women become the breadwinner. In this household, women have the double burden, to generate income as well to ensure that various domestic jobs are well completed. In the male-headed family, even though women have the responsibility in doing all the domestic jobs, but they are not the main income earner in the family. However, both types of families are likely to have a similar approach in dealing with flood and end with the same decision, i.e. stay instead of move to the saver areas. The findings much in line with a study in Nigeria, it does not necessarily to assume that female-headed families are poorer and therefore regard as more vulnerable even though they are likely to be low educated and earn less income than man [14].

4. Conclusion

The research resulted there are not any significant differences in the climate change adaptation strategies applied in male- and female-headed households. This shows that women in the female-headed households with their double burden, perform very well in managing their roles. Even though female-headed households are categorized as more vulnerable but the study has shown an evidence that women have good adaptation strategies to survive and even more become resilient. To end, in particular perspective, it may not be relevant to state that woman and female-headed households are likely to be more vulnerable compared to their counterparts.

References

[1] Takagi H, Esteban M and Thao N D 2014 *Introduction: Coastal Disasters and Climate Change in Vietnam* ed Thao N D, Takagi H and Esteban M (UK: Oxford) pp xxi-xxviii

[2] Wamsler C, Brink E and Rivera C 2013 Planning for climate change in urban areas: from theory to practice *Journal of Cleaner Production* **50** pp 68-81

[3] McGranahan G, Balk D and Anderson B 2007 The rising tide: assessing the risks of climate change and human settlements in low elevation coastal zones *Environment and Urbanization* **19**(**1**) pp 17-37

[4] Chandrappa R, Gupta S and Kulshrestha U C 2011 Coping with climate change: principles *and Asian context* (Heidelberg: Springer Science & Business Media) pp. 117-46

[5] Alston M 2013 Women and adaptation *Wiley Interdisciplinary Reviews: Climate Change* **4**(**5**) pp 351-8

[6] Terry G 2009 *Climate change and gender justice* (Warwickshire, UK: Oxfam GB,)

[7] Arora-Jonsson S 2011 Virtue and vulnerability: Discourses on women, gender, and climate change *Global Environmental Change* **21**(**2**) pp 744-51

[8] Alston M 2014 Gender mainstreaming and climate change *Women's Studies International Forum* **47** pp 287-94

[9] Van Aelst K and Holvoet N 2016 Intersections of gender and marital status in accessing climate change adaptation: Evidence from rural Tanzania *World Development* **79** pp 40-50

[10] Bhattarai B, Beilin R and Ford R 2015 Gender, agrobiodiversity, and climate change: A study of adaptation practices in the Nepal Himalayas *World Development* **70** pp 122-32

[11] Huynh P T and Resurreccion B P 2014 Women's differentiated vulnerability and adaptations to climate-related agricultural water scarcity in rural Central Vietnam *Climate and Development* **6**(**3**) pp 226-37

[12] Liu C, Esteve A and Treviño R 2017 Female-headed households and living conditions in Latin America *World Development* **90** pp 311-28

[13] Flatø M, Muttarak R and Pelser A 2017 Women, weather, and woes: the triangular dynamics of female-headed households, economic vulnerability, and climate variability in South Africa *World Development* **90** pp 41-62

[14] Oginni A, Ahonsi B and Ukwuije F 2013 Are female-headed households typically poorer than male-headed households in Nigeria? *The Journal of Socio-Economics* **45** pp 132-7

[15] Buvinić M and Gupta G R 1997 Female-headed households and female-maintained families: are they worth targeting to reduce poverty in developing countries? *Economic development and cultural change* **45**(**2**) pp 259-80

[16] Handayani W and Kumalasari N R 2015 *Migration as future adaptive capacity: The case of Java – Indonesia* ed Hillmann F, Pahl M *et al* (London: Environmental Change, Adaptation and Migration: Bringing in the Region) pp 117–38

[17] CBS (Central Bureau of Statistics) 2016 Tanjung Mas Village Monograph Semarang

[18] Singarimbun M 1982 *Survey Research Method* (in Indonesia) (Jakarta: Lembaga Penelitian Pendidikan dan Penerangan Ekonomi dan Sosial)

[19] Rochmayanto Y and Kurniasih P 2013 Gender role in climate change adaptation of mountainous ecosystem in Solok Regency, West Sumatra (in Indonesia) *Jurnal Analisis Kebijakan Kehutanan* **10**(**3**) 203-213.

[20] Fakih M 1996 Gender as a Tool in Social Field of Analysis (in Indonesia) *Jurnal Analisis Sosial* ed 4

Improving Indonesian cinnamon (c. burmannii (Nees & t. nees) Blume) value chains for Greater Farmers Incomes

S R Menggala[1], P V Damme[1]

[1] Department of Plant Production, Faculty of BioscienceEngineering, Ghent University, Coupure Links 653, Gent 9000, Belgium
[2]Corresponding author: sidirana.menggalasusanto@ugent.be, patrick.vandamme@ugent.be

Abstract. Genus *Cinnamomum (Lauraceae)* regroups some species whose stem bark are harvested, conditioned and traded as cinnamon in an international market. Over the centuries, the species have been domesticated so that now at least six different ones are grown in Southeast Asia countries. One of the species is *Cinnamomum burmannii*, also known as Korintje Cinnamon, which generates income for most smallholder farmers in Kerinci district, Jambi, Indonesia. Most cinnamon consumed in the world originates from this Korintje Cinnamon products. It is recognized for its unparalleled quality that comes with its sharp and sweet flavor, with a slightly bitter edge. However, international market requirements for product certification and quality standards make it difficult for a farmer to comply. Our research will address issues related to (improvement of) productivity, sustainability and value chains faced by cinnamon producers in Kerinci, to strengthen their product's value chains. Smallholder farmers are very vulnerable to climate change impacts, and thus empowering the value chains of agricultural products will increase farmers resilience to climate change. The research will analyze the development of agricultural value chains, certification & standards on trade mechanism to help farmers earn a better income and future prospects.

1. Introduction

Indonesia is a tropical country crossed by the equator. It has the second largest biodiversity and the third largest natural resource reserves for oil, natural gas, gold, copper and other minerals in the world. The country is also rich in various types of ecosystems: aquatic ecosystems, freshwater ecosystems, tropical rainforests, peat swamps, mangroves, coral reefs and coastal ecosystems. The highest biodiversity in Indonesia is to be found in the tropical forest environment. Not only serving as a source of commercial and industrial wood products, it also provides people's daily necessities, such as lumber, pulp, and paper. However, Indonesia is now facing multiple problems such as illegal logging, chaotic urbanization, unsustainable agriculture, and forest conversion (converting forests into large-scale plantations). As a result, Indonesia faces the great challenge of combining poverty alleviation and economic growth with sustainable use and conservation of biodiversity. For this reason, Indonesia needs to find solutions for the long-term sustainable use of biodiversity that will improve the social welfare of local communities.

Indonesia has several native natural products that have a potential value in local and global markets, such as *Cinnamomum burmannii Nees ex Blume*. This is one of the four types of cinnamon categorized as high economic value cinnamon besides *Cinnamomum verum, Cinnamomum cassia* (*C. aromaticaum*, also called Chinese cinnamon), and *Cinnamomum loureiroi* (also known as Vietnamese or Saigon cinnamon) [1]. Genus *Cinnamomum (Lauraceae)* regroups some species whose stem bark

Content from this work may be used under the terms of the Creative Commons Attribution 3.0 licence. Any further distribution of this work must maintain attribution to the author(s) and the title of the work, journal citation and DOI.
Published under licence by IOP Publishing Ltd

are harvested, conditioned and traded as cinnamon. It is categorized as spice products in an international market [2]. In addition, during the harvesting process, other commodities can also be obtained from a cinnamon tree, i.e., stem, leaves, root, and twigs, because these parts also contain beneficial constituents [3]. Cinnamon is considered as a high-value plant because every part of the plant, besides functioning as a spice, can also be used for pharmaceutical and perfume [4]. Cinnamon wood can be used as raw materials for particle board, while the leaves and branches can be distilled to obtain oil. Over the centuries, the species have been domesticated so that now at least six different ones are grown in Southeast Asia. One of the native cinnamon species in Indonesia is *Cinnamomum burmannii*, also known as Korintje Cinnamon, which generates income for most small growers in Kerinci district, Jambi, Indonesia. The largest cinnamon plantation is situated in West Sumatra in the region known as Kerinci, near the city of Padang. This native plant is a prime commodity, especially in West Sumatra and Kerinci regency, as the center of cinnamon production in Indonesia. Most of the cinnamon consumed worldwide is the Korintje Cinnamon products. It is recognized for its unrivaled quality that comes with its sharp and sweet flavor, with a slightly bitter edge.

Although cinnamon bark belongs to one of the Indonesia's export commodities, the prices obtained by farmers for its transactions are still low. Thus, this small income affects the revenue of cinnamon farmers and their families. Basically, the price of cinnamon received by farmers adapted to the type of cinnamon bark. However, international market demands of product certification and quality standards make it difficult for small growers to meet the request. The international consumer protection agency requires safe products that are free from chemical elements harmful to human health. It causes the exporters to be careful in providing quality commodities, as well as requiring farmers to conduct the cultivation process based on the operational standards.

Another factor is related to the gate prices received by small growers for the transactions that are still low.Besides, the price of the cinnamon in a producer level does not correlate with the export corporation rate [5]. Although cinnamon is not a primary revenue for the producer, its function is critical as a reserve fund for the farmers to meet the needs of their daily life. Most cinnamon small growers have a small area of land that decreases the number of crops. One cinnamon tree can produce approximately 20 - 25 kg of cinnamon. Korintje Cinnamon, one of the highest cinnamons consumed worldwide has a trouble in this value chain scheme. Small growers must wait for the harvesting time for up to 15 years. This long wait will certainly affect the total revenue of the farmers, causing them to have a low-income and lead to poverty. There should be particular efforts such as product derivation and diversification to create its added value. Accordingly, this research is aimed at finding the variables that can improve the welfare of cinnamon farmers.

2. Methods

The research was conducted in Talang Kemuning Village, Bukit Kermai Subdistrict, Jambi Province, Indonesia in February 2017. The object of this study were the relevant stakeholders, including small growers, cooperative initiatives, cinnamon collectors, local government and the other trading channels of Korintje cinnamon. The place for collecting the sample was chosen purposely by selecting the location where the producer and buyer work closely together in the concept of sustainable agriculture. The sample farmer population is a farmer who produces and sells cinnamon. A 10% sample is taken from each of the heads of households in Talang Kemuning Village. Collecting samples and other trading agents involved in the value chain channel using the purposive sampling method, which is traced to the producer to end market.

The research has addressed some of the issues on value chains improvement faced by cinnamon producers in Talang Kemuning village. The community lives in Bukit Kerman sub-district. It is located in the district of Kerinci, Jambi, Indonesia. Having an area of 1,600 Ha, there are totally 1,200 people living in this community, spreading in 520 families. The majority of people's livelihoods are farmers because the area has enough land to produce agricultural products such as cinnamon, coffee, and cocoa.

There are two types of data in this research, namely primary and secondary data. The primary data were collected through in-depth interviews with key informants. From this type of data, the researcher obtained the information of general economic actors, economic potentials, marketing channels, production and trading costs, and the purchase price and selling price. The secondary data were garnered by recording the related institutions or Non-Governmental Organizations (NGOs) and searching from the relevant sources of literature.

Based on a field study at Talang Kemuning in February 2017, the yields of cinnamon accounted for 10% of the total fixed revenue of a small cinnamon grower, where the other 90% is obtained from other crops. This little contribution of cinnamon income is caused by its low price and extended harvest period.

3. Result and Discussion

Nearly 90% of small growers in Talang Kemuning, Kerinci have the other sources of income to support their livelihoods. They work as a labor in a rubber plantation, housemaids in Malaysia, and many other types of hard work. The findings showed that there are two major constraints identified, namely on-farm and off-farm problems.

3.1. On-Farm Issues in Kerinci

Based on the value chain stream, in the first phase, there is a source of bark obtained by small growers that are lacking competence and knowledge on agricultural practices. It has been proven that they are the only clearcutting without replanting new trees. Small growers also use traditional tools to cultivate the bark from the cinnamon tree. Then, the second constraint deals with the ownership of land, most of 1 Ha wide (birding) are rented. The yearly production of cinnamon bark is increasingly high because of the external factors, such as seeding, fertilizers and labor cost. Due to this concern, the third constraints appear, there is a lack of industry to support products and research development in Kerinci. Furthermore, there is also a lack of investment in tree rejuvenation, replanting and further development of technologies for processing plant into essential oils and oleoresin either from government or buyers (business). The most important constraint is the lack of infrastructure to support cinnamon transportation from plantation to warehouse. Small growers used only water buffalo to carry their raw bark, from plantation to drying of the bark location.

3.2. Off-Farm Issues in Kerinci

Constraints in post-harvest cinnamon process arise because of regulations, standards, laws and also informal rules & norms that are not supporting the value chains improvement. Jambi provincial government should issue a policy stating that cinnamon should not be sold as raw materials (bark) but it should be processed through the cinnamon grinding mill to improve its value so that it can be sold both in local and international markets. Jambi provincial government can set and enforce rules to initiate and finance a grinding machine that can make cinnamon powder, essential oils, and oleoresin that has a higher sale value in the market. Another constraint is related to the direct supply chain. Because of the absence of harbor facility in Jambi, the product should be delivered to another city takes another process of trading. The last constraint in the off-farm issues deals with the international standards for agriculture practices that are difficult for small growers to comply. All of those issues eventually ended up in small growers as price takers on the basic gate-price of the cinnamon market.

Even though the income earned from cinnamon plays an important role for small growers, but the primary concern of cinnamon farmers deals with a long period of harvest time. One cinnamon tree produces about 20 kg of bark with the range of age up to 20 years for production with a variant of quality and price. Cinnamon tree is cultivated three times for its bark. The first harvest occured when the tree reaches the age of 6 years, and the second harvest is at ten years and the last harvest time is at 15 years old. During that harvest time, the farmers are looking for other solutions to generate income, such as intercropping and planting another product that is more productive [6]. Therefore, it is important to find a solution for the farmers to produce derivative products and create a product diversification that can give added value and improve the value chains for greater income. There are four types of upgrading [7]:

- Process improvement, where the transforming production process will reorganize or improve processing technology;
- Product development, where natural products will develop into diverse and more sophisticated product lines, with higher values per unit volume;
- Functional improvement, which refers to cases where new and superior functions will draw up in the value chains;
- Inter-sectoral upgrading, which occurs when new research or technology enables a product to shift from one sector to a different "new area."

The involvement of other parties to improve the value chains for greater farmer income is also important in the outcome, such as described below:

Business, producer associations, universities, NGO's and local government through policy support;

Establishment of research and development (R&D) capabilities of national or regional universities, or R&D facilities of large firms with whom partnerships will form;

Strategic use of labeling, branding, trademarks, and certification.

Majority of small growers in developing countries, including those who are in Korintje area, are facing series of constraints that often limit their ability to participate competitively in a value chain improvement model including supporting functions. The following are the model of four major constraints that limit the competitiveness of small and medium-sized manufacturers and their entry into value chains [8]:

- Access to end-market
- Access to skills and capacity improvement
- Collaboration and cooperative building
- Access to finance and incentives

First, access to the end-market is relevant to improve value chains for smallholders. In the context of this model, it refers specifically to the presence of value chain connections between small growers and buyers and how they can be established. It is also important for the consumer to be informed about the origin of the products using traceability tools and sustainable measurements. In this case, neither the small-grower organic or chemical pesticide can reduce the climate change impacts. Second, while smallholders work at the farm from their childhood, a specific training is often required to improve the productivity and product quality. Such training can include the introduction of new technologies and plant varieties, not only explaining how to comply with food safety and other certification requirements but also how the value chains works [8]. Nevertheless, there are also new agricultural practice adaptations including new cutting that impact on biodiversity loss and land-slide.

Third, building coordination and collaboration at two coordination levels that can trigger R&D and infrastructure improvement by any condition. Finally, the last part is an access to finance that can support for product diversification and technology investment including more environmentally friendly agricultural machinery.

To enhance cinnamon value chains for greater farmer income, it can be conducted on several models such as educating farmers about the sustainable practices for the environment, helping them to access local and international markets, supporting their productivity improvement, promoting good agricultural practices, providing organic fertilizer, and so forth. However, it also important for the farmers to increase their knowledge about the regulation standard so they can access EU market. Below are the issues dealing with the quality and standard affecting the performance of cinnamon export value chains according to the interview conducted with Cassia Coop and VECO.

- Marketing: Understanding about EU markets due to complying with food safety standards.
- Production and processing: Poor of the application on good agricultural practices and lack of traceability system in the supply chain.

IOP Conf. Series: Earth and Environmental Science **129** (2018) 012026 doi:10.1088/1755-1315/129/1/012026

- Technology and training: Poor of R&D work in value addition, new technologies, and mechanization (traditional cultivating), applied agricultural training programs for farmers by NGO's.

Branding the Korintje Cinnamon products by putting trademarks of traceability and certification can have a significant role in the future of Indonesia's spices export markets. Putting labels, marks and food safety certifications can influence the customers to pay more for the product that is certified and meet the standard regulations and categorized as a premium product. Consumers in Europe are increasingly interested in consuming a product that has a 'clean & green' label because they are aware of the health and security of the product that will be consumed. When the products have claimed for its sustainability and traceability campaign, it is crucial to ensure that those claims can be traced. It requires a system of traceability to be implemented in Korintje Cinnamon products. However, if the cinnamon farmers in Kerinci can achieve the demanding quality standards, they surely can get more benefit and gain brighter future prospects. Therefore, for the purpose of livelihoods improvement, the farmers should choose and decide whether or not a particular set of standards is good for their cinnamon through-out sustainability objectives.

4. Conclusion

Implementing several models to improve the farming system and marketing is associated with enhancing farmers' knowledge on the standard regulation to access EU market so that it can benefit cinnamon farmers in Kerinci in the future. Improving the cinnamon commodity is still prospective in Kerinci. However, farmer's traditional ways of cultivating cinnamon cause them to receive less income than the actual price. Therefore, farmers are expected to decide the standard of Korintje Cinnamon products in order to improve their income and livelihoods.

References

[1] Chen P, Sun J, and Ford P 2014 Differentiation of the four major species of cinnamons (*C. burmannii*, *C. verum*, *C. cassia*, and *C. loureiroi*) using a Flow Injection Mass Spectrometric (FIMS) fingerprinting method *J Agric Food Chem* **62** pp 2516–2521

[2] Al-Dhubiab BE 2012 Pharmaceutical applications and phytochemical profile of *Cinnamomum burmannii Pharmacogn Rev* **6** pp 125–131

[3] Kaul PN, Bhattacharya AK, Rajeswara Rao BR, Syamasundar KV and Ramesh S 2003 Volatile constituents of essential oils isolated from different parts of cinnamon (*Cinnamomum zeylanicum* Blume) *J Sci Food Agric* **83** pp 53–55

[4] Thangaselvabai T, Joshua JP and Jayasekar M 2009 Cinnamon (Cinnamomum verum presl) - the sweet bark spice for flavor and fragrance- a review *Agric. Rev.* **30** pp167-175

[5] Jaya A, Rustiadi E, Gonarsyah I, Bratakusumah DS, and Juanda B 2009 The cinnamon commodity development effect to regional economic: case study of Kerinci regency Jambi *Forum Pascasarjana* **32** pp 67-79

[6] Adlaida, Nainggolan S, and Murdy S 2015 *Jurnal Penelitian Universitas Jambi Seri Sains* **17** 50-67

[7] Humphrey and Schmitz 2001 *Governance in Global Value Chains* [IDS Bulletin 32.3] (Brighton, UK: Institute of Development Studies) pp 1-16

[8] Lie H, Rich KM, Kurwijila LR, and Jervell AM 2012 Improving smallholder livelihoods through local value chain development: a case study of goat milk yogurt in Tanzania *International Food and Agribusiness Management Review* **15** 55-86

[9] Tonsor GT 2011 Consumer inferences of food safety and quality *European Review of Agricultural Economics* **38** pp 213–235

A climate risk assessment of clean water supply in an urban area: A case study of South Tangerang city, Indonesia

S I W Nastiti[1], H Kusnoputranto[2], R Boer[3] and S W Utomo[4]

[1] Graduate School of Environmental Science, University of Indonesia
[2]Department of Environmental Health, Faculty of Public Health and Graduate School of Environmental Science, University of Indonesia
[3]Department of Geophysics and Meteorology, Faculty of Mathematics and Natural Sciences, Bogor Agricultural University, Indonesia
[4]Department of Environmental Health, Faculty of Public Health and Graduate School of Environmental Science, University of Indonesia

Abstract. The demand for clean water in South Tangerang, Indonesia, is very high. At present, this demand is mostly met by groundwater that is much influenced by climate variability, land cover change, and human activities. The local company on water services (PDAM) provides clean water services for only about 9% of the population. The climate risk assessment conducted by South Tangerang Government in 2016 indicates that several areas are potentially exposed to a high risk of climate change. Survey and in-depth interview with communities and sectoral officers suggest that a risk to clean water supply in this city is increasing. This study aims to assess climate potential risks on clean water supply based on the 2016 study. We adopted the method of that study by modifying some of the vulnerability indicators that can represent clean water access and supply. The results of the study demonstrate that many wards in South Tangerang would be exposed to high climate risks of clean water supply. By 2021, about 54% of wards would be exposed from high to the very very high risk of clean water supply. These results signify the tangible need of adaptation actions, to prevent the worsening impacts of climate on clean water supply.

1. Introduction

South Tangerang City is a neighbor and buffer zone to Jakarta Capital City. It is located downstream of three watersheds (Cisadane, Angke, and Pesanggrahan watershed) and is potentially vulnerable to climate disasters such as flood and drought. Based on interviews with the wards, sub-district officials and communities in several areas in South Tangerang from January to September 2017, each ward has drought-prone and flood-prone areas. The risk of drought and flood is felt increasing. The increasing risk of drought is mainly caused by the rapid increase of population and rapid growth of development process leading to the exceptionally high need for clean water, while the level of clean water supply does not much increase but becomes more uncertain. The increasing risk of flood is caused by increasing rainfall, narrowing and blocking of river's body by domestic solid waste leading to more frequent overflowing of the river of the upstream. It is reported that flood also causes the problem to clean water supply as many of community wells in the flood-prone areas are being contaminated by the flood water. In addition, the flood also causes the spread of various diseases, such as dengue fever in the city.

Content from this work may be used under the terms of the Creative Commons Attribution 3.0 licence. Any further distribution of this work must maintain attribution to the author(s) and the title of the work, journal citation and DOI.
Published under licence by IOP Publishing Ltd

In the future, the supply of clean water in South Tangerang City may pose a greater problem. It is not only due to climate change, but also high population growth. South Tangerang City has the highest population growth rate in 2010-2011 in Banten Province, which recorded 3.6% growth per year [1]. Thus, the demand for clean water will increase rapidly in the coming years. On the other hand, infrastructure for clean water services has not much improved. At present, only about 9% of the population in this city has access to clean water supply system or PDAM (Local Company of Drinking Water), while the rest depends on ground surface water (wells). Previous studies found that water shortage is a major problem in the South Tangerang City [2]. Boer *et al.* [3]indicate that Tangerang City will expose to the increased risk of water shortage even without climate change. By 2080, a water deficit will be more severe.

Ways of communities to cope with limited clean water supply during flood and drought occurrence are by deepening the groundwater wells, buy clean water, and ask for help from neighbors or from other parties such as the government. Considering the existing rapid increase of population and development in the urban areas of the South Tangerang, adequate availability of clean water becomes essential for sustainable growth of the city. However, the evidence of rapid growth and facts obtained in the field suggest that the future availability of clean water in South Tangerang is at a threatened level. To anticipate this condition, it is necessary to establish a clean water management policy considering climate change.

Based on the previous study, several of the city's environmental pillars are even categorized in vulnerable conditions [4]. On the other hand, spatial policies in the Local Spatial Plan (RTRW) regulation are not well defined in term of its relation to water management [5]. The South Tangerang City Government has had A Master Plan of Drinking Water Service System (RISPAM) in 2013. However, it is still in the preparation phase of its action plan [6]. The Assessment of Vulnerability and Risk for Climate Change in South Tangerang City has been prepared in 2016 by local government, but this does not include clean water aspects [7].

2. Objective

This study aims to assess the climate risks of water supply in South Tangerang City at present and in the near future. The result of this study is expected to be used as a reference for policymakers in designing strategy for sustainable water supply and avoiding greater losses in the future.

3. Method

Climate risk assessment of clean water supply is conducted to evaluate the level of risk of wards to clean water supply at South Tangerang City. The risk is measured by possible damage toward that can occur when threats become reality, including the anticipated severity of consequences. The City is situated in the western part of Java Islands, Indonesia (Figure 1a) consisting of 7 sub-districts and 54 wards (Figure 1b).

The assessment looks at to what extent awards are and going to be affected by the impacts of climate hazard (floods and drought) regarding clean water availability. The extent of the impact will be determined by the level of vulnerability of the wards, the degree to which award is unable to cope with the hazards.The occurrence of drought will limit the availability of clean water for household, particularly when the household relies very much on groundwater and is not connected to PDAM. Flood occurrence may also limit the availability of clean water as the flood may contaminate the quality of water in wells. Thus, the potential impact of the climate hazards on the household in the ward dealing with clean water will depend on the ward vulnerability or, in other words, the risk is defined as a function of hazard (H) and vulnerability (V) [8].

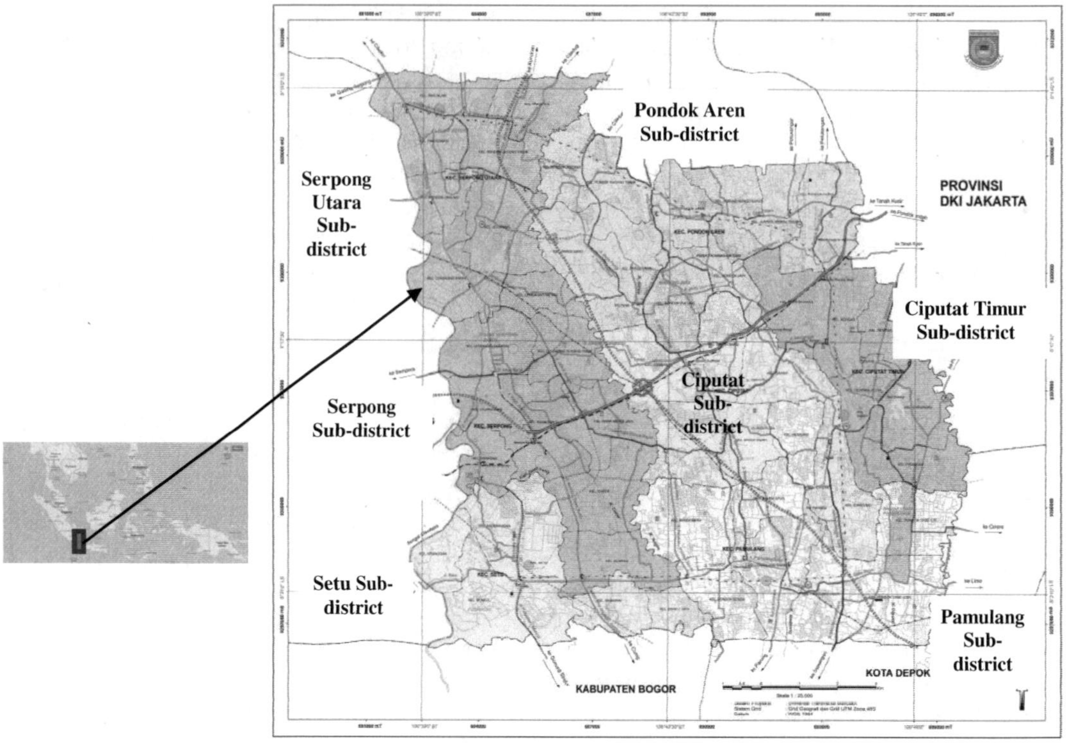

Figure 1. Map of South Tangerang city is divided into 7 sub-districts

Intergovernmental Panel on Climate Change (IPCC) formulates the vulnerability [9]as the following:

$$\text{Vulnerability} = \text{f(exposure, sensitivity, adaptive capacity)} \quad (1)$$

Exposure is the condition or level at which a system is exposed to significant climatic variations [10]. Sensitivity is defined as the degree/level of an affected system, either adversely or favorably, by climate-related stimuli, including all elements of climate change such as average climatic characteristics, climate variability, and extreme frequency and magnitude [10]. Adaptive capacity is defined as the ability of the system to adapt to climate change, to reduce potential damage from it, and to benefit or overcome its consequences [10].

Many studies developed approaches and methods to quantify the vulnerability as index [11][12]13]. A set of indicators have been used to represent exposure, sensitivity and adaptive capacity. In general, the index is developed by summing the multiplication of indicators values with their corresponding weights as the following:

$$\text{Index} = \Sigma w_i{*}I_i \quad (2)$$

where w_i and I_i is the weight and indicator value. The weight of an indicator is determined based on the importance level of the indicator in explaining the vulnerability which relies on expert opinion. However, there are many other approaches that can be used to define the weight. In this study, we used some indicators from the Government Study on the vulnerability of the Tangerang City conducted in 2016 [7] and introduced two new indicators that represent sensitivity and adaptive capacity related to clean water (Table 1a and 1b). These include clean water needs representing the sensitivity and PDAM's customers representing adaptive capacity. Clean water needs are classified as

sensitivity indicator as RWs (association of community) with high water needs will be more sensitive to clean water shortage than those with low water needs. Meanwhile, the number of PDAM customers is classified as adaptive capacity as the RW with the higher number of PDAM customers will have a higher capacity to adapt to water shortage than those with a lower number of PDAM customers.
The clean water needs of RW (CWN) are calculated using the following equation:

$$CWN = \text{population size} * \text{water consumption} \tag{3}$$

Water consumption is assumed to be about 175 liters/person/day (by using average amount based on the standard 150-200 liter/person/day) [14]. The value of CWN of an RW is normalized to make the value range between 0 and 1 by dividing the CWM with the CWM maximum. While the number of PDAM customer is also normalized by dividing the number of the customer with the population size of the RW.

Table 1a. New indicator of exposure and sensitivity

No	Data	Weighting
1	Population density	0.1
2	Clean water needs	0.25
3	Sensitive population	0.15
4	Low income family	0.15
5	Slump area	0.05
6	Developed area	0.1
7	Functional area	0.05
8	Unemployment	0.15
	Total weighting	**1**

Table 1b. New indicator of adaptive capacity

No	Data	Weighting
1	PDAM's customer	0.25
2	Education facilities capacity	0.20
3	Health facilities capacity	0.15
4	Source of income	0.15
5	Housing types	0.06
6	Sanitation types per family	0.10
7	Waste disposal method per family	0.05
8	Source of Energy	0.04
	Total weighting	**1**

Using the equation 1, we define two indices namely sensitivity and exposure index (IKS) and adaptive capacity index (IKA). The vulnerability index of the ward is defined using the quadrant system as suggested by the Permen LHK 33.2016 [15](Table 2).

Table 2. Category of vulnerability index

REQUIREMENT	QUADRANT	CATEGORY	COLOR
If Ano.IKS<0, Ano.IKA>0, Ano IKS-IKA<-0,25	1	Very Low	
If Ano.IKS>0, Ano.IKA>0, Ano IKS+IKA>0,25	2	Low	
If Ano IKS+IKA<0,25, Ano.IKS+IKA>-0,25, Ano IKS-IKA<0,25, Ano.IKS-IKA>-0,25	3	Medium	
If Ano.IKS<0, Ano.IKA<0, Ano IKS+IKA<-0,25	4	High	
If Ano.IKS>0, Ano.IKA<0, Ano IKS-IKA>0,25	5	Very High	

Source: CCROM, 2013 in PermenLHK 33, 2016[15]

While hazard index (HI) is also calculated following equation (1) in which the hazards include drought and floods. The level of flood and drought hazards is obtained based on the perception of respondent on the intensity of the hazards occurrence. The scores are 1, 2, 3, 4 or 5 representing very low, low, medium, high and very high intensity. The weight of the drought is 0.6, and the flood is 0.4. The drought has a higher weight, as the impact of the drought is regional wise while that of floods is a quite site-specific. The index of hazard is then classified into five, as shown in Table 3. The risk index is defined using the risk matrix as shown in Table 4.

Table 3. Category of hazard index

REQUIREMENT	CATEGORY	COLOR
IfHI< 1	Very Low	
If1< HI <=2	Low	
If2< HI <=3	Medium	
If3< HI <=4	High	
IfHI>4	Very High	

Source: Author's calculation

Table 4. Category of risk level

PROBABILITY		HAZARD				
		Very High	High	Medium	Low	Very Low
VULNERABILITY	Very High	VVH	VVH	VH	H	M
	High	VVH	VH	H	M	L
	Medium	VH	H	M	L	VL
	Low	H	M	L	VL	VVL
	Very Low	M	L	VL	VVL	VVL

Source: Permen LHK 33, 2016[15]

Note:

VVH : Very Very High
VH : Very High
H :High
M :Medium
L :Low
VL : Very Low
VVL : Very Very Low

This study simulated 6 scenarios, which are:

- Scenario 1 describes the current situation;
- Scenario 2 describes projected situation in 2021 in which the number of PDAM costumer per year was increased 0.5% in every ward;
- Scenario 3 is similar to scenario 2, but ward that has no PDAM customer will start to have new PDAM access in 2017 for every ward;
- Scenario 4 describes projected situation in 2021 in which the number of wards affecting by flood was increased by 2% per year
- Scenario 5 describes projected situation in 2021 in which the number of RW as PDAM costumer were increased by 2% per year for every ward;
- Scenario 6 projected situation in 2021 with a combination of scenario 3, scenario 4, and scenario 5.

4. Results

At the current condition, the level of vulnerability of wards of South Tangerang City is mostly dominated by medium to very high level. Factors causing the vulnerability of the wards are influenced mainly by clean water needs, unemployment, education level and the percentage of households that have access to clean water system (PDAM customers). Wards that have very high vulnerability level are Paku Alam, Pakulonan, Serpong, Ciater, Pondok Kacang Timur, Pamulang Barat, Pamulang Timur, Pondok Benda, Kedaung, Benda Baru, Sawah, Jombang, Pondok, and Ranji, otherwise, award that has low vulnerability index is Cilenggang. It is found that the wards with very high vulnerability level are highly populated with high clean water demand, high unemployment rate, low education

IOP Conf. Series: Earth and Environmental Science **129** (2018) 012027 doi:10.1088/1755-1315/129/1/012027

level and limited access of the household to clean water services (PDAM). These wards include Paku Alam, Ciater, Pondok Kacang Timur, Pamulang Timur, Sawah, Jombang and Pondok Ranji (Figure 2). In term of climate hazard, wards that are exposing to the high intensity of flood and drought are mostly situated in the north part of the city and a few in the south (Figure 2). Wards with very high vulnerability level and very high hazard index are considered to be very high risk. It is found that there are more than 20% of wards exposed to very high climate risk to the water supply. These wards include Lengkong Karya, Pondok Kacang Timur, Pondok Aren, Jurang Mangu Barat, Muncul, Kranggan, Pamulang Barat, Pamulang Timur, Kedaung, Cipayung, and Pondok Ranji (Figure 2).

With the scenario of increasing number of PDAM customer (scenario 2) and ward with PDAM access (scenario 3) with a low rate as well as increasing number of the ward of flood area (Scenario 4) will not change much the level risk (Figure 3). However, by increasing number of the ward with clean water service system (PDAM) at a high rate (2%; scenario 5 and 6), many of wards that currently have high risks will have lower risks. Under scenario 5 and 6, in the next five years (2021) the wards that have very low risks will increase significantly (Figure 3). The maps showing distribution of vulnerability index, hazards index, and risk index of wards under scenarios 2, 3, 4, 5 and 6 are presented in Figure 4.

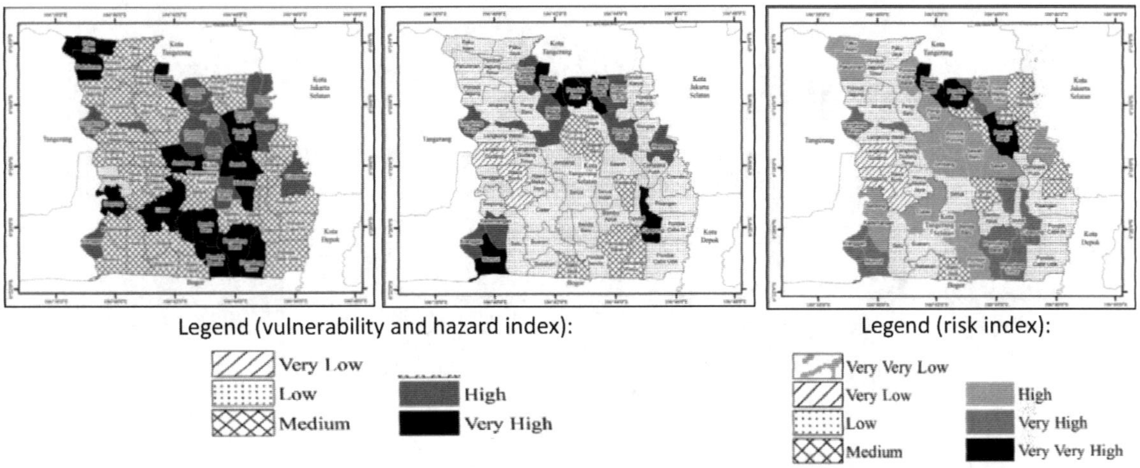

Figure 2. Map of vulnerability index (left), hazard index (middle) and risk index (right) of award under current conditions.

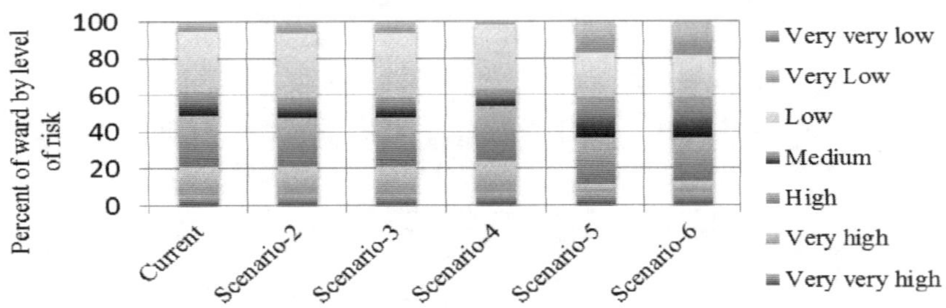

Figure 3. Percentage of ward based on risk level at current situation and under the scenarios

5. Discussion

Based on this study, it is shown that South Tangerang City has a potentially high risk of clean water supply caused by climate hazard, land cover changes, and human activities. It needs an adaptation action to reduce the worse impact by flood or drought that will potentially happen in the future. There is an option to build water reservoir as the water storage, as well as alternative water resources. Moreover, several river bodies are needed to be normalized, since these are occupied by buildings. The river normalization program may reduce the flood incidents in some areas. For preventing the worse drought condition in the future, South Tangerang City Government should also create many environments and water conservation by greening, bio pore hole, or rain harvesting. There should be a policy on sustainable groundwater management for preventing the water clean supply problems.

Besides climate hazards, the adaptation action should work on increasing the PDAM access throughout the city. The source of PDAM is not always from Cisadane River only, there are many other potential water resources in South Tangerang City since they have several lakes in the south and east part of the city. In 2013, RISPAM of South Tangerang is thought to be sufficient to provide the basis for the planning of water supply system from a variety of water sources according to the existing resources in South Tangerang. The action plan in the RISPAM can be improved by climate hazard consideration. Therefore, hopefully, the worse impact of climate hazards, population and development growth can be alleviated for more sustainable life.

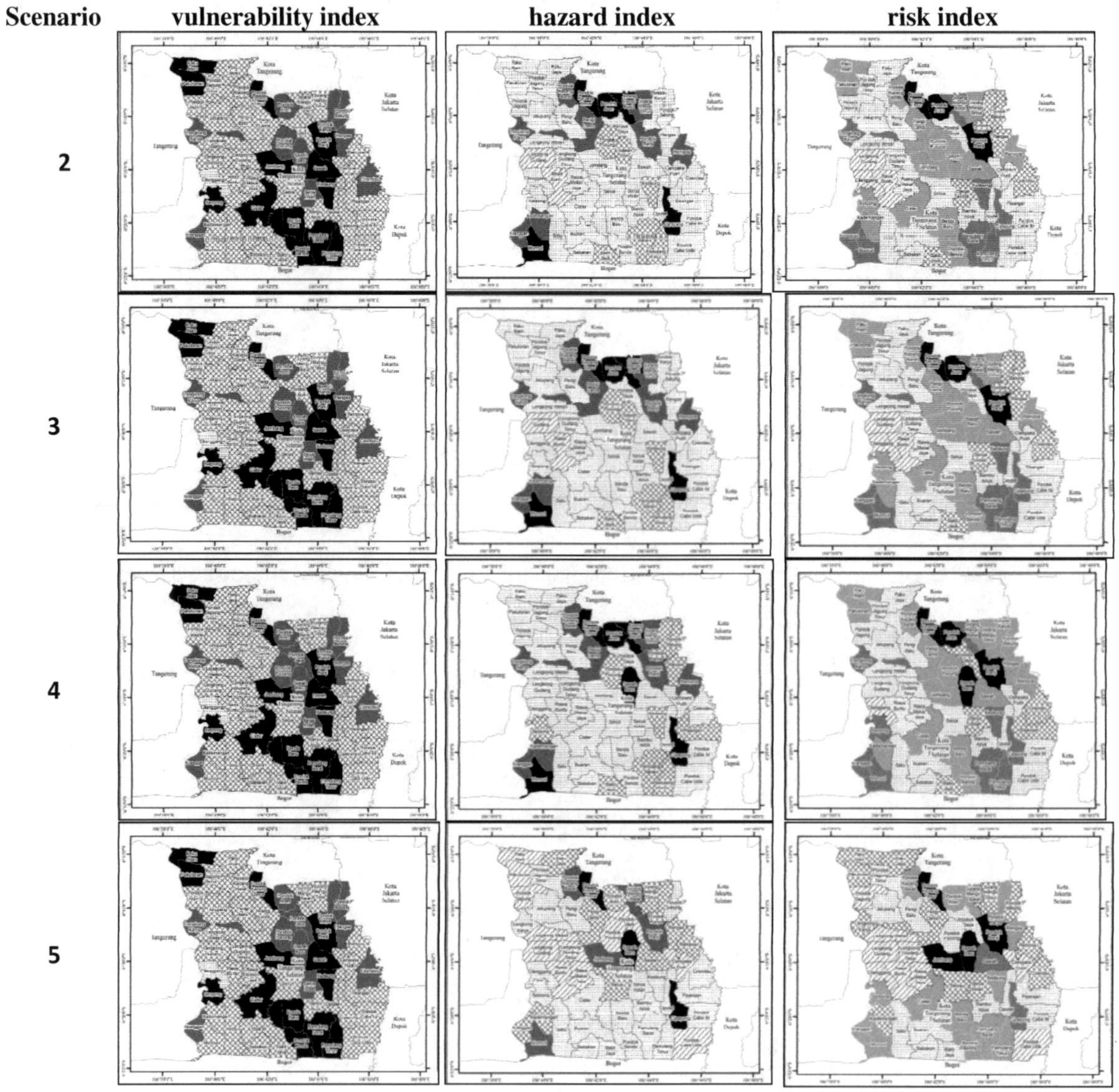

6

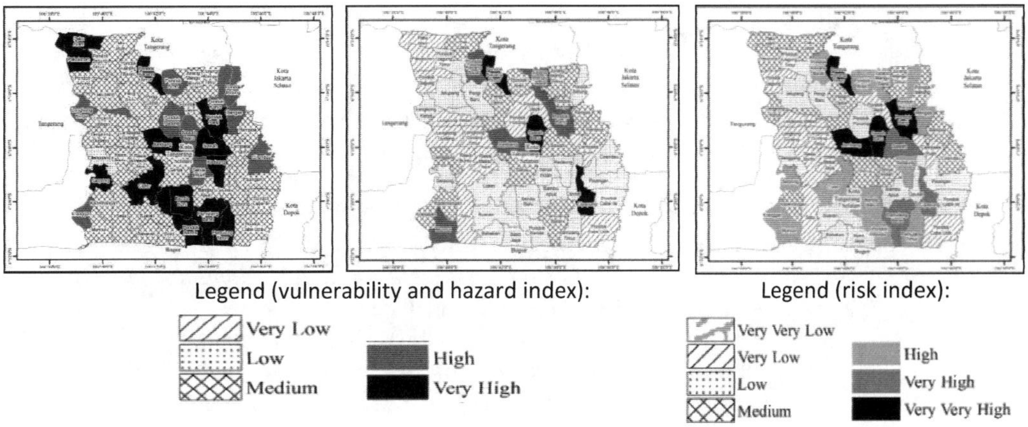

Figure 4. Map of vulnerability index (left), hazard index (middle) and risk index (right) of award under scenarios 2, 3, 4, 5 and 6.

6. Conclusion

South Tangerang City with its high population density and high clean water demand is also vulnerable to climate change. Since only about 9% population serviced by PDAM, most of them are highly dependent on groundwater that is very much influenced by climate variability, land cover changes, and human activities.

Based on this study, approximately 54% of the wards in South Tangerang have high risk until very very high-risk level and can be worse in the future if the climate hazards are not being anticipated as well as PDAM access is not being increased for all residents. It is necessary to develop a long-term adaptation strategy especially for clean water provision that takes into account aspects of climate change. Such strategy will enable the proponent to anticipate the possibility of worse risk and minimize the perceived loss.

Acknowledgement

Acknowledgments are submitted to all parties who have been willing to be interviewed and fill out the questionnaire and provide data and information for this study, as well as for 2017 PITTA grant of the University of Indonesia for supporting this study. Hopefully, the results of this study can be beneficial for the community and the South Tangerang City Government in achieving a sustainable water supply in the future.

References

[1] Provincial Government of Banten 2013 *Report of Status of Environment of Provincial of Banten*(Serang: Provincial Government of Banten)

[2] Rachmawati N 2012 Distribution and determination of water facilities and infrastructures in South Tangerang city *Jurnal Tata Kota Dan Daerah* **(4)2:**111-8

[3] Boer R, Rakhman A, Faqih A, Perdinan and Situmorang AP 2015 *Index of Climate Vulnerability and Risk of East Nusa Tenggara UNDP-SPARC Project Ministry of Environment and Forestry* (Jakarta: Ministry of Environment and Forestry)

[4] Apriyanto H, Eriyatno, Rustiadi E and Mardi I 2015 Sustainability status of South Tangerang city-Banten using key performance indicators *J. Manusia dan Lingkungan***22 (2)**: 260-70

[5] Sjarifudin A 2013 Implementation of regional spatial planning of South Tangerang city and future threat Bandung *Bulletin KontribusIllmiah***11(3)**: 140-52

[6] City Government of South Tangerang 2013 *Master Plan of System of Drinking Water Services (RISPAM) of South Tangerang City* (Tangerang: City Government of South Tangerang)

[7] City Government of South Tangerang 2016 *Assessment of Vulnerability and Risk of Climate*

Change of South Tangerang City (Tangerang: City Government of South Tangerang)

[8] IPCC 2014 Climate Change 2014: *Synthesis Report. Contribution of Working Groups I, II and III to the Fifth Assessment Report of the Intergovernmental Panel on Climate Change* [Core Writing Team, R.K. Pachauri and L.A. Meyer (eds.)] (Geneva:IPCC)p 151

[9] IPCC 2007 Climate Change 2007: *Impacts, Adaptation and Vulnerability - Contribution of Working Group II to the Fourth Assessment Report of the Intergovernmental Panel on Climate Change* [Parry M L, Canziani O F, Palutikof J P, Van Der Linde P J, and Hanson C E (eds.)](Cambridge: Cambridge University Press) pp 7-22.

[10] Arif AM 2013 *ToResponseA Climate Change Threat: An Adaptation An Urgent and Priority Choice*(Jakarta: National Board of Climate Change)

[11] Balica SF, Wright NG and van der Meulen F 2012 A flood vulnerability index for coastal cities and its use in assessing climate change impacts *natural Hazards***64**:73-105

[12] Boer R, Dasanto B D, Perdinan and Marthinus D 2012*Hydrologic Balance of CitarumWatershed under Current and Future Climate,* W. Leal Filho (ed.), *Climate Change and the Sustainable Use of Water Resources Climate Change Management* DOI: 10.1007/978-3-642-22266-5-3

[13] Prakasa R B S, Murty K S R and Amminedu E 2005 *Estimation of Flood Vulnerability Index for Delta Areas Through RS and GIS*(Seoul: Geoscience and Remote Sensing Symposium)pp 3611–14

[14] Ministry of Public Works 2006 *Services Unit Material of Training of Refreshing of Human Resources of Drinking Water Sector*(Jakarta: Ministry of Public Works)

[15] Ministry of Environment and Forestry 2016 *Ministry of Environment and Forestry Regulation Number P.33/MenLHK/Setjen/Kum.1/3/2016 on Guidance of Establishment of Climate Change AdaptationAction* (Jakarta: Ministry of Environment and Forestry)

[16] "South Tangerang city in number 2016". *tangselkota.bps.go.id*, 2016. [Online], Available:http://tangselkota.bps.go.id/. [Accessed: 27-Mar-2017].

[17] "Sub-district in number 2014". *tangselkota.bps.go.id*, 2014. [Online], Available:http://tangselkota.bps.go.id/. [Accessed: 27-Mar-2017].

[18] "Sub-district in number 2015". *tangselkota.bps.go.id*, 2015. [Online], Available:http://tangselkota.bps.go.id/. [Accessed: 27-Mar-2017].

[19] "Sub-district in number 2016". *tangselkota.bps.go.id*, 2016. [Online], Available:http://tangselkota.bps.go.id/. [Accessed: 27-Mar-2017].

IOP Conf. Series: Earth and Environmental Science **129** (2018) 012028 doi:10.1088/1755-1315/129/1/012028

A Group in Urban: The Social Capital of Ciliwung Depok Community (KCD)

D M Tampi[1,2], J Sumabrata[1], A Zubair[1] and N H Kinan[1]

[1]Urban Development Studies: School of Strategic and Global, Universitas Indonesia, Salemba Raya Street Number 4, Central Jakarta, Indonesia.
[2]Corresponding author: danielmambo131293@gmail.com

Abstract. Climate change is a phenomenon that not only affects physical aspect of the environment, but also the social aspect. As a consequence, the urban communities are also its group which is highly vulnerable to climate change impact. Depok Ciliwung Community or *Komunitas Ciliwung Depok* (KCD) is located under Grand Depok City Bridge to keep an environment such as land conversion and garbage. This study aims to determine the social capital of KCD's group related to the potential of social resilience in Ciliwung riverbank along Depok City. Data were gathered from in-depth interviews, observation, and documentation within 4 months. KCD invited the local people to keep and maintain an environment of Ciliwung riverbank. The result of this study lies on the strengths and weaknesses of KCD's Group and involvement of human resources, with the support of public and private sectors facilities. Their commitments and actions on behalf of social interests protecting the rights of individuals in society. The conclusions of this research are KCD group have several aspects of social capital that consists of strengths and weaknesses. The presence of social capital and the potential of social resilience helped KCD and the surrounding of Ciliwung riverbank in facing of climate change/ threat of natural disruption caused by the humans itself.

1. Introduction

Climate change is a global phenomenon triggered by human activities. These changes have some impacts in urban areas on the physical aspect of the environment, health, politics, and social aspect. According to Evers [1], the sustainability of the city is currently dependent on the rules and regulations imposed by the government. These rules and regulations are the political decisions of the country and the city government. The compilers and their enactment are based on political struggles in which alliances, winners, and losers often change. Evers in such situations (the absence of common city concepts and meanings) the dominant city may try to imprint their seal on the city through buildings built for the purpose of dominating urban spaces, such as conference buildings, high-rise office buildings, shopping centers, apartments, toll roads that cross the city and so on.

If that happens then no wonder various social inequalities occur in urban areas. Ife [2] argues that urban community structures are much weaker, the boundaries are difficult to see or do not exist. There are people who usually contact groups and structures that are substantially detached from their local community. These things seem to be a challenge to the city community in witnessing the state of space that increasingly exclude the interests and rights of urban communities. These local communities are important to trace because they are located in urban areas which according to the characteristics of Patembayan urban life patterns (GESELLSCHAFT) [3] among others: Individualism, Egocentric, Prioritizing money, Competition in the struggle for sustenance and the importance of property rights.

Content from this work may be used under the terms of the Creative Commons Attribution 3.0 licence. Any further distribution of this work must maintain attribution to the author(s) and the title of the work, journal citation and DOI.
Published under licence by IOP Publishing Ltd

Depok Ciliwung Community (KCD) is one of the communities in urban areas that have the potential to possess such traits. The first Ciliwung community in Depok City is a community that was established in 2010 under the leadership of Mr. Taufik and Mr. Sahroel. This community began when the existence of irregularities of housing development in the village of Kalimulya as well as small industries on the riverbanks caused water pollution and landslides in Ciliwung.

Humans are social beings, who cannot live alone but have relationships/relations with other humans as individuals or groups. The Ife States that Social Capital can be seen as 'adhesive' that unites society, relationships between people, who do what they do to each other because of the social obligations of reciprocity, social solidarity, and community. Part of building social capital is the strengthening of 'civil society'. Civil society is a term used for formal and semiformal structures that society voluntarily forms, on their own initiative, not as a consequence of certain government programs or directives. According to Djohan [4] When discussing the capital in Indonesia, especially in order to improve the superiority of society, says that we need to see and pay attention to things that have happened in the past. Soeharto had changed the way of thinking of Indonesian people who previously are more tied to the problems in the country (bonded) to become a nation that would relate to other countries in the world (bridging). By paying attention to global events, Soeharto transformed Indonesia's previously agriculture-oriented economy to industrial economy. In the era of reform after Soeharto's downfall, the wind of change directed the Indonesian people to the real practice of democracy. Since then democracy is considered important.

Unfortunately, the government failed to do its homework to expand employment and improve the welfare of the people. As a result, unemployment and poverty increased. Considering that unemployment and poverty greatly affect the social capital of a nation, the Indonesian government took these matters seriously. What happened then was, the Indonesian economy was growing remarkably. The GDP growth averaged 7% per year. It was no surprise that Indonesia was then classified as an economic market that could compete at the global level, especially in the fields of mining, textiles, and electronics. However, the dominant role of the government actually weakened the Indonesian nation. Inevitably, the economy went into recession and Indonesia's role in the world became very weak. From the background, the purpose of this study is to analyze the strengths and weaknesses of Depok Ciliwung Community and social capital in relation to the potential of social resilience in facing the threat of disaster caused by human activities.

2. Materials and Method

2.1. Materials

According to Gottdiener [5] City is a limited space due to its densely populated and relatively large population and heterogeneous cultural population. The city is also defined as a social stage which emphasizes the interaction and social role of cities that are nurtured in behavior (Mumford in Gottdiener). In addition, there are two understandings of the city; first, it is the physical aspect (built with nature) as a form of space with its elements; second, it is a human aspect as the subject of development and city space user.

The community is the ability of social units to reduce hazards, the impact of disasters they feel, and carry out recovery activities by minimizing social disruption and reducing the impact of future disasters [6]. A community is a group that thinks of itself as having a strong and lasting bond, especially when the groups share location geographically. One way to measure the Community is the regular participation of individuals in their activities. Then, the Community is also a specific physical space and location that is generally understood as a group territory. This space provides and organizes its own materials in which its members have strong emotional ties [5].

Social capital is considered important in individual and group level. Then, the following are aspects of social capital in the dimension of reconciliation and time that build and decrease the quality of social capital according to Dale in Darmajanti [7]:

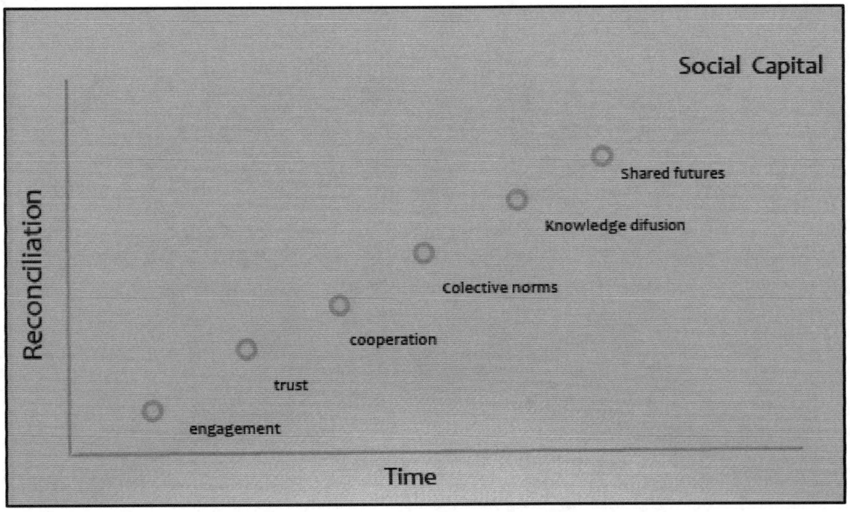

Figure 1. Aspects of Social Capital Builders

In building a social capital, some real actions are required in the dimensions of reconciliation and time; first with engagement, trust, collaboration, collective norms, knowledge diffusion, and shared futures, while the decline in social capital can be described as follows:

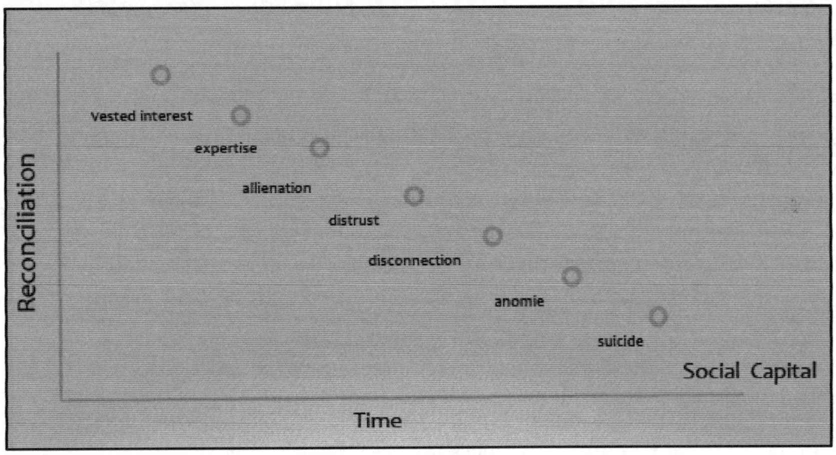

Figure 2. Aspects of Reduced Social Capital

Figure 2 indicates that the aspects that degrade the quality of social capital are: Vested interest, expertise, alienation, distrust, disconnection, disorder (anomie), and suicide. According to Dale Social capital in relation to Community development such as a Social capital, are identified by the following indicators: diversity, stakeholders, commitment, leadership, and voice, facilitate decision-making and key actions that will provide a reconciliation of three imperatives, and thus contribute to sustainable development (Figure 3). Here is a model of a social capital formation that contributes to sustainable development:

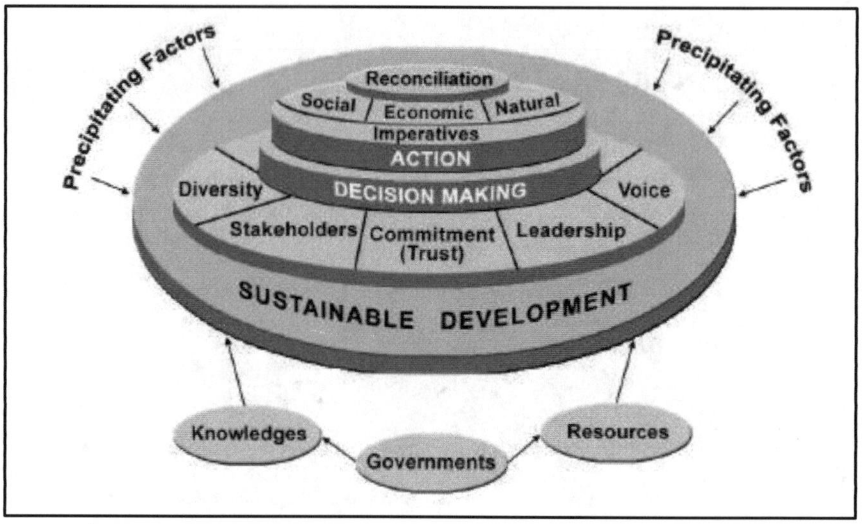

Figure 3. Model of Social Capital Formation [8]

A good combination of bonding and bridge networks will, as suggested by Tompkins and Adger in Dale [8] results in greater resilience and more increased adaptability. Adaptation is only half of the profit earned by such social networks; Large strategic impulse lies in the collective capacity for creative innovation that increases the importance. We would like to emphasize that the creation of this importance is proactive exercise. We distinguish the proactive side of managing environmental change as a "development" component of sustainable development.

2.2. Method

The method used in this study is a qualitative method. Through qualitative approach, the researcher can study the process and gain perspective and meaning from key informants to everyday life that goes on in the region, or to the symptom they experience either in that place or elsewhere. Qualitative research is a study that intends to understand the phenomenon of what is experienced by research subjects such as behavior, perception, motivation, action, etc., holistically, and by way of description in the form of words and language, in a special context that is natural and by utilizing various natural methods [9].

This study uses a case study approach. The case study focuses on the status of a research subject that is concerned with a specific or distinctive phase of the whole personality. Research subjects can be individuals, groups, institutions, and society. The purpose of the case study is to provide a detailed overview of the background, traits, and characteristics of the case, or the status of the individual, which then of the above characteristics will be made general [10]. The case study approach seeks to look at the historical, physical setting and relates the context to the theories to be used [11]. This study has limitations on the role of actors, groups, social capital and community-based at Depok City's Ciliwung riverbank. Stages of data analysis in this study include field findings, matrix, categorization, abstraction associated with the theory, the results of analysis and triangulation.

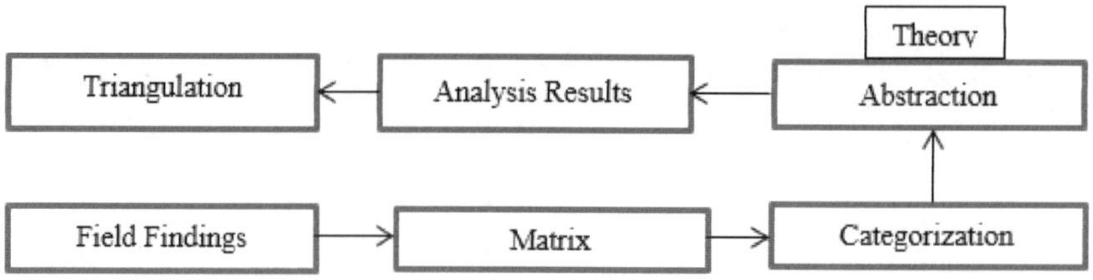

Figure 4. Data Analysis Stage

In analyzing the data, the researchers also have a measurement of analysis of the findings and theories expressed in scale. This scale in the form of scoring numbers from 1-10 with low, medium and high quality.

Table 1. Qualitative Analysis Rating Scale

Scale	**Quality**
1-3	Low
4-6	Medium
7-10	High

Source: Researchers Processed Results, 2017

This scale is made by the first author in conducting a qualitative assessment. This assessment is made to make measurements on the analysis that discusses the findings and concepts referred to. It is realized that ideally, the concept that the researchers concerned with depends on the extent of the researcher's interpretation of the findings in the field.

3. Results and Discussions

In this research, the researcher started the research by doing the initial observation at Ciliwung Depok Community basecamp. Researchers searched for key informants determining the informants to be interviewed. They divided the informant into two groups; the core and supporting informants. The informants were given initials which are the core informants as A1-A6 and Supporting Informants as B1-B4, it is an ethic in academic writing to avoid the conflict among actors. The location of Ciliwung Depok Community (KCD) is located under the Grand Depok City Bridge, Boulevard Street, Depok Village, Pancoran Mas District, Depok City, West Java Province. The Ciliwung Depok Community Basecamp (KCD) is often called *"Kolong"* by its members as a place where they conduct their meetings and activities. Here is a map of the location of Ciliwung Depok Community:

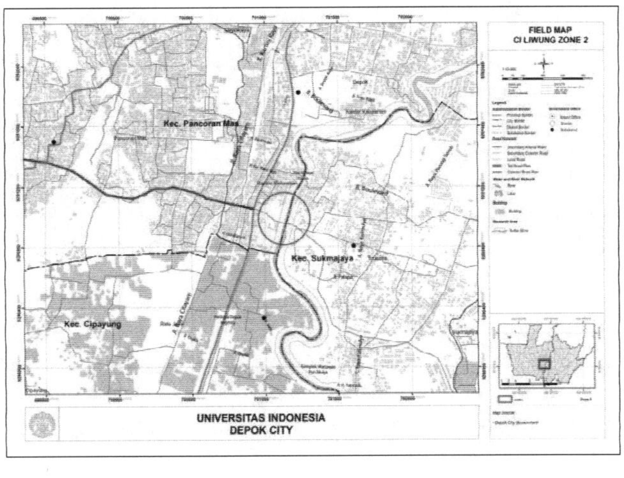

Figure 5. KCD Location Map (Personal Documentation, 2017)

Started by getting a boat, Informant A1 and his friends thought about the idea of their gathering place. They agreed to crush under the GDC bridge. Informant A1 said that the activities under Kolong GDC began in 2014 by holding a flag ceremony on 17 August 2014. At first, they cleaned the place and the Ciliwung River and implemented *halal bi halal* at the time of fasting month.

Here is the documentation when observing Kolong GDC Basecamp (KCD Basecamp, April 8, 2017) :

Figure 6. Kolong GDC (Basecamp KCD)

In 2010, Informant A1 became familiar with Informant A2. Then they started their communication through facebook message. Informant A1 explained that Informant A2 is also a Depok resident. Informant A2 is the person who first had the access to the government, especially those in the environmental service. Informant A2 also helped Informant A1 to tag the latest information to the parties involved in environmental services in Depok at that time. Since the initial communications of the time, their meetings began to intensify and became partners in advocacy. They began to work together to fight for all forms of development misappropriation in the Ciliwung border. The case they fought for was refusing the construction of Taman Anyelir III which had not been licensed.

Figure 7. Evidence of Rejection of Taman Anyelir III

KCD is a fluid group bound by their emotions and interests. KCD has contributed a lot in fighting for clean Ciliwung and brought benefits for the Ciliwung riverbanks community. These contributions will not be realized without the internal and external support of the KCD in facing various problems and realizing the social interests in Ciliwung. The internal and external factors that directly affect the development of powerful communities and can organize their groups (*How they organize themselves to solve their own problems*) :

Internally, KCD has a membership of 50 people, but the only ¼ of the total membership is actively involved in basecamp activity.

"Only 10, at most 15 people are active and it keeps on changing. Informant A2 and I are the ones remain since the beginning. (Informant A1)

In terms of quantity, the members of KCD are quite a lot. But in terms of routine involvement, they are very little. This happens because only one-fourth of KCD members actually spend their spare time doing activities in Kolong. Most of the others only work on events that are temporary. Members of KCD are engaged in their respective interests. Apart from that, Depok Ciliwung Community always shows its existence with the existence of social capital that binds the members of the group which is also part of the society along the riverbanks of Ciliwung and Depok City in general. Here are the results of the analysis that has been made by researchers. This is the social capital analysis findings of KCD.

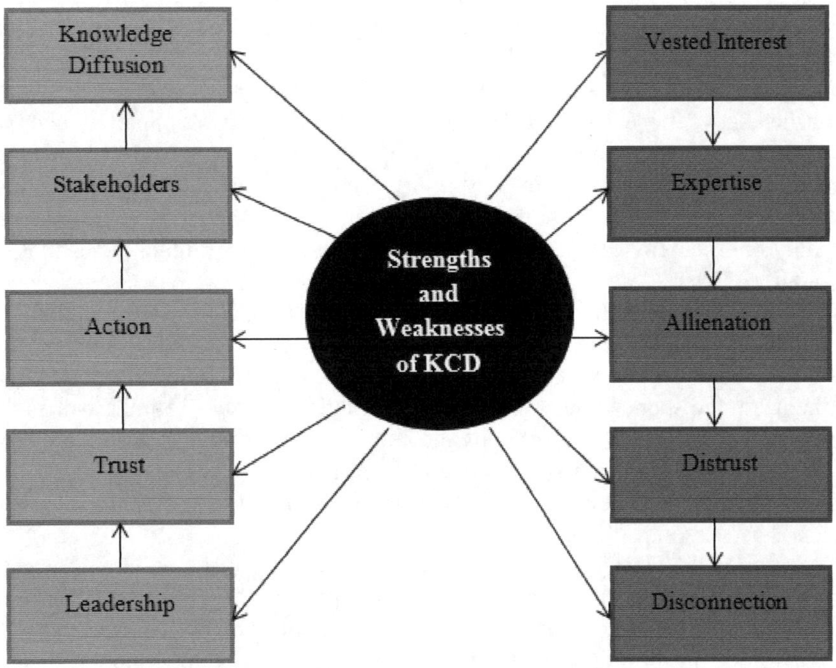

Figure 8. Chart of Strength and Weakness of KCD Analysis

The strengths of KCD actors lie in leadership, the mutual trust among actors, their faculty, the existence of the rules and complement each other as a resource that has its own expertise. While the weaknesses of KCD actors are the existence of personal interest, the competition because each has a specialty of expertise, the existence of actors who are alienated, the occurrence of distrust of the actors and the separation caused by eliminating themselves from KCD. The first strength of KCD is Leadership. This group was initiated by Informant A1 who has a lot of experience in his life. He gained the experiences from his workplace and conditions around his neighborhood. This what makes Informant A1 brave. His courage in advocating and protecting rights makes him known by many people. Informant A1 was able to organize KCD as a group. Not only that, Informant A1 also a wise leader in KCD members' point of view.

In addition to Informant A1, Informants A2, A5, and Mr. A are individuals who have great influence in KCD. Informant A2 has a good relationship with several related agencies in Depok City. Informant A2 is able to collaborate with Informant A1. They advocate and monitor every form of frauds found in Ciliwung riverbank. They are able to bridge the society and government. Informants A1 and A2 are captured the volunteers and their aspirations through facebook social media. Not a few volunteers come to the KCD basecamp. They come with a variety of purposes. KCD facilitates volunteers to work in Ciliwung.

Because of the Courage, Wisdom, and Knowledge of Informants A1 and A2, they work together with the community at some SPA (Saung Pustaka Air) point, they succeeded in getting the community represented by RT / RW to participate as KCD members. Through them, the KCD is considered as the mover and an open space for children. Through them as well, KCD is considered as a facilitator in training activities to the community, especially those on the riverbanks of Ciliwung. The strength that lies in leadership is also shown by Informant A5 and Mr. A. Their professions and experiences led Informant A5 and Mr. A to participate and merge to become part of KCD. The participation, exemplary and knowledge of informant A5 and Mr. A become the KCD builder. They not only become facilitators but they are also capable of being reliable partners. Their involvement as coaches makes KCD empowered. They sparked the concept of Ciliwung as Ecotourism and RT (Rukun Tetangga) friendly water in Ciliwung Depok community. So the researchers gave a value of 7 on the aspects of leadership.

The second power in KCD is Trust. In social capital, the element of trust is considered important. Trust is established between members and builders of KCD. Such trust is apparent when they gather in meetings and activities. These activities are carried out together. Although these activities are temporal, but KCD is able to organize every activity. The trust that was built was also seen from the participation of Ciliwung citizens in the KCD. They view the KCD as a useful community and make significant changes in their environment. In the aspect of trust, the researcher gives an 8 value because it has a high quality of trust.

The next KCD's power is Action. Actions held by actors in KCD invited members as part of the community. Thanks to the awareness and concern of actors, they also work with their intuition. By using intuition then other people can witness something else that can not be done by others because it comes from the heart and spontaneous movement. The intuition comes from the values they hold with regard to where they live. This is supported by the cooperation with the government. The government supports KCD in their actions. Such support can be seen from various contributions and grants given to KCD and vice versa. The researcher gives the value of 6 with medium quality because there are still many obstacles as barriers in KCD actions.

The next strength is the existence of Stakeholders. KCD maintains a good communication among stakeholders. They have partners such as Depok City Government officials and other groups who take on the task of keeping Ciliwung together. In addition, to support their activities, KCD is working with Corporate Social Responsibility (CSR) and some private sectors to conduct their activities in Ciliwung. The stakeholders have involved and contributed to the existence of KCD for seven years. The researchers gave a value of 6 with moderate quality. The last strength is the dissemination of knowledge. The lack of knowledge possessed by KCD members require some resources. In addition to the coaches who directly train their members, they also involve several Institutions to train them. Institutional contributions are very large in KCD. KCD cannot exist without the help of knowledge held within the Institutions. In this aspect, the researcher gave a value of 7 because of openness and willingness among members in sharing knowledge/skill.

Of the five components of KCD strength, there are weaknesses that make social capital in KCD declining; the Personal Interest. Since KCD members are diverse, they depart from the motivation and goals of each member. Goals with different orientations make each of them focus and work on the things they demand. It has the potential for emerging expertise that considers itself knows better and best. Even because of this, the potential for conflict and friction occurred. As people living in urban areas, selfish attitudes often arise. Egocentric attitudes and putting money first are proof that there are many pressures and motives that are indirectly seen and witnessed. In this aspect, the researcher gave 6 with medium quality.

The next thing is Partition. After these attitudes, then those who lead to conflict begin to eliminate themselves. Attitudes are nothing but the separation/distance to one person and to the KCD as a group, the aspect of separation is rated 3 because the potential is still low. It then brings disbelief to the person who eliminates himself. The trust to the person is gone. KCD distrust of the person resulted in the breaking of the previously constructed relationship. The researcher assigned a score of 3 on the aspect of distrust. The disconnected relationship and the distrust lead to disconnection. Relationships are not only limited to mistrust but may also have an, impact on termination of relationships between individuals as well as to KCD as a group. This breaking aspect is rated 2 because of its low quality.

The five components of the decreased social capital are analyzed by researchers as a weakness of KCD.

From the strengths and weaknesses, Social Capital is related to Micro Social Resilience Potential. It should be realized that social interests in society certainly have different levels.

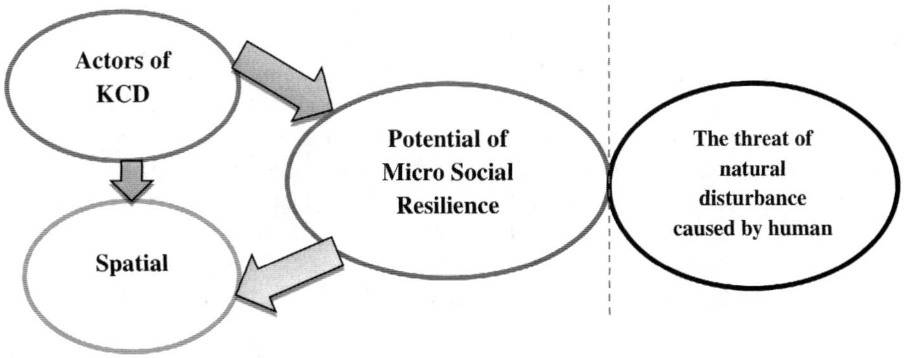

Figure 9. Chart of Analysis of Social Capital Linkage and Potential of Micro Social Resilience

Based on social capital owned by actors who are Individuals in groups then have a common goal. In fact, the effort of Individuals in Social Capital is closely related to the concept of micro social resilience; how the actors/individuals have commitments and actions on behalf of social interests in protecting the rights of individuals in Ciliwung community, even society in general in facing stressful situation / threat of natural disturbance caused by humans themselves. Thus, in realizing the macro (National) interests, we should start from the realization of micro social interest that prioritizes individual actions in fighting for the rights of people at the local level. The power of the actors is closely related to the potential of micro social resilience. KCD actors demonstrate their commitment and ability in society to act on behalf of common social resilience. The actors have also protected the rights of the community and have the ability to address the problems in their environment. This potential is seen in several locations that have been mapped as a spatial form of the distribution of KCD activity locations. The analysis of the linkage of social capital and the potential of micro CS social resilience is actually their effort in dealing with the threat of natural disturbance caused by man-made.

4. Conclusions

Ciliwung Depok Community (KCD) is an urban community that is actively fighting for not only social rights but also the river environmental sustainability as part of the society life. KCD is also a group facing the climate change impact and the potential disaster in Ciliwung watershed. This research is a historical record of the KCD trips that have fought in protecting the rights of the Ciliwung community from the threats of natural hazards caused by human actions. Ciliwung Depok community is bound by a sense of togetherness and trust that is not limited by the standard rules. The leadership figure owned by Informants A1 and A2 is able to lead its members and invite the aspirations of many volunteers and institutions to engage in its activities.

The social capital that becomes adhesive between members and stakeholders helps them to show the spirit of the community and its existence as a community in urban areas. Researchers also concluded that in the urban community there are citizens who care about the preservation watershed (DAS). The social capital of Ciliwung Depok Community is divided into two parts: social capital builder: Leadership, Trust, Action, Stakeholders, and Knowledge Diffusion. While the second part is the social capital-lowering aspects include: Vested Interested, Expertise, Alienation, Distrust and Disconnection. The strength of KCD lies in the aspect of social capital builders, but in its development, KCD has constraints or weaknesses. The weakness is found in the social capital-lowering aspect as analyzed by the researcher. The results of the researchers' analysis in this study also found that social capital is closely related to the potential of micro social resilience. The potential is analyzed by looking at the aspect of the actors and the community itself as well as spatial because it involves the location of community activities on the riverbanks, precisely under the bridge of GDC Depok City. Finally, the analysis of social capital and the potential of micro resilience and the factors found in the community are potential to response climate change phenomenon as part of realizing sustainable development.

The recommendation of this research is that the Government should close any development and industrial activities that may damage the Ciliwung watershed (DAS), potentially bringing a landslide disaster and increasing climate change impacts. The government needs to be a liaison between stakeholders and non-partisan groups with particular interests. People should be aware that they should not throw garbage into the river. The community should support the programs implemented by KCD and other organizations that are working to maintain the Ciliwung watershed (DAS). For the sake of the development of science then the Academics should be able to identify the cause of Ciliwung watershed (DAS) damage and the arrival of landslide disaster in the process of planning Depok City. Ultimately, the development of this research is to increase the insight of science, contribute to the development of the Indonesian Nation related to the strength of urban communities in the maintenance of the environment in the watershed (DAS) and facing the climate change impacts. Learning from KCD cases as part of the community, participating and having social concerns in sustainable urban development.

Acknowledgment
The authors gratefully acknowledged that the present research is supported by Higher Education Science and Technology Development Grant funded by Indonesia Ministry of Research and Higher Education.

References
[1] Evers, Hans-Dieter, Korff and Rudger 2002 *Urbanism In Southeast Asia: Meaning And Power In Social Spaces* (Jakarta: Yayasan Obor Indonesia)
[2] Ife, Jim, Tesoriero, Frank 2008 *Alternative to Community Development in the Era of Globalization* (Yogyakarta: Pustaka Pelajar)
[3] Suparlan P 2004 *Urban Society and Culture: Urban Anthropology Perspective* (Jakarta: Yayasan Pengembangan Kajian Ilmu Kepolisian)
[4] Djohan R 2007 *Lead to Togetherness Leaders & Social Capital. (Jakarta: Fund Asia Education*)
[5] Gottdiener, M. Budd and Leslie 2010 *Key Concepts in Urban Studies*. (Sage)
[6] Wilbanks T J 2009 *How Geographic Scale Matters in Seeking Community Resilience* (United States: CARRI Reports 7)
[7] Darmajanti L 2017 *The Principal Of Community Development Course* (Jakarta: Urban Development Studies, Universitas Indonesia)
[8] Dale A 2005 *Social Capital and Sustainable Development* (Accessed on June 3, 2017: https://www.crcresearch.org/files-crcresearch/File/proposal.pdf)
[9] Moleong L J 2006 *Qualitative Research Methodology Revised Edition* (Bandung: Remaja Rosdakarya Offset)
[10] Nazir M 1988 *Research Methodology* (Jakarta: Ghalia Indonesia)
[11] Septiady Y 2016 *Urban Society and Culture Course* (Jakarta: Urban Development Studies, Universitas Indonesia)

IOP Conf. Series: Earth and Environmental Science **129** (2018) 012029 doi:10.1088/1755-1315/129/1/012029

Land management on soil physical properties and maize (*Zea mays* L. var. BIMA) growth (An adaptation strategy of climate change)

M K Zaki[1,5], Komariah[2], B Pujiasmanto[3] and K Noda[4]

[1]Agronomy Dept., Graduate School Program, Sebelas Maret University, Jl. Ir. Sutami 36A, Kentingan, Surakarta, Central Java 57126 Indonesia.
[2]Soil Science Dept., Agriculture Faculty, Sebelas Maret University, Jl. Ir. Sutami 36A, Kentingan, Surakarta, Central Java 57126 Indonesia.
[3]Agrotechnology Dept., Agriculture Faculty, Sebelas Maret University, Jl. Ir. Sutami 36A, Kentingan, Surakarta, Central Java 57126 Indonesia.
[4]Applied Biological Sciences Faculty, Gifu University, Yanagido Street 1-1, Gifu City, Japan.
[5] Corresponding author: zakimuhamad30@gmail.com

Abstract. Water deficit is a problem on rainfed maize production but can be solved by proper land management. The objective of the study to determine the soil physical properties and maize yield affected by land management to adapt to drought. The experimental design was a randomized complete block using 5 treatments with 4 repetitions, including: (i) Control (KO), (ii) Rice Straw Mulched (MC), (iii) Compost Fertilizer (CF), (iv) In-Organic Fertilizer (AF), (v) Legume Cover crop (CC). Soil physical and maize growth properties namely soil moisture, soil texture, soil bulk density, plant height, biomass, and yield were investigated. The results showed that composting land increased soil water availability and provided nutrient to crops and thus increase soil physical properties, maize growth and yield. Although inorganic fertilizer also increased plant growth and yield, but it did not improve soil physical properties.

1. Introduction

Climate change is a natural phenomenon occurs nowadays due to global warming. In natural, it occurs due to the increase of solar intensity and the rotation of the Earth's orbits approaching the sun [1]. In addition, climate change caused by global warming occurs due to the increased carbon dioxide concentration in the air. One of the global climate changes is the increasing frequency and intensity of extreme climates such as storms, floods, and droughts [2].

Drought is a natural phenomenon related to the climate especially insufficiency of rainfall that affects agricultural, environmental, and socioeconomic conditions [3]. Drought is divided into meteorological, hydrological, agronomic, and socio-economic droughts. Meteorological droughts are defined as rainfall deficit; agronomic drought is a drought occurs due to low groundwater, high evapotranspiration and water scarcity [4]. Drought is an important issue for rain-fed agriculture, because of the dependency of water availability to rainfall. This leads to drought threat during the dry season.

Various efforts have been made to anticipate the occurrence of drought on the rain-fed land, due to rainfall irregularity such as the utilization of Nino 3.4 surface temperature data to predict monthly rainfall [2], as well as the implementation of various land management, monitoring soil moisture to

Content from this work may be used under the terms of the Creative Commons Attribution 3.0 licence. Any further distribution of this work must maintain attribution to the author(s) and the title of the work, journal citation and DOI.
Published under licence by IOP Publishing Ltd

determine cropping pattern [5]. This study aims to investigate the effects of land management to anticipate drought by employing organic, inorganic fertilizer and cover crop on soil physical, properties, maize growth and yield.

2. Method

This research was conducted in the experimental field of Faculty of Agriculture, Sebelas Maret University, Surakarta. Administratively the research area is located in Central Java, Indonesia (110°56'51 "S and 07°37'48" E) with elevation 191 meters above sea level. Soil type at research site is Alfisols. The condition of research land as shown in Figure 1.

Figure 1. Land preparation condition in Jumantono district with Alfisol soil.

The research was designed using Randomized Block Design with 4 treatments (single factor), including 1) Control or no treatment (K0), 2) Composting (CF), 3) Straw mulch giving (MC), 4) An-organic fertilizer (AF), and 5) Use of cover crop (CC) with 4 repetitions. Each treatment consists of 4 repetitions. Soil moisture was observed with the gravimetric method, soil bulk density with pycnometer method and soil texture with hydrometer method. The observation of maize growth and yield occupied volumetric method. Analysis of variance (F-test) and advance test (Duncan Multiple Range Test) were employed to analyze the data, at α =0.05.

3. Results and discussions

3.1. Soil physical properties

3.1.1. Soil moisture. Water in soil acts as a solvent and a binder element between soil particles that will affect the structure and strength of the soil. Chemically, water acts as a transporting agent of solutes involved in soil development and degradation. Soil moisture at each treatment is shown in Figure 2.

Figure 2 shows that soil moisture content under all treatments are similar to control, but managing rain-fed farmland with compost and inorganic fertilizer resulted in significant different soil moisture. Applying compost on rain-fed farmland can significantly maintain soil moisture remains high (18.1%) than implementing just inorganic fertilizer (14.5%). Thus it is important to control inorganic fertilizer usage in order to minimize drought risk at rain-fed farmland.

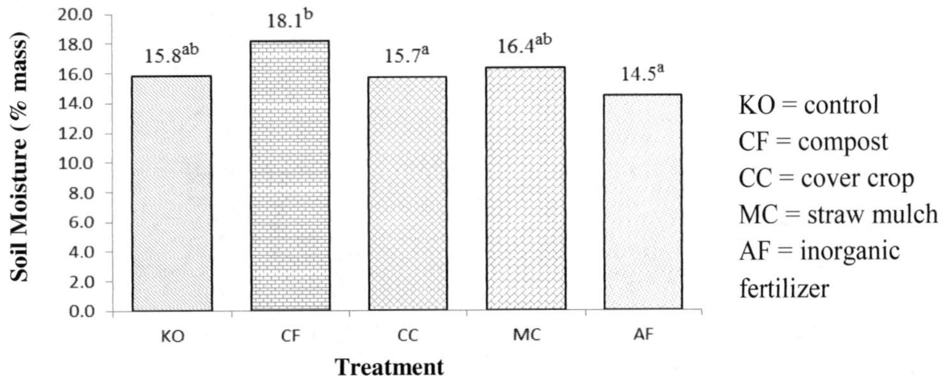

Figure 2. Soil moisture at each treatment

3.1.2. Soil texture. The soil texture shows the fraction comparison between sand (2mm - 50μ), silt (2μ-50 μ), and clay (<2μ) in the fraction of fine soil [6]. The relative size of soil particles is expressed in terms of texture that refers to the fineness or roughness of the soil [7]. The soil texture result under treatments as presented in Table 1.

Table 1. Soil texture of each treatment (0-30 cm)

Treatments	Texture (%)		
	Sand	Silt	Clay
control	29.07	7.98	62.93
compost	28.55	5.26	66.63
cover crop	34.05	11.72	56.97
straw mulch	26.68	6.68	66.17
inorganic fertilizer	27.94	20.99	48.30

Table 1 shows that the sand content ranges from 26.68 to 35.05%, silt ranges 5.26 – 20.99% and clay ranges from 48.30 to 66.63%. Soil fraction which strongly correlates with soil moisture is clay. The clay content at each treatment is shown in Figure 3.

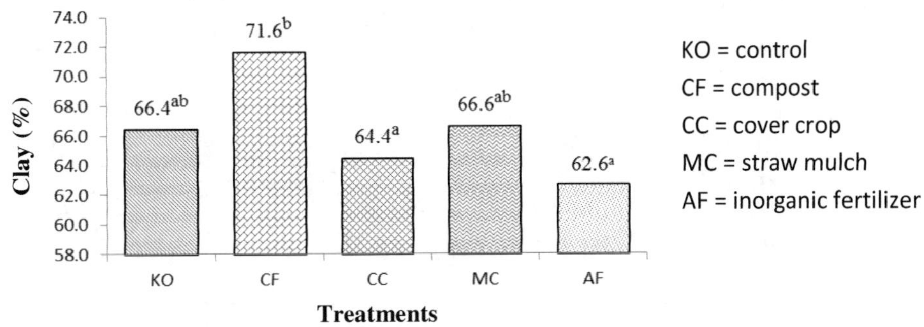

Figure 3. Clay content at each treatment

Figure 3 shows that clay content under all treatment were not significantly different with control. This is the same result as shown by soil moisture in Figure 2 that the soil moisture under all treatments were not different as that performed by control. It is shown in Figure 3 that compost treatment had higher clay content than that under cover crop and inorganic fertilizer treatments. That is probably because compost donated clay to the soil, but not much compare to control. Soil texture requires very long time to be affected by treatments, and thus it will be too early to determine the effects of this study on soil texture. However, compost played role in contributing clay than inorganic fertilizer.

3.2. Maize growth and yield

3.2.1. Plant height. Plant height is an indicator of plant growth that is easily observable and measured. The result of maize height measurement is presented in Figure 4.

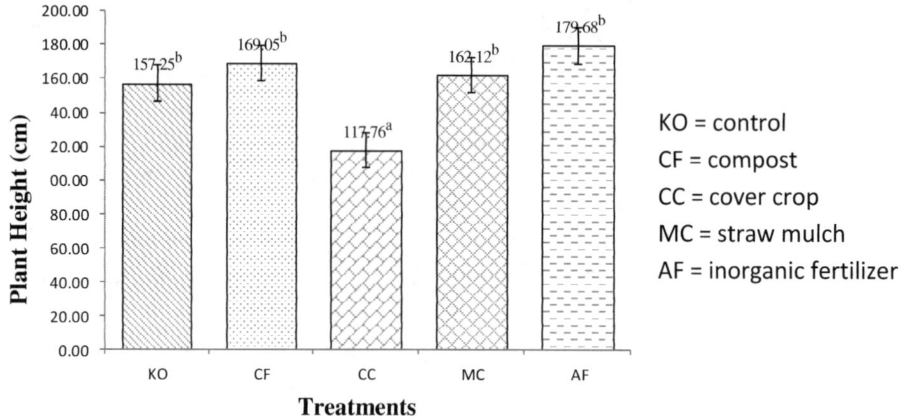

Figure 4. Plant growth at each treatment

Figure 4 shows that the cover crop treatment significantly resulted in the lowest plant height (117.76 cm), while other treatments are nearly similar to control (157.25 cm). That is may be due to the nutrient competitiveness of maize with cover crop during the growth. This result shows that plant height is regardless of soil moisture, nutrient plays more role in determining plant height than water availability.

3.2.2. Leaf width. Leaf width is an indicator of plant growth that is easily observed. Leaf width is one component of leaf area. Sitompul and Guritno [8], said that the leaves function organ that plays a role in photosynthesis. Leaf width is the main parameter for determining leaf area and rate of photosynthesis of plant unity. The result of leaf width observation shown in Figure 5.

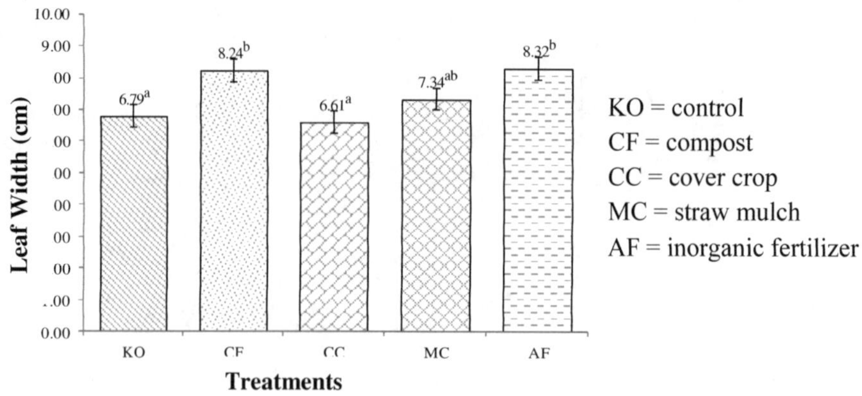

Figure 5. Leaf width at each treatment

Figure 5 shows that leaf width under compost (8.24 cm) and inorganic fertilizer (8.32 cm) were significantly higher than control (6.79 cm). That is probably because compost and inorganic fertilizer provide nutrient than other treatments, and thus leaf width growth seemed to in correspondence with nutrient supply from the soil.

3.2.3. Cobs weight. The result of cobs weight measurement is performed in Figure 6.

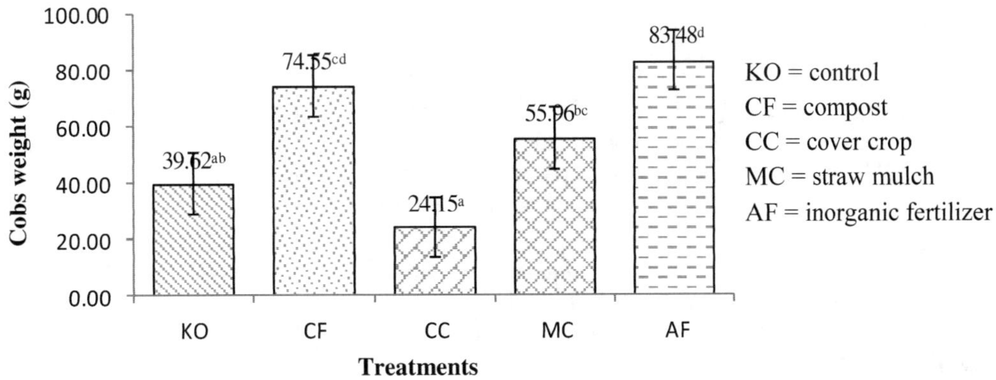

Figure 6. Cobs weight at each treatment

Based on Figure 6, it can be seen that the cobs weight were significantly higher under compost (74.55g) and inorganic fertilizer treatment (83.48 g), compared with control (39.62gr). This is similar pattern as shown in leaf width growth, that nutrient availability probably plays more role in determining the cob weight.

3.2.4. Biomass. Figure 7 shows biomass at each treatment.

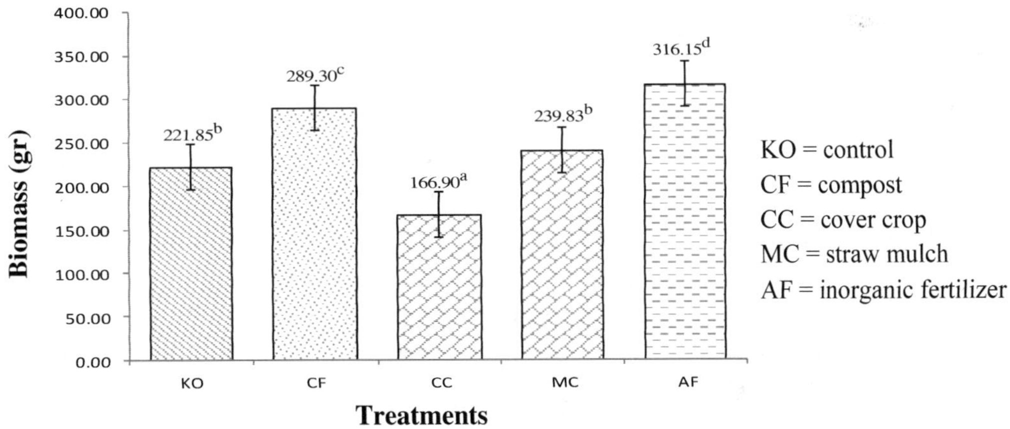

Figure 7. Biomass at each treatment

Figure 7 displays that the highest biomass reached by inorganic fertilizer (316.15g), followed by compost treatment (221.85 g), with the lowest is undercover crop treatment (166.90 g). It shows that biomass determined by nutrient availability and competitiveness with other crops included cover crop.

4. Conclusion

Rain-fed land management to anticipate climate change adverse effect on drought by conditioning soil with compost, cover crop, mulching and conventionally (only implementing chemical fertilizer), resulted in different effects. Composting land will increase soil water availability and provide nutrient to crops and thus increase soil physical properties, maize growth and yield. Although inorganic fertilizer also increased plant growth and yield, but it did not improve soil physical properties.

Acknowledgment
The authors gratefully acknowledged that the present research is supported by Laboratory Station of Gifu University, Japan at Sebelas Maret University, and Ministry of Higher Education, Republic of Indonesia.

References
[1] Adedeji O, Reuben O and Olatoye O 2014 Global climate change *Journal of Geoscience and Environment Protection* **2** pp 114-22
[2] Nurjanah A, Sumani, Komariah, Ariyanto D P and Zaki M K 2016 Analysis of agricultural drought based on daily water balanced at rainfed farmland *Proceeding of International Conference on Climate Change* pp 121 -7
[3] Golian S, Madziyasni O and Aghakouchak 2015 Trends meteorological and agricultural droughts in Iran *Theor Appl Climatol* **119** pp 679–88
[4] Wilhite D 2000 *Drought: a global assessment* (New York: Routledge) pp 3-18
[6] Glantz M H 2001 *Currents of Change: Impacts of El Niño and La Niña on Climate and Society* (United Kingdom: Cambridge University Press) pp 52-3
[6] Hardjowigeno 2007 *Soil Science* (Jakarta: Pustaka Utama Publishing) pp 1-5
[7] Foth H D 1943 *Fundamental of soil science* (United States of America: Arcata Graphics Company) pp 256-7
[8] Sitompul S M and Guritno B 1995 *Plant Growth Analysis* (Yogyakarta: Gajah Mada University Press) p 412

IOP Conf. Series: Earth and Environmental Science **129** (2018) 012030 doi:10.1088/1755-1315/129/1/012030

The Dutch colonial architecture of buildings in Manado's Old City: A response to the coastal tropical climate

V A Kumurur[1,3] and D M Tampi[2]

[1] Architecture Department, Faculty of Engineering, Sam Ratulangi University, Jl. Kampus Unsrat – 95115, Malalayang Manado, Indonesia
[2] Urban Development Studies: School of Strategic and Global Studies, Universitas Indonesia, Salemba Raya Street Number 4, Central Jakarta, Indonesia
[3] Corresponding author: vkumurur@gmail.com

Abstract. The late 19th and early 20th centuries was an era when the phenomenon of global warming began, as did the development of cities in Indonesia. In that era, cities in Indonesia functioned as colonial cities. The city of Manado is one of the coastal cities, written in the Dutch Royal Act of 1814 as the territory of Dutch sovereignty, was amended in 1848, 1872 and 1922. Dutch colonial art and architecture in Indonesia are not only influenced by culture but also the climate. For the purpose of physical comfort in the tropical environments, architects began to use local building materials, since the early 19th century, and the building began to be replaced by a customizing architecture. Descriptive analysis was employed as the method in this study. The result found that the Dutch Colonial Architecture emphasized the physical aspects, the royal style adapted to local conditions, and the local building emphasis on function. The tropical climate of Manado City influences the shape of the building with Dutch colonial architectural style in this area. As climate change is shown by rising temperatures, further observations on the design of colonial architecture will be important.

1. Introduction

The last ten years has become the hottest decade of the earth since the last 100 years. It is possibly the hottest year in the last millennium century [1]. The city is well researched and illustrated that since there is a city, there is also development, either in whole or in part, to both positive and negative directions. Since there are a close relationship between a city and people living in different times, the city likely tends to be dynamic. As one of the discoveries in the history, the city is a point of maximum concentration showing the strength and culture of a society [2]. Many Dutch city managers and architects applied the concept of local or traditional values in planning and developing the city, settlements, and buildings [3]. According to Ardiyanto *et al* [4], Dutch colonial architecture in the tropical area in the early 20th century adapted to the tropical climate. In addition to wind, the climate depends on the amount of total solar radiation received at a site. It is also influenced by the length of the daytime. The duration of the daytime will be longer than a night in the cities located in the north of the equator [5]. Climate and environmental aspects are factors affecting the architectural products [6]. Humans have a large enough capacity to adapt to a variety of climates and environments [1], including on the design of the building. Roaf *et al* [1] said that vernacular or traditional building designs are climate-sensitive, using a number of climate-friendly building techniques, and providing appropriate passive or isolation cooling techniques. Tropical climates of coastal cities on the island of Sulawesi, such as Makasar and Manado, have buildings with facades that have adapted to the tropical climate.

Content from this work may be used under the terms of the Creative Commons Attribution 3.0 licence. Any further distribution of this work must maintain attribution to the author(s) and the title of the work, journal citation and DOI.
Published under licence by IOP Publishing Ltd

Those buildings have the characteristic of the art deco style of European architecture that has been adapted into an Indonesian neo-vernacular style of tropical architecture [7]. According to Purwanto, a building that has a hall as a building protector tends to be cooler than the buildings that directly receive the heat of the sun on the walls of its facade. Other researchers prove that the roof is the most important building element in the process of forming the passive cooling in humid tropics [8]. The building style of Dutch colonial architecture in Indonesia has adapted to local traditional architecture [3], as well as the colonial buildings in Manado.

Manado is one of the coastal cities in Eastern Indonesia that was established by the Dutch in the 1700-1900 period [9]. Meanwhile, the Residency of Manado was one of the cities built in 1824 [10]. The city of Manado has 18.7 km of coastline and coordinates 1.4748°LU, 124.8421°BT (tropical cancer). Here is the map of the old city of Manado.

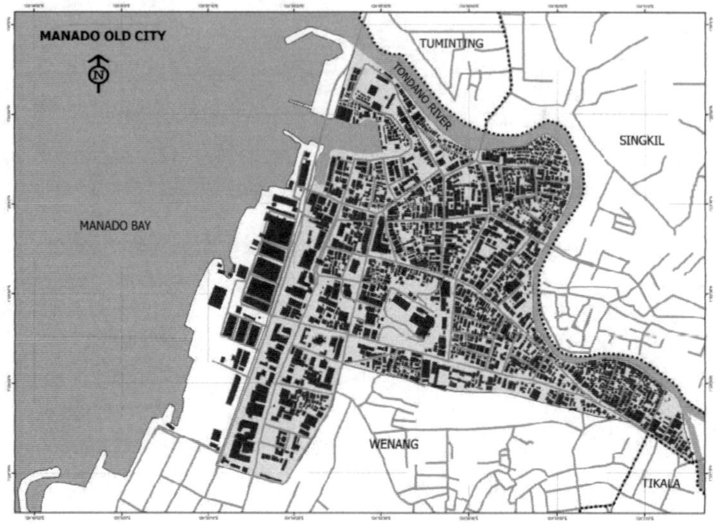

Figure 1. Map of the Old City of Manado (Personal Process 2016)

Old buildings in the old city of Manado are identified as having Dutch colonial architectural style [10]. Dutch colonial architecture in Indonesia is a unique cultural phenomenon. It is a mixture of Dutch colonial and Indonesian culture. It does not exist elsewhere, even in the other former colonized countries [11]. The uniqueness of these buildings can be seen in their colonial relics. According to the results of identification and analysis of Handinoto [12], the architectural style in that the period is divided into three, namely Indische Empire Style (18th-19th century); Transitional Architecture Style (1890-1915); and Indo-Europe Style (1920-1930). In the mid-20th century, global warming resulted in climate change and increased the frequency and intensity of extreme weather events. The IPCC states that global warming can cause significant changes in physical and biological systems (KLH) [13]. The purpose of this study is to analyze the colonial architectural style of Dutch architecture in the old city of Manado that is influenced by a local culture and climate change. This study accommodates the perspective of architects, environmentalists, and urban planners. Their collaboration is important in this study to get multidisciplinary perspectives rather than a monodisciplinary viewpoint. This study will be useful in understanding the relationship between the dynamic climate change that occurs in Indonesia and the architectural style of buildings in this country.

2. Materials and Method

2.1. Materials

Based on the results of identification, there are three architectural styles of Dutch colonial architecture in the old city of Manado, namely Minahasa Raad (1930), St. Ignatius Church (1953), and Benteng Cinema (1953). Three colonial architectural styles that influenced colonial buildings in the city of Manado are the style of Indische Empire, intermediate, and modern colonial architecture [10].

Indische Empire architecture is a style that developed in the 19th century in the Dutch East Indies. This style was popularized by the Governor-General HW Daendles (1808-1811) [12], while the Transitional Architecture Style was a Romanic plagiarism of European architectural style [12]. Transitional architecture style buildings are mostly designed by building inspectors who work in the Dutch government development department [12]. The Indo-European Architecture style is the combination between the buildings that have the shape of a mixture of the archipelago and modern architectures adapted to the climate, building materials, and technologies developed at that time [12].

The elements of architectural style can be divided into these four aspects, including building plan, building view, building materials and construction system [12]. According to Ardyanto et.al [4], the style of Dutch colonial architecture in Indonesia have anticipated the condition of humid-tropical climate. Therefore, the buildings in Indonesia must adjust to the conditions of tropical climate and the dynamics of the changes. The following building elements can be used in the concept of colonial architectural style presented in the table.

Table 1. Elements of Building Materials

Indische empire (18th-19th Century)	Transitional architecture (1890-1915)	Modern colonial architecture (1915-1940)
Bricks on columns and walls	Bricks on columns and walls	Concrete building materials
The main material of wood on truss, sills, and doors	The main material of wood on the truss. Frames and doors	The use of glass material is wide enough (especially on windows)
Not much use of glass	The use of glass on windows is still very limited	

According to Handinoto in Wulur [14], the architectural style of the Indische Empire (Ages 18-19) is determined by the main building construction materials such as bricks (both columns and walls), wooden building materials (mainly for the truss), board and doors. It also does not use a lot of glass. In the style of Transitional Architectural building (1890-1915), the building materials are bricks on columns and walls, the main material of wood on truss, board, and doors, and the limited use of glass on windows. Then, the materials in Modern Colonial Architecture style (1915-1940) are mainly consisting of concrete building materials and quite wide glass materials (especially on windows).

The element of a construction system in an Indische Empire-style building (1819) is on the bearwalls, with rows of columns on the front and back porch. Then, the use of columns and beams construction system, and a roof shield construction with roof tile coverings. The Transitional Architectural Buildings (1890-1915) used a bearish wall construction system with prominent front levels, using a saddle shape and a shield using roof tiles. There was also an attempt to use the additional construction as a vent on the roof. For the style of Modern Colonial architecture (1915-1940), the frame construction system is used, so the wall only serves as a cover. It is still dominated by a saddle roof with roof tile or shingle roof, and there is a section of the building using concrete construction, using a flat roof of concrete (Table 2).

Table 2. Elements of Construction Systems

Indische empire (18th-19th Century)	Transitional architecture (1890-1915)	Modern colonial architecture (1915-1940)
Bearing wall, with column rows on the front and back porch	Bearing wall with prominent front gevels	The frame construction system, so the wall only serves as a cover
Using a column and beam construction system	Roof: the shape of the saddle roof and the shield by using roof tile coverings	Roof: still dominated by gable roof with roof tile or shingle roof material
Construction of roof shield with roof tile coverings.	There is an effort to use additional construction as a vent on the roof	There are parts of the building using concrete construction, using a flat roof of concrete material, which has never existed

2.2. Method

In this study, the data were collected through field observation, photo documentation, identification of buildings that become the object of research, and analysis between the findings and the concept of colonial-style buildings in the tropical areas. The data were then analyzed using descriptive analysis. It is used to analyze the colonial architectural building style elements in Manado in response to the tropical climate of this region.

3. Results and Discussions

The characteristics of European architecture are still clearly seen in the building. Nonetheless, the existence of surrounding corridors whose the rooms have many doors depicts an enormous adaptation to the tropical climate [4]. Hadinoto stated that the colonial architecture in Malang during the years of 1990-1915 has adapted the design to the local climate. Having a good ventilation, it is especially designed to adjust to the heat of the sun and rain [4].

Minahasa Raad is one of the buildings representing the government buildings during the Dutch colonial period. Minahasa Raad has a symmetrical floor plan with a central space surrounded by the main hall. It uses the light beam elements on the back porch and windows but not on the entire window. Some windows do not use light-resistant elements, especially on windows in the western part of the building (Figure 2). From the floor plan, the analysis shows that 50% is influenced by the transitional architecture style (1890-1915). The buildings look symmetrical and do not use columns, appeared as simple buildings that do not use many complicated ornaments. The Minahasa Raad building was built in 1930 (entered in the 19th century). According to Soekiman in Kumurur (2015) [10], the 19th century is known as the eclectic period. It was a period when practical perspectives lifestyles were applied. At that time, people were more concerned with function and work. Instead of presenting the beauty, they were more aware of the usefulness of the work. Similar to Minahasa Raad building, the building is very simple but there is still an ornament attached to the facade (gevel). The building envelope can lose the heat by infiltration and transmission through thermal conduction, convection, and radiation.

It is visible that the design of Minahasa Raad building is clean, without using many ornaments. From the analysis of building-looking elements, it shows that 67% is influenced by the transitional architecture style (1890-1915). Bricks are the important materials in building Minahasa Raad, especially for constructing the walls. While wood materials are used to make roof and door frames. The use of glass is still very limited, used only on the windows. From the analysis of elements of building materials, it is found that Minahasa Raad building is influenced by two colonial architectural styles with the same percentage. They are Indische Empire style (33%) and transitional architecture style (33%).

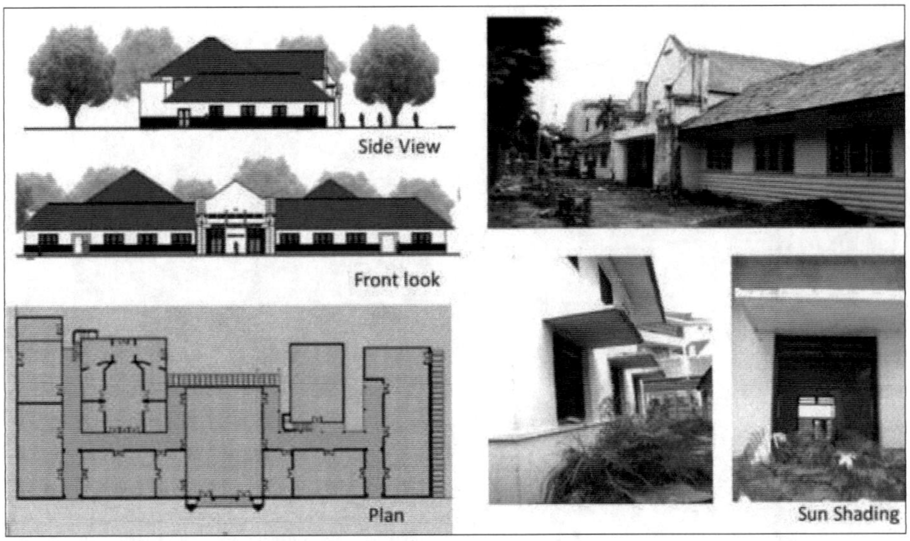

Figure 2. Minahasa Raad Building (Personal Processed, 2017)

According to Goulding and Lewis [16], solar energy can make a major contribution to the heating requirements of a building. Minahasa Raad building is dominated by a shingle roof and combined with a Tuitgevel gevel type. The analysis of construction system elements shows that 22% of it is a transition architectural style. Overall, Minahasa Raad building appears as a clean design. It has a sun-shading (overhang) on every window, high main door (3.20 m), and a canopy (overhang) on the roof with the width of 1.127m. Meanwhile, the sun-shading width is 0.50-0.8 m. It is made of wood materials and brick wall fillers.

Next is the St. Ignatius Catholic Church Building (Fig. 3). The Catholic Church of St. Ignatius represents a religious building built in Dutch colonial times. This church is located in the Don Bosco Manado school complex. The St. Ignatius Church plan consists of altars, the sitting room of the congregation, the confessional space, and its full symmetrical form. Although there is no terrace surrounding it, the entrance uses a light beam element or canopy.

The materials of the building consist of bricks (as the main construction material in both column and wall), wood (especially on truss), boards and doors. It uses natural stone materials on the outer walls, which serves to protect the walls from the effects of sunlight and rain. The use of glass is still very limited. It is only used on doors and windows with a narrow size. Serving as a natural lighting, glass is used only in the windows. According to Goulding and Lewis [15], the optimal use of natural daylight (especially in buildings used mainly in a daytime) can, by replacing the artificial light, make a significant contribution to energy efficiency, visual comfort, and the well-being of occupants.

Wood materials are used in window frames, as well as indoors. Meanwhile, a truss uses a concrete construction with a frame system, in which the building wall serves as a wall cover. The roof is dominated by a saddle roof with a shingle roof cover.

There is a section of the building that uses a concrete construction. It is a flat roof of concrete material, which has never existed, where the flat roof becomes an additional decoration on the building. The roof serves as a balcony on the church of St. Ignatius. There are windows to enter the light and natural air to the church room, using lightweight building materials; bricks as the main construction materials for both the column and the wall. It uses wood materials, especially on truss and doors and also natural stone materials on the outer wall that serve to protect the walls from the sunlight and rain. These buildings were not designed for an air-conditioner and electricity fan usage. All of the physical environment aspects have been considered and applied in the building design.

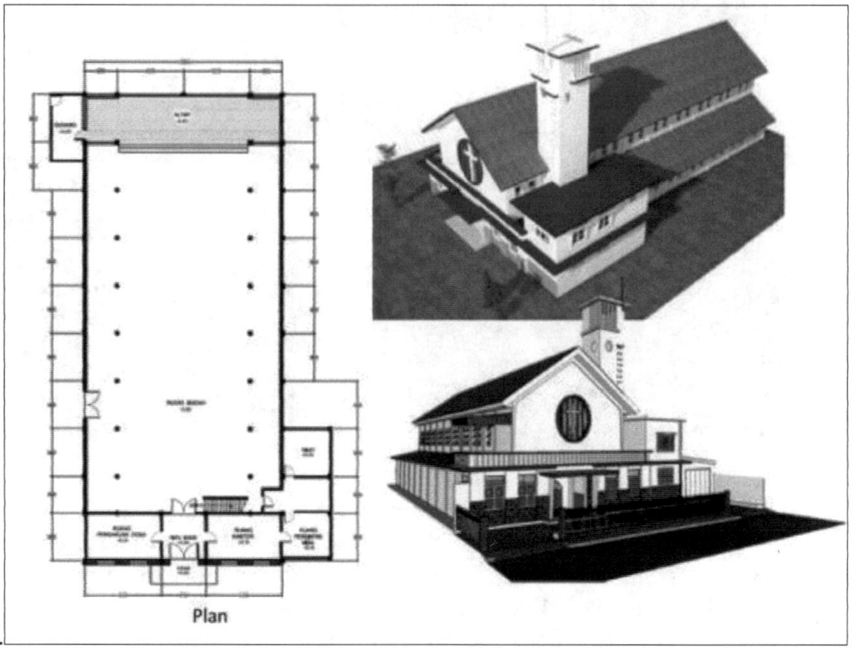

Figure 3. St. Ignatius Church Building (Personal Process, 2015)

The next building, which becomes the focus of the author's study, is the former Benteng Cinema. The current cinema building plan is not symmetrical but more varied (Figure 4). However, when viewed in the initial form, the balance point of the former Benteng Cinema building is located on the left and right side of the building. Since about 1952-1954, the building had been rebuilt and changed its shape by putting a balance point in the center of the building. A two-story building with no terrace surrounds the building and uses sunbathers, though not on all windows. From the analysis of floor plan elements of the building, it can be known that the former "Benteng" Cinema is 67% influenced by a modern colonial architectural style.

The use of openings on the walls in the form of windows indicates that this building uses natural exposure and lighting, although it does not use the roof as a passive coolant. In addition to the amount of contrast and the lighting window, the penetration of natural light can be controlled by reducing the flow of events [15]. Thus, the possibility of building without a saddle roof or shield is to store heat, so it is necessary to add the electrical energy to cool the room.

Among several buildings in the old city of Manado, the castle theater is one of the buildings that has a unique architectural style. This is one of the Dutch heritage buildings that has an art deco style. Now the Benteng Cinema is converted into small shops, restaurants, and salons. In this building, bricks are used as wall fillers whereas wood materials are used only on its windows and doors. There is not much use of glass on windows.

Figure 4. The former building of Benteng Cinema in 1990 and 2000 (Personal Process, 2015)

4. Conclusions
Dutch colonial building style was designed by utilizing elements that were adapted to the tropical climate. The buildings were designed to reduce heat through large overhangs and cross ventilation. As the author's recommendation, there are some things that could be a concern and follow-up later on in the future. First, the government should review the policy on the development of the city, especially in the old city of Manado. It must be done by rejuvenating old buildings in the old city. The government is willing to discuss with development actors, academics, industry, and society as well as the concept of "fourth helix". For academician and researchers, there is a need for further studies to explore the other issues on the Dutch Colonial architecture style buildings in the city of Manado. Finally, this study was conducted for the purpose of reducing the impact of climate change in the city.

Acknowledgement
This research is supported by Directorate of Research and Community Service, Directorate General of Research and Development Strengthening Ministry of Research, Technology and Higher Education Subject to Contract Research of Fiscal Year 2017. We thank our colleagues from Sam Ratulangi University who provide insight and expertise, which greatly helped research, although it may not be in accordance with all the interpretations/conclusions of this paper.

References
[1] Roaf S, Crichton D and F Nicol 2005 *Adapting Buildings and Cities for Climate Change* (Italy: Elsevier LTD) pp 51-149
[2] Mumford L 1953 *The Culture of Cities* (London: Lund Rumfhries) pp 479-81
[3] Keling G 2016 Typology of Dutch colonial building in Singaraja *Archaeological Forum* **29(2)** pp 65-80
[4] Ardiyanto A, Djunaedi A, Ikaputra and J A Suryabrata 2015 The Architecture of Dutch colonial office in Indonesia and the adaptation to tropical climate *International Journal of Scientific and Research Publications* **5(4)** pp 1-6
[5] Yuliatmaja M R 2009 *Sun Radiation and Solar Radiation Intensity to Sun Movement During Solstice In Semarang (Case Study of Meteorology and Geophysics Agency Climatology Station Semarang On June and September 2005 Until 2007)* (Semarang: Faculty of Mathematics and Natural Sciences Department of Physics Unnes) p 11
[6] Rapoport A 1969 *House Form and Culture* (Prentice Hall: Englewood Cliffs, N) pp 104-5
[7] Pujantara R 2013 Characteristics of the building faceade colonialism heritage and spatial distribution in Makassar City *Building Forum Journal* **11(2)**
[8] Malama A and Sharples S 1996 Thermal and economic implications of passive cooling strategies in low-cost housing in tropical upland climate *Architectural Science Review* **39(2),** pp 95-106

[9] Rutz W 1987 *Cities and Town in Indonesia* (Berlin Stuttgart: Gebruder Borntraeger) pp 50-3
[10] Kumurur V A 2015 The Effect of Dutch colonial architecture style on historic buildings in the old town manado area *Proceedings of IPLBI Scientific Meeting 2015* pp 167-172
[11] Sumalyo Y 1995 *Dutch Colonial Architecture in Indonesia* (Yogyakarta: Gadjah Mada University Press) pp 2-3
[12] Hadinoto 2010 *Architecture and Cities in Java in Colonial Period* (Yogyakarta: Graha Science) pp 145-7
[13] Ministry of Environment 2007 *National Action Plan for Climate Change* (Jakarta: Ministry of Environment) p 1
[14] Wulur A, Kumurur V A and I R B Kaunang 2015 Colonial architecture building style at historic buildings in Manado City *Sabua Journal* **7(1)** pp 371-82
[15] Goulding J R and J O Lewis 1997 *Bioclimatic Architecture* (LIOR E.E.I.G. Panoramalaan 7, B-1560-HOEILAART)

IOP Conf. Series: Earth and Environmental Science **129** (2018) 012031 doi:10.1088/1755-1315/129/1/012031

Stakeholders' perspectives towards effective climate change adaptation on the Mongolian livestock sector

A Batbaatar[1,3], P Apichayakul[2] and S Tantanee[2]

[1] Department of Civil Engineering, Faculty of Engineering, Naresuan University, 99 Moo 9 Tambon Tha Pho, Muang Phitsanulok 65000, Thailand
[2] Department of Electrical and Computer Engineering, Faculty of Engineering, Naresuan University, 99 Moo 9 Tambon Tha Pho, Muang Phitsanulok 65000, Thailand
[3] Corresponding author: amarjargalb58@email.nu.ac.th

Abstract. Climate change is one of the greatest threats that world is facing today, and having significant deleterious effects on natural and human systems. Recent climate-induced extreme events and their impacts demand timely adaptation actions to the changing odds of their occurrence. The great phenomenon is already being felt in the Mongolian plateau, especially on the livestock sector. The sector provides the main income and livelihood for one-third of the population of about three million people. A high number of livestock is lost due to a unique phenomenon is known as a "dzud". This paper examines the key stakeholders' perspectives in the implementation of climate change adaptation and identifies its barriers, with a focus on the livestock sector. In order to meet the objectives, this research used a semi-structured interview with organizations related to the livestock sector and climate change. The extent of stakeholders' perspectives might be depending on the way they share information, stakeholder engagement, and their experiences with extreme events, as well as their location and level in government. The research findings will indicate an understanding of climate change perspectives, adaptation, and level of capacity of organizations, which can be used as a guideline for organizations to develop climate change adaptation policies related to the livestock sector in Mongolia.

1. Introduction

The Earth's climate is changing in profound ways, while there has been growing debates on, that have addressed the risk and vulnerability of affected systems, and have issued Declarations that have been endorsed by most countries. Among the climate science community, there is a growing consensus that climate change is not just an unfortunate phenomenon, but indeed it is an anthropogenic tragedy by emitting greenhouse gas (GHG) into the atmosphere. Nevertheless, this trend is still not completely approved [1]. However, it is certain that the phenomenon has created uncertainty in, and a threat to, the future of sustainable development. The increasing number of extreme events has gained massive attention from a scientist, questioning how and what is the connection of those events to climate change. The following are several extreme events occurred over the past few years; 1) the 2013 typhoon Haiyan in Southeast Asia (especially in the Philippines), considered as one of the deadliest cyclones ever recorded [2], 2) the disastrous flooding in 2011 in Thailand [3], and countries bounding the Bay of Bengal is the most exposed to cyclones and flooding that have resulted in the loss of thousands of lives, displacement, damaged infrastructures and economic crisis, and 3) the 2008

Content from this work may be used under the terms of the Creative Commons Attribution 3.0 licence. Any further distribution of this work must maintain attribution to the author(s) and the title of the work, journal citation and DOI.
Published under licence by IOP Publishing Ltd

IOP Conf. Series: Earth and Environmental Science **129** (2018) 012031 doi:10.1088/1755-1315/129/1/012031

catastrophic cyclone Nargis that resulted in enormous destruction of socio-economic sectors and more than 130.000 mortalities [4].

As a result of climate change, millions of people are suffering from scarce natural resources, unsanitary living conditions, insufficient health services, and insufficient or non-existent social services [5]. Most tragically, affluent industrialised nations have the biggest role to release GHG into the atmosphere, whereas the world's poorest communities mostly located in the Asian and African continent, who solely depend on agriculture and livestock as their livelihood and income, are suffering most from harmful effects of climate change, yet they are least responsible for causing it [6]. The recent science tells us that impacts of climate change will become more harmful in a myriad of ways including more frequent and widespread of extreme events, accelerating the extinction of biodiversity and destroying ecosystems, infrastructures, economic losses, and diminishing natural resources. Therefore, many countries have been trying to diminish such adverse effects through developing climate change adaptation approaches and related policies at all levels and sectors. The livestock sector is now and will inevitably be affected by climate change. Adverse effects felt in the livestock sector can threaten other sectors, including the economic sector, through the loss of livestock for herders. It means that they have less income to buy goods, which impacts the sales and income of providers, or the shortage of produce; meat, milk, milk products that lead to increasing prices of these foodstuffs. Therefore, it will cause economic hardship and hunger, even starvation, to many, as well as affecting the food security of countries.

Since no country is immune to climate change, Mongolia has not been left from the harmful effects of climate change. Developing adaptation strategies to climate change with the focus on the livestock sector is important in Mongolia. Thus, this paper aims to explore the key stakeholders' perspectives on climate change impacts on the livestock sector, identify adaptation approaches and barriers, identify stakeholders' capacity to overcome those barriers, and conclude with a summary of the discussion and some suggestions for further work.

1.1. Climate change adaptation

The higher average global temperature and variations of precipitation are projected to have a wide range of impacts on natural resources including fresh water, agricultural and grazing areas, and such natural resources likely to become more limited so that livestock, of significant importance in economies, will find it harder to survive in certain areas with harsh and deteriorating conditions [7]. The livestock sector provides livelihoods for approximately 2 billion people and uses the largest land surface areas of drylands that covers more than 40% of the total surface areas of the Earth. Since the sustainability of the livestock sector is a subject susceptible to climate change, it is crucial to implement an appropriate form of adaptation [8].

According to the Intergovernmental Panel on Climate Change, adaptation is "the process of adjustment to actual or expected climate and its effects" [5]. A variety of adaptation has been developed by stakeholders; such as capacity building (e.g., education and training), management (e.g., livelihood diversification), policy (e.g., implementation or revision of regulations and policies, integrating adaptation into development policies), financing (e.g., livestock insurance, contingency fund) and technology (e.g., enhance water access technologies) [9]. Numerous factors are being and will influence the effectiveness of adaptation, which can create sudden or dramatic changes in adaptation. Consequently, it is important that adaptation should appropriately target a certain level, location, and duration. Sometimes, success in the short-term adaptation may turn out to have little or no benefit in the long-term process and vice versa. Thus, we need more viable adaptation to the increased risk and vulnerability of the future.

2. Climate and the livestock sector in Mongolia

2.1. Mongolia's climate

Mongolia is a landlocked country situated between Russia to the north and China to the south, which experiences a continental climate. A long-lasting harsh cold winter, dry hot summer, with an annual average temperature of minus 0.8°C, with the temperature dropping below -45°C in winter and reach up to more than 40°C in the summer season for each year. The significant changes have been observed on the Mongolian plateau. Climate change brings ever drier and hotter conditions, and average temperature has been risen by 2.1°C in the 1940-2007, and future climate scenarios indicating that annual average temperature will be increased by 0.4-1.6°C in the coming next 30 years [7 10 11]. The increased average temperature in the winter season is estimated as 3.6°C. The trend of increasing temperatures in winter has been accompanied by several adverse events such as strong wind storm and heavy snowfall. This temperature is sometimes rapidly rising over a short period of 3-7 days and then abruptly dropping again to sub-zero temperatures. This situation has led to the ice-sheet coverage of grassland areas in some years, making it harder for livestock to find and access feed [12]. Snow is one of the main sources of water for livestock during winter, but an excessive amount of snowfall that accompanies low temperatures creates unfavorable conditions for herders and livestock.This unique phenomenon is usually called as a "dzud". According to Mongolian Second Assessment Report on Climate Change in 2014, drought and dzud are the most catastrophic extreme events due to their impacts on socio-economics [10].

2.1.1. Dzud. It is a Mongolian term for a complex and long-lasting natural disaster consisting of summer drought followed by cold and harsh winter that result in a high number of livestock losses due to hunger, exhaustion, and exposure to freezing air temperatures. There is also a potential for loss of human life and damaging impacts on herder livelihoods and infrastructures [13]. Currently, Mongolia has experienced 5 types of dzud: tsagaan, khar, tumer, khuiten and khavsargan (table 1). The duration of a dzud continues from the beginning of November; possibly little earlier, lasting to the end of March [14]. In the last two decades, Mongolia has faced with four devastating dzuds. First, three consecutive dzuds in 1999-2000, 2000-2001 and 2001-2002 that resulted in the impoverishment of thousands of herder households and livestock losses estimated at 30% of total livestock. Then, a severe dzud in 2009-2010, with temperatures dropped below -53°C in some areas in January, resulted in widespread livestock losses of some 20% of the total livestock, with an estimated 28% of the Mongolian population being harshly affected in 17 out of 21 provinces [13].

Table 1. Typologies of dzud.

Types of dzud	Description
Tsagaan dzud or White dzud	The higher amount of snowfall with low air temperature.
Khar dzud or Black dzud	Lack of snowfall which leads to insufficient water availability for livestock and herders.
Tumer dzud or Iron dzud	Mixed temperatures, sudden temperature drop after a short-term warming resulted in ice sheet on grazing areas.
Khuiten dzud or Cold dzud	Low precipitation with strong stormwind.
Khavsargan dzud or Combined dzud	Combination of two or more of the above-mentioned dzuds.

2.1.2. Drought. In Mongolia, drought occurs ever longer periods, and its intensity and frequency are significantly increased since the end of the last century [11]. As a result, more than 880 out of 5138 rivers and approximately 1200 out of 3747 lakes are dried up at the beginning of 21st Century [15]. Along with the three consecutive dzuds, there were three consecutive summer droughts considered as the worst drought that Mongolia ever faced with, which affected 50-70% of the total territory. The growing season for livestock forage and agricultural crops in Mongolia is relatively short, between May and September, and the growth of grasses is heavily dependent on the weather condition. If there has been insufficient forage and hay, livestock will fail to fully fatten and gain the necessary strength for the coming winter [12]. When the drought is followed by dzud, a huge number of livestock losses are experienced. Particularly, a combination of drought and white dzud has a higher risk and impact than drought and other types of dzud. Thus, drought is considered as one of the triggers that intensify dzud [16].

2.2. The livestock sector in Mongolia
Mongolia has a traditionally nomadic society and the livestock sector or animal husbandry as a principal component of culture and identity, a primary means of livelihood and income for herders, and the main food for the rest of the population. Accordingly, climatic conditions are of primary importance in the success of that significant sector of the Mongolian economy and society [12]. The total number of livestock is approximately 61.5 million, consisting of horse, cattle, camel, sheep, and goat. The sector is dependent almost solely on forage productions, hence, is highly vulnerable to natural hazards. Mongolia is divided into six ecological zones that experiences more localized conditions such as high mountains, steppe, forest-steppe, desert-steppe, desert, and taiga.The herding in the high mountains, forest-steppe, steppe and arid grassland areas are highly sensitive to extreme climatic conditions that pose a higher risk of dzud and drought [17].

3. Research method

This research conducted semi-structured interviews with the key stakeholders from public and private, and international organizations at all levels, to gather their experiences with, and perspectives of climate change, impacts and adaptation, and identify barriers encountered in developing adaptation approaches. Stakeholders for the interview were identified firstly, through eliciting the organizations related to disaster management and climate change and the livestock sector. Secondly, in order to collect reliable information and select rightful stakeholders, this research is based on the possible criteria for selection of stakeholder participation. The selected stakeholders are those who have the most appropriate information, knowledge, and expertise, the capability to influence adaptation process, and an envisaged pluralistic of perspectives. In addition, they should be willing to give information and available for the interview [18].

3.1. Developing interview questions
The interview, with pre-developed 21 questions, is divided into four sections: (i) stakeholders' perspective on climate change, (ii) adaptation approaches (iii) barriers of adaptation (iv) organizational capacity.

3.1.1. Section 1: Stakeholders' perspectives on climate change. This section aims to draw out stakeholders' perspectives and understanding on impacts of climate change and identify the main drivers of changes in the livestock sector. The number of drivers of climate change to the livestock sector is identified: quantity and quality of feeds; heat stress; water availability; livestock diseases; and biodiversity [6].

3.1.2. Section 2: Identification of adaptation approach. This section is aimed to identify specific examples of adaptations and potential responses to climate change based on stakeholders' experience and knowledge. First, this section will analyze the efficiency of previously implemented adaptations related to the livestock sector. Then, the selection of adaptation approaches, which they consider as the most necessary or required, is implemented. All identified distinct adaptations will be grouped into ten categories including capacity building, management and planning, practice and behavior, policy, information, physical infrastructure, warning and observing systems, green infrastructure, financing, and technology [9].

3.1.3. Section 3: Identification of barriers. This section is aimed to identify common barriers that exist within the development of adaptations and how stakeholders overcome those barriers. Every adaptation approach has a challenge of how to handle the diverse barriers, so defining the adaptation scope must include the identification of the barriers that are likely to be encountered. Several researchers categorized adaptation barriers such as institutional governance issues, financial, politics, leadership, and communication [19]. Nevertheless, this research focuses primarily on politics, and institutional and governance related to the barriers so that it can explore the strategies that they used to overcome such barriers.

3.1.4. Section 4: Capacity of organizations. The aim of the section is to assess the organizational capacity of stakeholder. The purpose of this assessment is to recognize the capacity and resources that they have in order to develop effective adaptation approaches. The capacity of organizations includes highly educated human resources, sufficient and reliable scientific information, sufficient financial resources, and effective cooperation with other related organizations [20]. As well, this section is also used to identify their enhancement strategies for improving their capacities.

4. Discussion

As a global phenomenon, the effects of climate change are experienced differently in every region. Since climate change is persistently broadening its negative effects, stakeholders' involvement in the adaptation to climate change will be more important to reduce the risk of vulnerable systems. The frequency and intensity of dzud and drought are predicted to increase the more scarce availability of natural resources and threaten more population, mostly in local areas, properties and other sectors across Mongolia over the next coming decades.

The sustainability of the livestock sector is directly influenced by herders' livelihoods. It is interesting to note that herders are not only become vulnerable due to loss of their livestock but also vulnerable to some other non-climatic factors. For instance, most of the herders have little or no formal education, somehow its influence being not able to access loans from the bank to diversify their income or source of livelihoods. The diversification of income source of herder households is important as a safety net in times of dzud and drought occurrence. There is a growing number of adaptation on the sustainability of livestock sector and herders' livelihood in Mongolia, yet it is not clear that all of these adaptations would be successful due to underlying barriers. Sharing scientific information, partnership and qualified perspectives of stakeholder has the potential influence to the successfulness of adaptation approaches.

5. Conclusion

Mongolia is the nomadic society and has the dependency on natural resources. It makes the country more vulnerable to extreme events. Thereby, this research intends to explore the most appropriate adaptation approaches to the impacts of climate change, which is inspired by the series of the historic dzud and drought events. A broad perspective of stakeholders is important to enhance the quality and quantity of adaptation that can provide sustainability of herders' livelihood, of the livestock sector and pastoralism.

For further research, this research recommends that the focus is on the climate change communication and how information sharing and engagement of stakeholders could be improved for the rational decision making, for possible actions in response to climate change. As well, one of the most important needs is mainstreaming climate change adaptation into sustainable development policy. The ultimate climate change adaptation needs to involve all potential stakeholders and be viable targeted at specific context and spatial so that it can increase the capacity of affected sectors and sustainable development. Overall, the Government of Mongolia's support should focus on formulating climate change adaptation policies and programmes related to the livestock sector that improve the capacity of livestock and herder community.

References

[1] Cook J, *et al.* 2013 Quantifying the consensus on anthropogenic global warming in the scientific literature *Environ. Res .Lett.* **8(2)** p 7

[2] Dai K, Chen S E, Luo M and Loflin G 2017 A framework for holistic designs of power line systems based on lessons learned from Super Typhoon Haiyan *Sustain. Cities Soc.* **35** pp 350-64

[3] Saito N 2014 Challenges for adapting Bangkok's flood management systems to climate change *Urban Clim.* **9** pp 89-100

[4] Tasnim K M, Shibayama T, Esteban M, Takagi H, Ohira K and Nakamura R 2014 Field observation and numerical simulation of past and future storm surges in the Bay of Bengal: a case study of cyclone Nargis *Nat. Haz.* **75(2)** pp 1619-47

[5] IPCC 2014 *Climate Change 2014: Synthesis Report* (Geneva, Switzerland: IPCC Press)

[6] Thornton P K, van de Steeg J, Notenbaert A and Herrero M 2009 The impacts of climate change on livestock and livestock systems in developing countries: A review of what we know and what we need to know *Agric. Sys.* **101(3)** pp 113-27

[7] Wang J, Brown D G and Agrawal A 2013 Climate adaptation, local institutions, and rural livelihoods: A comparative study of herder communities in Mongolia and Inner Mongolia, China *Glob. Environ. Change* **23(6)** pp 1673-83

[8] Martin R, Müller B, Linstädter A and Frank K 2014 How much climate change can pastoral livelihoods tolerate? Modelling rangeland use and evaluating risk *Glob. Environ. Change* **24** pp 183-92

[9] Biagini B, Bierbaum R, Stults M, Dobardzic S and McNeeley S M 2014 A typology of adaptation actions: A global look at climate adaptation actions financed through the Global Environment Facility *Glob. Environ. Change* **25** pp 97-108

[10] Ministry of Environment and Green Development of Mongolia "Mongolia second assessment report on climate change" *http://www.jcm-mongolia.com/* 2014 [Online] Available: http://www.jcm-mongolia.com/wp-content/uploads/2015/11/MARCC-Final-Bk-2014-book-lst.9.17-ilovepdf-compressed.pdf [Accessed: 04-10-2017]

[11] Batima P "Climate Change Vulnerability and Adaptation in the Livestock Sector of Mongolia" *http://start.org/* 2006 [Online] Available: http://sedac.ciesin.columbia.edu/aiacc/progress/FinalRept_AIACC_AS06.pdf [Accessed: 04-10-2017]

[12] Batima P, Bat B and Tserendorj T "Evaluation of adaptation measures for livestock sector in Mongolia" *http://start.org/* 2006 [Online] Available: http://www.start.org/Projects/AIACC_Project/working_papers/Working%20Papers/AIACC_WP41_Batima.pdf [Accessed: 04-10-2017]

[13] UN Mongolia Country Team "Mongolia: Dzud Appeal" *http://www.unocha.org/* 2010 [Online] Available: https://docs.unocha.org/sites/dms/CAP/2010_Mongolia_DzudAppeal_SCREEN.pdf [Accessed: 05-10-2017]

[14] Begzsuren S, Ellis J E, Ojima D S, Coughenour M B and Chuluun T 2004 Livestock responses to droughts and severe winter weather in the Gobi Three Beauty National Park, Mongolia *J Arid Environ.* **59**(4) p 785-96

[15] Szumińska D 2016 Changes in the surface area of the Böön Tsagaan and Orog lakes (Mongolia, Valley of the Lakes, 1974–2013) compared to climate and permafrost changes *Sediment.*

Geol. **340** pp 62-73

[16] Tachiiri K and Shinoda M 2011 Quantitative risk assessment for future meteorological disasters *Clim. Change* **113(3-4)** pp 867–82

[17] Sternberg T 2008 Environmental challenges in Mongolia's dryland pastoral landscape *J Arid Environ.* **72(7)** pp 1294-304

[18] Hage M, Leroy P and Petersen A C 2010 Stakeholder participation in environmental knowledge production *Futures* **42(3)** pp 254-64

[19] Ekstrom J A and Moser S C 2013 Institutions as key element to successful climate adaptation processes: results from the San Francisco Bay Area *Successful adaptation to climate change: linking science and policy in a rapidly changing world* pp 97-113

[20] Measham T G, *et al.* 2011 Adapting to climate change through local municipal planning: barriers and challenges *Mitig. Adapt. Strategies Glob. Chang.* **16(8)** pp 889-909

IOP Conf. Series: Earth and Environmental Science **129** (2018) 012032 doi:10.1088/1755-1315/129/1/012032

Land use change on climate parameters at Samin subwatershed in Central Java, Indonesia

Sutarno[1,4], Komariah[3], T Gunawan[2], D Purnomo[3] and Suntoro[3]

[1]Doctoral Program in Environmental Science, Graduate School, Universitas Sebelas Maret, Jl. Ir. Sutami No.36A, Kentingan, Surakarta, 57126, Indonesia
[2]Faculty of Geography, Gadjah Mada University, Bulaksumur, Depok, Sleman, Yogyakarta 55281, Indonesia
[3]Faculty of Agriculture, Universitas Sebelas Maret, Surakarta, Indonesia
[4]Correspondence author: sutarno@staff.uns.ac.id or sutarno.uns@gmail.com

Abstract. The Samin sub-watershed (SSW) is one of the critical watersheds in Indonesia which need conservation. The identification of land-use/ land-cover changes (LUCC) can help in deciding the priority of conservation areas as well as limiting the widespread of critical lands in the watershed, which can contribute to climate change. The purpose of this study is to determine the impact of land use change on climate parameters, i.e. precipitation, air temperature and relative air humidity. The method is by using the descriptive explorative. The study employed Indonesian topographic map and Landsat's imageries of 1996, 2001, 2006, 2011 and 2016. The climate data from 1996 to 2016 were obtained from surroundings weather station. Data were analyzed using Geographic Information System (GIS) and SPSS. The results showed that land use was dominated by rice fields 22,552.83 ha (69.20%), and converted to non-agricultural lands 165.05 hectares/year for the last 20 years. Forest area decreased 65.8 ha/year, and settlement (housing and industrial estates) increased 253.87 ha/year (11.07%). The statistical analysis resulted in a negative relationship between forest area and air temperature and, but no significant correlation with rainfall.

1. Introduction

Study of land use/ land cover change (LUCC) or land conversion of agricultural has received considerable attention [1]. There is debate as to whether agricultural land should be maintained or converted to other uses [2]. The LUCC plays an important role in the climate system [3]. LUCC on a local, regional, and global scale is one of the key drivers of global climate change [4,5]. LUCC is one of the main causes of global climate change. At the moment, the climate impact of LUCC has become a hotly discussed topic worldwide. According to the fourth assessment report of Intergovernmental Panel on Climate Change (IPCC) in 2007, human activities, such as urbanization and agriculture, accounted for 90% of global warming [6,7]. LUCC is the loss of major agricultural land, consequently vegetation lost, which has long been known to modify the climate near the soil. Geiger (1950) [8] said that canopy cover has a strong influence on extreme climatic conditions [9]. LUCC is an almost inevitable phenomenon during the period of economic growth and population growth [10]. In China, since 1980, LUCC of agricultural land to non-agricultural land has been running and widespread. The three main factors causing LUCC are high population density, urbanization processes and rapid economic growth [11].

Content from this work may be used under the terms of the Creative Commons Attribution 3.0 licence. Any further distribution of this work must maintain attribution to the author(s) and the title of the work, journal citation and DOI.
Published under licence by IOP Publishing Ltd

IOP Conf. Series: Earth and Environmental Science **129** (2018) 012032 doi:10.1088/1755-1315/129/1/012032

Human activities, such as deforestation, expanded use of non-agricultural vegetation and agricultural land, grassland degradation, urbanization and other large-scale land use change the natural quality of land cover [12]. According to the Intergovernmental Panel on Climate Change (IPCC), LUCC activity can result in an average temperature rise of 4°C in 2100. Cinar [12] explains this will change the physical characteristics of the ground surface and will affect the soil interaction of atmospheric energy. They also explain that urbanization and deforestation lead to increased albedo, with rising atmospheric temperature, which will affect the thermal comfort of buildings and urban life. The increase in land use change can provide different ecosystems that affect climate parameters such as air temperature, relative humidity, precipitation and wind speed. Therefore, it is necessary to improve water-use efficiency by adjusting to future precipitation reductions due to climate change [13].

LUCC can be grouped into three categories: urban encroachment to farmland, conversion of agricultural land into the non-agricultural land, and transition of agricultural land to the natural landscape. The transition differs greatly in their degree of reversibility as well as it's associated environmental and socioeconomic consequences [1]. There is a strong correlation between monthly average temperatures and land changes that resulted from the increased urban land use, changes in forest cover and the conversion of permanent lands into artificial greenhouses. Zhang *et al.* [14] which studied about a simulation of the effect of urbanization on climate in the Yangtze River Delta found that the conversion of rural land and most irrigated agricultural land to urban land cover produced significant changes to near-surface temperatures, humidity, wind speed, and precipitation.

Although many studies already discussed the effects of land use changes on climate, the discussion in the Samin subwatershed has not been found. Thus, the aim of this study is to investigate the influence of LUCC in Samin Subwatershed on climate parameters.

2. Material and methods

The research was conducted from March 2015 to August 2017. The study area is Samin Subwatershed, Central Java, Indonesia, located at the coordinates of 49 S UTM, 480000-520000 mT, and 9144000-9160000 mU. The total area is approximately 32,590.49 ha, with 108.282 km circumference and the total population in 2015 was approximately 675,900. Morphologically, Samin Sub watershed can be divided into 2 parts for the upstream located in Karanganyar Regency while downstream areas are located in Sukoharjo Regency. The boundaries of the northern area are Mungkung Subwatershed, southern area is Walikan subwatershed, eastern part is Magetan Regency, and the western side is Bengawan Solo river, Surakarta Municipality, and Sukoharjo Regency.

The river length is 410.319 km and seasonal river length is 320.7704875 km, and the drainage density is 1.28 km/km^2. High drainage densities can reduce peak discharge and surface runoff is rapidly [15]. The outlet of Samin Subwatershed is Bengawan Solo River.

The research was conducted by explorative descriptive. Land use changes were analyzed using Geographic Information System (GIS) by observing Landsat imageries of 1996, 2001, 2006, 2011 and 2016. Land use was classified as rice field, settlement (settlement and industry), forest, garden, grassland, and upland. Land use type analysis on Landsat imagery is done on the basis of hue, color, shape, density, and roughness of the image

Climate data of 20 years (1996-2016) were obtained from Jumantono weather station (7°37'47.1"S, 110°56'51.8"E), including rainfall, air temperature, air relative humidity, and evaporation. Rainfall was observed using ombrometer and the data is collected daily every 7 am. Air temperature is observed manually with a thermometer and the data is recorded every 7 am, 1 pm and 6 pm, respectively, every day. Air relative humidity is observed using psychrometer dry and wet bulb, and the data is recorded in a similar way with air temperature. Evaporation is measured from class A pan evaporation, and the data is recorded every morning at 7 am. Statistical analysis was employed to find the correlation between LUCC and the climate parameters.

IOP Conf. Series: Earth and Environmental Science **129** (2018) 012032 doi:10.1088/1755-1315/129/1/012032

3. Results and discussion

3.1. Results
The land use changes analyses are presented in Figure 1 and Figure 2.

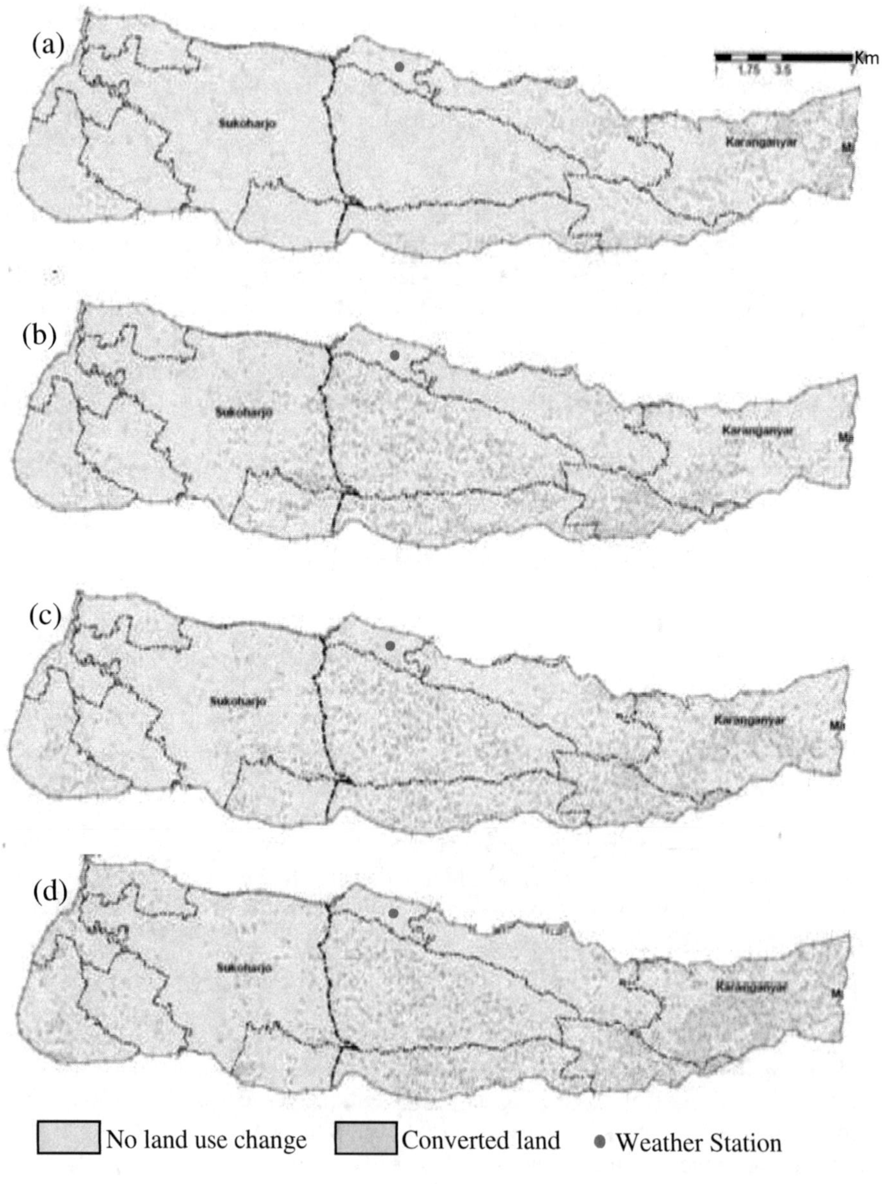

Figure 1. Land use change distribution during 1996-2001 (a), 2001-2006 (b), 2006-2010 (c), and 2010-2016 (d)

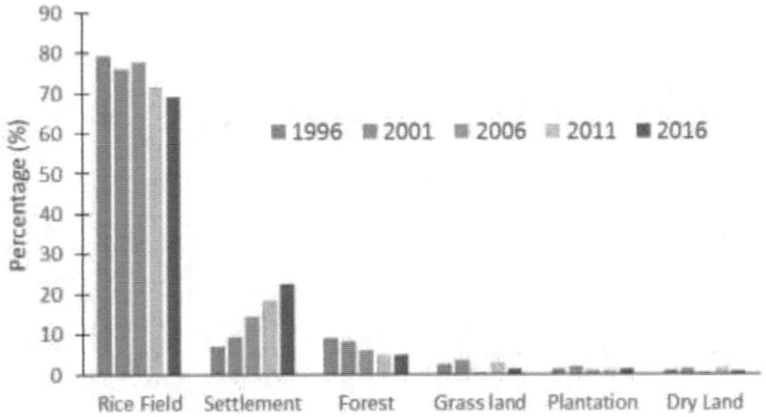

Figure 2. Land Use Changes in Percentage

Figure 1 and 2 show that during 20 years (1996-2016) land conversion occurred. Figure 2 presents rice field decreased from 79.33% (25,854.15 ha) to 69.20% (22,552.83 ha). This means there has been a decrease of 3,301.32 ha (10%) of rice field with the mean annual decrease is 165.05 hectares/ year. Rice field may have changed to non-agricultural functions i.e. settlement as it increased 221.46% over the period or approximately 253.87 ha per year. Figure 2 also shows that forest reduced 8.98% (2,926.74 ha) in 1996 to 4.94% (1,610.28 ha) in 2016, which means that 1,316 ha forest lost during 20 years or 65,8 ha per year. Grassland, plantation and dry land were changed over periods (fluctuating).

Annual rainfall, annual average air temperature, annual average relative air humidity, and annual average evaporation from pan during 20 years (1996 – 2016) are presented in Figure 3,4 5 and 6, respectively.

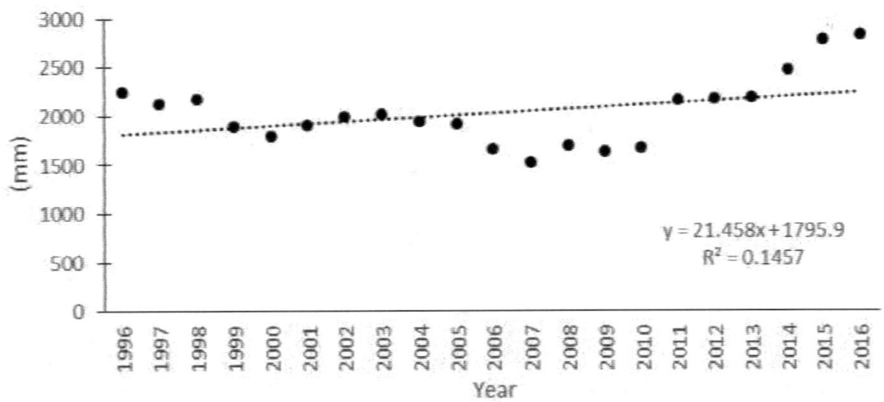

Figure 3. Annual Rainfall 1996 – 2016

Figure 3 shows fluctuating rainfall for 20 years with increasing trend. The lowest annual rainfall was in 2007 (1,523 mm) and the highest is in 2016 (2,825 mm), with coefficient of determination (R^2)=0.1457

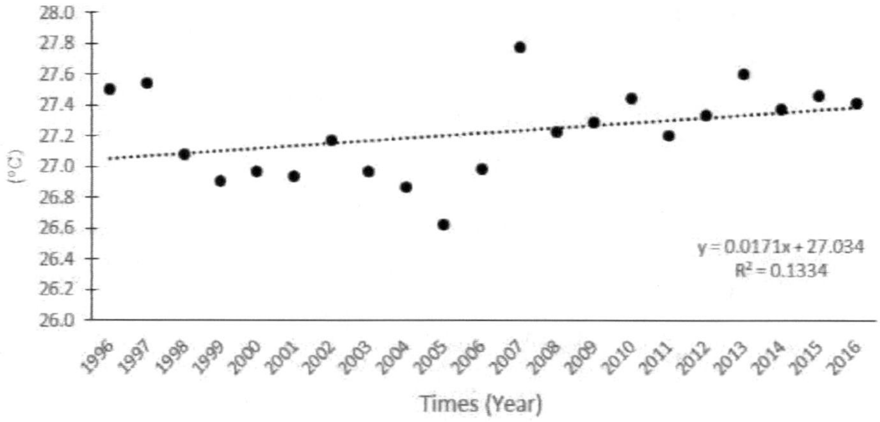

Figure 4. Annual average air temperature 1996-2016

Figure 4 presents the average air temperature of Samin Sub watershed from 1996 to 2016, which are fluctuating. The lowest and the highest annual average air temperature is in 2005 (26.6°C) and 2007 (27.8°C), respectively. It is also shown an increasing trend of annual average air temperature with R^2=0.1334.

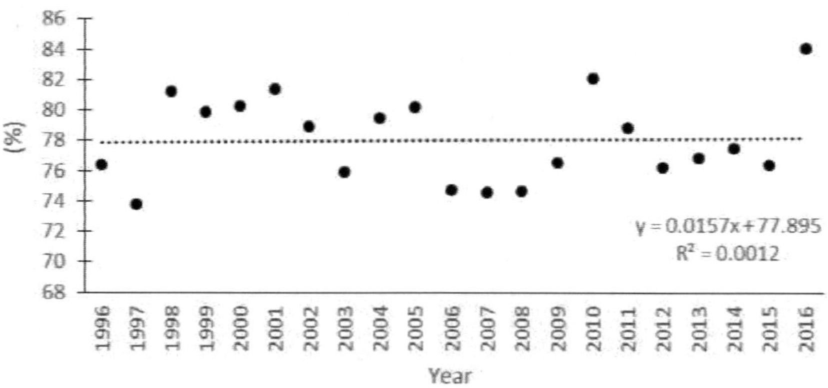

Figure 5. Annual average daily air relative humidity 1996-2016

Figure 5 figures out the annual average daily air relative humidity, which fluctuated over periods. The lowest and the highest average air relative humidity within 20 years period is 73.8% (1997) and 84.1% (2016), respectively. There is no trend shown in Figure 5, with very low R^2, which approached 0.

The annual average daily evaporation from evaporation pan shown in Figure 6 fluctuates with the lowest was in is 3.69 mm (2000) and the highest was 4.8 mm (2008). The evaporation performs a slightly increasing trend with R^2= 0.3533.

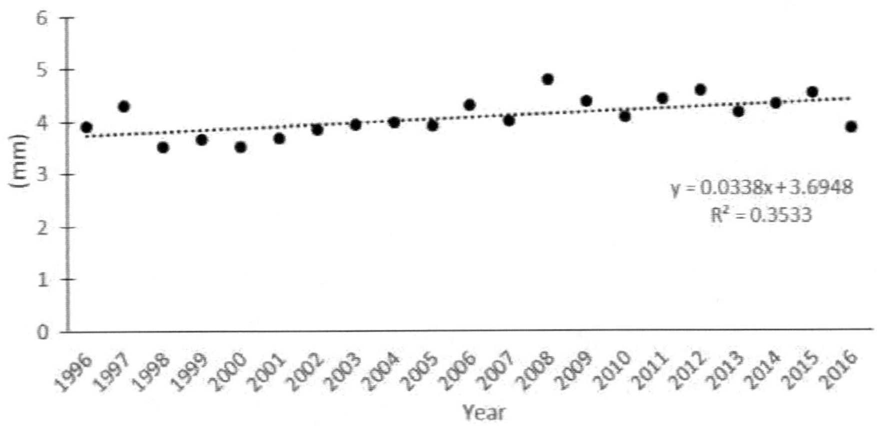

Figure 6. Annual Average Daily Evaporation 1996-2016 (evaporation pan)

3.2. Discussion

Spearman's correlation analysis resulted in non-significant correlations between land use changes and rainfall (P>0.05), that is probably because rainfall is a global hydrological cycle which covers a wide area. Thus the rainfall in the study site may be more influenced by the global rainfall process mainly Asian monsoonal system, which follows the global wind direction according to solar position. The land use change at research site maybe not that severe which directly impact the rainfall.

Correlation analyses obtained a significant correlation between forest-land use change with climate parameters significantly influenced the air temperature (P=0.044) with a negative correlation coefficient (r = -0.912). The regression analysis is presented in Figure 7.

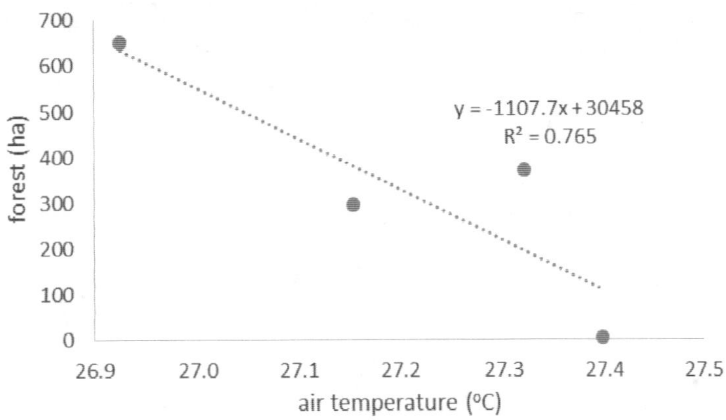

Figure 7. Linear regression of forest-land use change vs. air temperature

Figure 7 shows that land use change of forest area negatively correlates with air temperature. That means the reduction of forest area lead to air temperature increase, with a coefficient of determination (R^2) 0.765. The reduction of forest area means decreasing leaf area index (LAI) which will promote air temperature increase [16].

Spearman's analysis also resulted in a significant correlation between plantation-land use change with air temperature (P=0.05), also with a negative correlation coefficient (r = -0.9). The regression analysis is shown in Figure 8.

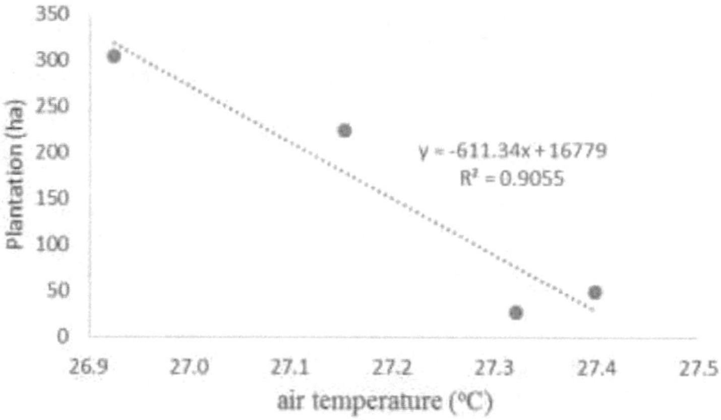

Figure 8. Linear regression of plantation – land use change vs. air temperature

It is shown in Figure 8 that similar to that of forest area change influenced on air temperature, the depletion of plantation area also consequently result in the decrease of air temperature with R^2 0.9055. The mechanism is explained by Gardiner and Miller) [17], that plants need at least 17 elements for metabolism system and the elements supplied from the air are hydrogen, carbon, and oxygen for photosynthesis. Light energy split off hydrogen becomes water molecules that were delivered to the leaf through the root and stem. Hydrogen is combined with carbon and oxygen from CO_2 molecules which are diffused into leaf through stomata. This process resulting in lower air temperature under canopy than surroundings. The other 14 essential elements are provided to plant directly from the soil. Plant tissue transports and absorbs water, mineral, and nutrient dissolved in water from the soil through the root.

Therefore, the conversion of forest and plantation indicated by plants removal will drive high solar radiation exposure on the land surface, low infiltration, high runoff, and very low oxygen supply from the plants. The circumstance leads to heat increase in the surrounding environment. A similar mechanism occurs in urban development which changes the landscape structures and thus resulting in a higher temperature at surroundings environment.

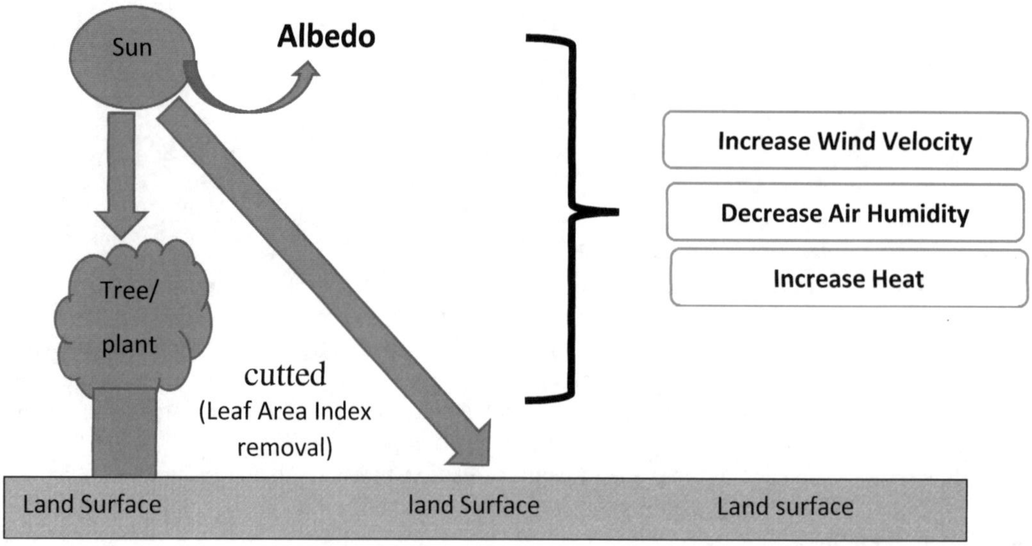

Figure 9. Driving factors of land use change to climate change

Figure 9 illustrates when the plants are cut or removed, the leaf of area index (LAI), or also called as canopy removal will consequently effects the evaporation of air particles near the land surface, and thus promote the heat to increase air temperature [16][17][18]. Therefore, forest and plantation conservation is required to prevent higher temperature rise, as well as to minimize the impact of climate change at Samin subwatershed.

Land use changes did not influence air relative humidity (P>0.05) due to humidity is more affected by rainfall than other climatic factors. Rainfall influences surroundings humidity, and thus humidity has no correlation with land use change as a result of global precipitation pattern. However, dryland correlates with evaporation (P=0.039) with positive correlation coefficient (r = 0.921). This means that the increase of dry land resulting in high evaporation.

4. Conclusion

Land use conversion from forest and plantation to other usages increased air temperature at surroundings environment of Samin subwatershed. Forest and plantation conservation is required to prevent higher temperature rise, as well as to minimize the impact of climate change at Samin subwatershed.

References

[1] Wang H, Feng Q and Ruan X 2016 Loss or gain: a spatial regression analysis of switching land conversions between agriculture and natural land. *Science Direct Agriculture, Ecosystems and Environment journal. Agriculture, Ecosystems and Environment* **221** 222–234

[2] Azadi H, Ho P and Hasfiati L 2010 Agricultural land conversion drivers: a comparison between less developed, developing and developed countries. *Land Degrad. Develop.* DOI: 10.1002/ldr.1037

[3] Verburg P H, Neumann and Nol L 2011 Challenges in using land use and land cover data for global change studies *Global Change Biology* **17(2)** 974–989

[4] Foley J A , Defries R and Asner G P 2005 Global consequences of land use *Science* **309(5734)** 570–574.

[5] Pielke R A 2005 Land use and climate change *Science* **310(5754)** 1625–1626.

[6] Pachauri R and Reisinger A 2007 *IPCC Fourth Assessment Report* (Geneva, Switzerland: IPCC)

[7] Yao X, Zhanqi W and Hua W 2015 Impact of urbanization and land use change in middle and lower reach the yangtze river, 1988-2008 *Advances in Meteorology* pp 10

[8] Geiger R 1950 *The Climate Near the Ground* (Cambridge: garvard University Press)

[9] Ashcroft M B, Gollan J R 2012 Fine resolution (25 m) topoclimatic grids of near surface (5 m) extreme temperature and humidity across various habitats in a large (200x300 km) and diverse region. *Int.J.Climatol* **32(14)** 2134-2148

[10] Tan R, Beckman V, Van D B L and Qu F 2009 Governing farmland conversion: comparing China with Netherland and Germany *Land Use Policy* **26** 961-974

[11] Ho S P S and Lin G C S 2004 Converting land to non-agricultural use in China Coastal Provinces, Eviden from Jiangsu *Modern China* **30** 81-112.

[12] Ismail C 2015 Assessing the correlation between land cover conversion and temporal climate change—a pilot study in Coastal Mediterranean City, Fethiye, Turkey *Atmosphere* **6** 1102-1118

[13] Hong N B and Yabe M 2016 Improvement in irrigation water use efficiency: a strategy for climate change adaptation and sustainable development of Vietnamese tea production *Environ. Dev. Sustain.* pp 1–17

[14] Zhang N, Gao Z Q, Wang X M and Chen Y 2010 Modeling the impact of urbanization on the local and regional climate in Yangtze River Delta, China *Theoretical and Applied Climatology* **102(3-4)** 331–342

[15] Rahayu N 2017 *Transfer of land function and rainfall to changes of hydrological condition of*

samin sub watershed (in Indonesian: Alih Fungsi Lahan dan Curah Hujan Terhadap Perubahan Kondisi Hidrologi Sub DAS Samin) (Surakarta: Universitas Sebelas Maret Press)
[16] Hardwick S R, Ralf T, Marion P, Eadgar T, Reuben N and Ewers R M 2015 The relationship between the leaf area index and microclimate in tropical forest and oil plant plantation: forest disturb anca drives changes in microclimate *Agricultural and Forest Meteorology Journal.* **201** 187-195
[17] Gardiner D T and Miller R W 2008 *Soil in our Environment* (New Jaesey Colombus, Ohio: Upper Saddle zriver)
[18] Kaufmann R and Cutler C 2008 *Environmental Science International Edition* (New York: Mc. Graw-Hill)

IOP Conf. Series: Earth and Environmental Science **129** (2018) 012033 doi:10.1088/1755-1315/129/1/012033

Climate change and farmers' cropping patterns in Cemoro watershed area, Central Java, Indonesia

Sugihardjo[1,3], J Sutrisno[1], P Setyono[2] and Suntoro[1]

[1]Agricultural Faculty, Universitas Sebelas Maret
[2]Mathematics and Natural Science Faculty, Universitas Sebelas Maret, Jl. Ir, Sutami 36A, Solo 57126, Indonesia
[3]Corresponding author: giek_bb@yahoo.com

Abstract. Cropping pattern applied by farmers is usually based on the availability of water. Farmers cultivate rice when water is available. If it is unavailable, farmers will choose to plant crops that need less water. Climate change greatly affects to farmers in determining the cropping pattern as it alters the rainfall pattern and distribution in the region. This condition requires farmers to adjust the cropping pattern so that they can do the farming successfully. This study aims to examine the application of cropping patterns applied by the farmers in the Cemoro Watershed, Central Java, Indonesia. Descriptive analysis approach is employed in this research. The results showed that farmers' cropping pattern is not based on the availability of water. However, it adopts a habit that has been practiced since long time ago or just adopt others farmer's habit. The cropping pattern applied by irrigated paddy farmers in Cemoro watershed area consists of two types: rice-rice-rice and rice-rice-secondary crops. Among those two types, most farmers apply the rice-rice-rice pattern. Meanwhile, there are three cropping patterns applied in the rain-land, namely rice-rice-rice, rice-rice-secondary crop, and rice-rice-fallow. The majority of farmers apply the second pattern (rice-rice-secondary crops). It was also found that farmers' cropping pattern was not in accordance with the recommendation of the local government.

1. Introduction

Climate change is a change in climate status that is identified with the average change and/or variability on climate-related factors and occurs for a longer period [1]. Generally, climate change is characterized by the increase of air temperature on the earth's surface and the increase of sea surface, which in the next step, affects to human's health, agriculture, water resource, and many other life sectors [2].

The agricultural sector is the most prone sector against climate change due to its high dependency to climate and weather, and also because the persons involving in the sector, likely tend to be poor, compared to those working in the industrial sector and service sector. In addition, the agricultural sector is prone to climate change, because it affects to cropping pattern, planting time, production, and the quality of harvest products. The global climate change will affect the occurrence of the change of rainfall's pattern and distribution, which leads to the occurrence of the shift of early season and planting time. Recently, many farmers experience a hard time to decide the right time of cropping or crop failures due to a long drought or unexpected rains [3].

A sufficient amount of water supply is a condition to reach optimum plants' production. The expected rains that occur at the right time with the right amount of water will give positive impacts for the agricultural sector. Climate change may cause floods or drought [3] that can cause the decrease in

Content from this work may be used under the terms of the Creative Commons Attribution 3.0 licence. Any further distribution of this work must maintain attribution to the author(s) and the title of the work, journal citation and DOI.
Published under licence by IOP Publishing Ltd

the amount of production [4]. Thus, farmers should act to face climate change, so that the farming production does not experience a decreased productivity. The continuity of adaptation conducted by the farmers, in fact, may overcome the climate change and the decrease in crops productivity [5].

Farmers may practice adaptation in several ways. One of them is by changing the planting time, by adjusting the time in the beginning of the rainy season [6]. The change of planting time, by itself, means conducting the change of cropping pattern, which should be applied in their farming. Farmers are supposed to change their cropping pattern in facing the climate change. The change of cropping pattern is meant for keeping the plants' optimum growing because the water supply is expected to be enough. This research examines the usual cropping pattern conducted by farmers in Cemoro watershed area, Central Java, Indonesia. The cropping pattern will also be compared with the one suggested by the government and cropping pattern prediction supposed to be done by paying attention to the rainfall.

2. Research method

This research is conducted in Cemoro watershed area, Central Java, Indonesia. It is located in the southeast of Central Java. Cemoro watershed is situated in several regencies including Semarang, Boyolali, Sragen, and Karanganyar. This research uses 120 farmers cultivating irrigation rice fields and 120 farmers cultivating rain-fed rice fields as the samples. The information collected includes the characteristics and habits of farmers in carrying out their farming. The data are then analyzed descriptively using a cross-tabulation assistance.

3. Result and discussion

Land utilization for farming activities needs to pay attention to the water supply to gain the optimum result. In general, farmers will decide the kind of cultivated plants based on their experience and the recommendation from the government. Kinds of plants cultivated by the farmers based on the time sequence are called cropping pattern. The appropriate and suitable cropping pattern is an expected way to guarantee enough water supply to fulfill water needs for the growing plants. Cropping pattern applied by farmers in Cemoro watershed area is shown in Figure 1.

Farmers in Cemoro watershed area, apply two kinds of cropping pattern on irrigation rice fields, the first is rice-rice-rice and the second is rice-rice-secondary crops, with 72% of farmers apply the first kind of pattern. Meanwhile, three cropping patterns are applied for rain-fed rice fields by the farmers, they are rice-rice-rice, rice-rice secondary crops, and rice-rice-fallow. The applied cropping pattern is commonly based on the farmers' experiences. Farmers pay attention to the surrounding and other farmers' habits.

Cropping pattern practiced by the farmers in Cemoro watershed area seems not suitable for the one recommended by their local government. The rice-rice-rice cropping pattern is not recommended by the government due to the high demands of water supply. Boyolali regency is a region that has issued a resolution related to the appropriate cropping pattern for both rain-fed and irrigation rice fields. The cropping pattern suggested for rice fields in Boyolali Regency, based on the water supply factor to fulfill the needs of growing plants are seen in Figure 2.

The Cropping Pattern 1 is suggested for irrigation rice fields with more than enough water supply. The Cropping Pattern 2 is suggested for irrigation rice fields with enough water supply. Meanwhile, the Cropping Pattern 3 is suggested for irrigation rice fields with lack of water supply. The government recommended farmers to apply the rice-secondary crops-fallow cropping pattern for rain-fed rice fields. This reality shows that farmers do not practice government's suggestion due to the economic problem. Farmers assume that rice has more economic value compared to other commodities. The unsuitable pattern causes a higher risk of crop failure that leads to the decrease in income.

Irrigation Rice Fields												
Cropping Pattern	MT I				MT II				MT III			
	Nov.	Dec.	Jan.	Feb.	Mar.	Apr.	May.	June	July	Aug.	Sept.	Oct
Type 1 (72 %)		Rice					Rice			Rice		
Type 2 (28 %)		Rice					Rice			Secd crop		

Rain-fed Rice Fields												
Cropping Pattern	MT I				MT II				MT III			
	Nov.	Dec.	Jan.	Feb.	Mar.	Apr.	May.	June	July	Aug.	Sept.	Oct
Type 1 (12 %)		Rice				Rice				Rice		
Type 2 (70 %)		Rice				Rice				Secd crop		
Type 3 (18 %)		Rice				Rice				Fallow		

Figure 1. Cropping plant on irrigation and rain-fed rice fields in Cemoro watershed area

Irrigation Rice Fields												
Cropping Pattern	MT I				MT II				MT III			
	Nov.	Dec.	Jan.	Feb.	Mar.	Apr.	May.	June	July	Aug.	Sept.	Oct.
Type 1		Rice				Rice				Secd crop		
Type 2a		Rice				Rice				Fallow		
Type 2b		Rice				Secd crop				Secd crop		
Type 3a		Rice				Secd crop				Fallow		
Type 3b			Secd crop			Rice				Fallow		

Figure 2. Suggested cropping pattern on irrigation rice fields in Boyolali regency

The application of cropping pattern actually should adjust with water supply for the growing plants. Thus, farmers should pay attention to the rainfall. Farmers can refer to the monthly average of rainfall, as a reference to start cultivating rice and then other plants. Figure 3 and Figure 4 shows the monthly average of rainfall recorded in Simo District, Boyolali Regency, and Gondangrejo district, Karanganyar regency, which may be a reference for farmers to apply the suitable applied cropping pattern.

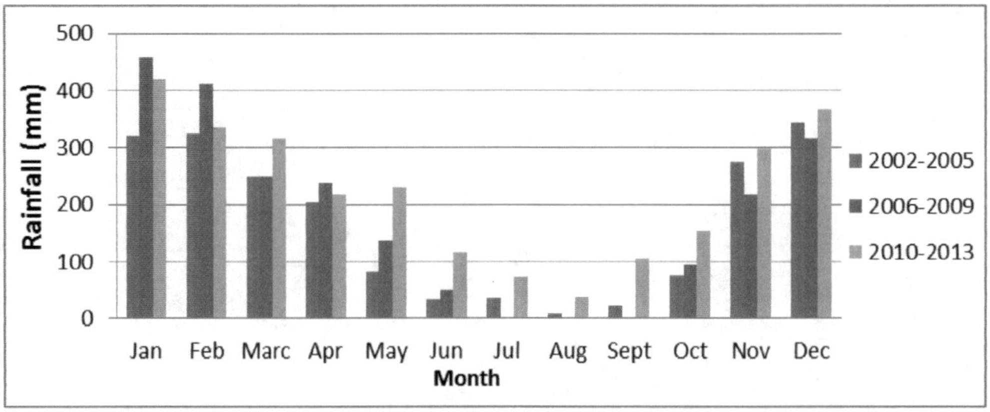

Figure 3. Monthly rainfall pattern in Simo District between 2002 and 2013

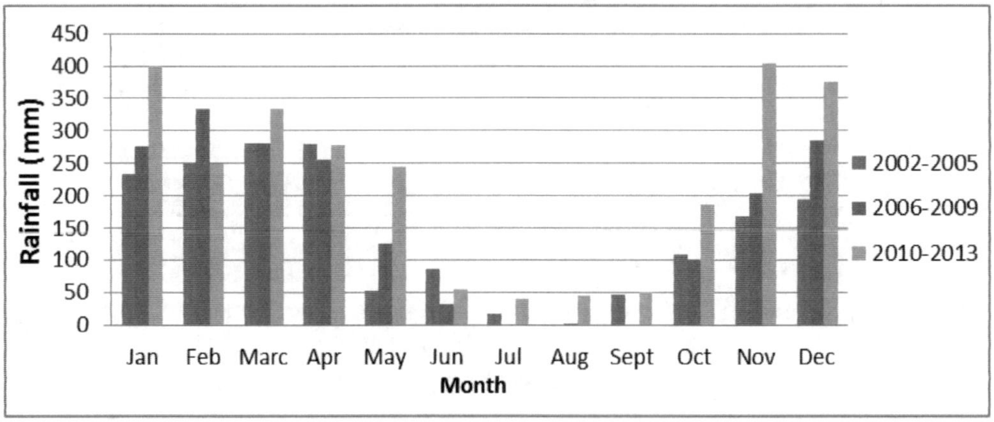

Figure 4. Monthly rainfall pattern in Gondangrejo District between 2002 and 2013

The difference of rainfall pattern between 2002 and 2013 in Simo District and Gondangrejo district may be identified by Figure 3 and Figure 4. The wet month, when the rainfall reaches 200 mm or more, occurs for six months in November, December, January, February, March, and April in Simo district between 2002 and 2005. Meanwhile, in Gondangrejo district, it occurs for four months in January, February, March, and April. Wet month occurs for six months in November, December, January, February, March, and April between 2006 and 2009, and seven months in November, December, January, February, March, April and May between 2010 and 2013. This condition can be an indicator of the change of climate pattern in Gondangrejo district. The wet month is characterized by 200 mm or more of rainfall while a humid month is characterized with 100 mm – 200 mm of rainfall, and dry month with less than 100 mm of rainfall. The development of rainfall in Simo District and Gondangrejo district can be seen in Table 1.

Cropping pattern can be formulated according to the rainfall data mentioned in Figure 3 and Figure 4, based on rainwater supply, which may be utilized for cultivating rice and other plants. Generally, rice may be started to plant when monthly average of rainfall reaches at least 175 mm, not less. According to the rainfall data seen at Figure 3 and Figure 4, the ideal cropping pattern formula, based on rainwater supply in Simo district between 2002 and 2005, 2006 and 2009, is rice-secondary crops-fallow and between 2010 and 2013 is rice-rice-secondary crops. Meanwhile, in Gondangrejo district, the ideal cropping pattern for 2002-2005 is rice-rice-secondary crops and rice-rice-fallow for 2006-2009.

Table 1. Numbers of wet month, humid and dry in Simo and Gondangrejo between 2002 and 2013

No	Period of Time	Wet Month (month)	Humid Month (month)	Dry Month (month)
1	Simo			
	2002-2005	6	0	6
	2006-2009	6	1	5
	2010-2013	7	3	2
2	Gondangrejo			
	2002-2005	4	3	5
	2006-2009	6	2	4
	2010-2013	7	1	4

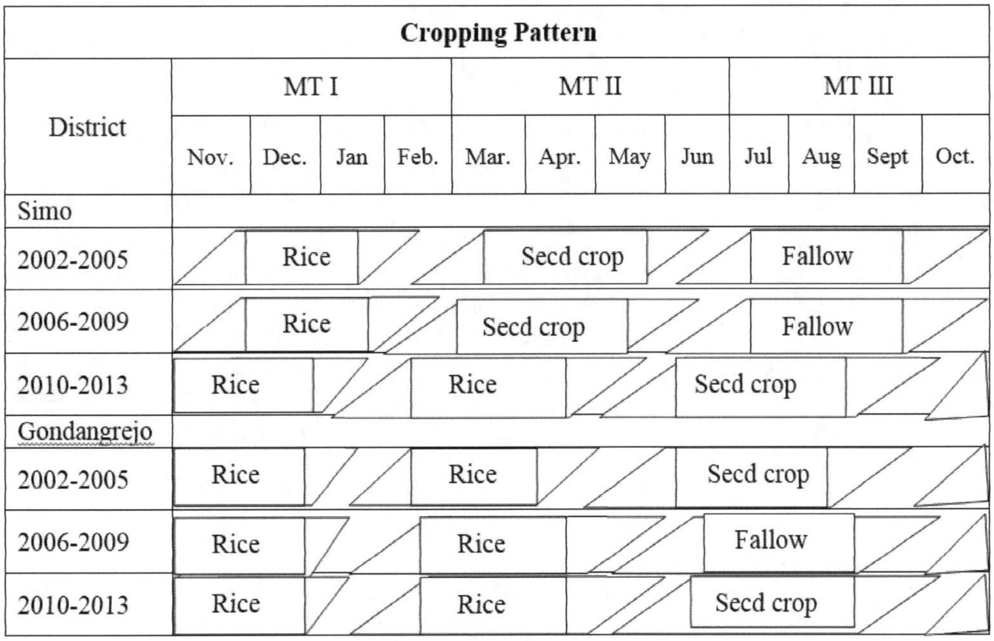

Figure 5. The suggested cropping pattern based on rainwater supply in Simo district and Gondangrejo district

These findings show that there are no farmers applying the ideal cropping pattern based on monthly rainfall that occurs in the area of Cemoro Watershed. They prefer to consider their desires and habits in deciding the pattern, although it may cause higher crop failures.

4. Conclusion

The data analysis shows that farmers apply rice-rice-rice cropping pattern on irrigation rice fields. Meanwhile, rice-rice-secondary crops pattern is applied on rain-fed rice fields. Farmers do not apply the cropping pattern suggested by the local government. This unsuitable cropping pattern leads to the higher risk of crop failures.

Acknowledgement
Thanks to the enumerators who helped with data collection and processing.

References

[1] Angles, Chinnadurai and Sundar 2011 Awareness on impact of climate change on dry land agriculture and coping mechanisms of dry land farmers *Indian Journal of Agricultural Economics* **66(3)** 65-72

[2] Ayunwuy, Kuponiyi, Ogunlade and Oyetoro 2010 Farmers perception of impact of climate changes on food crop production in Ogbomoso, agricultural zone of Oyo State, Nigeria *Continental Journal Agricultural Economics* **4** 19-25

[3] IPCC 2001 *Climate Change* 2001: *Impacts, Adaptation, and Vulnerability-Intergovermental Panel on Climate Change* (Cambridge: Cambridge University Press)

[4] Mburu B K, Kung'u J B and Muriuki J N 2014 Effects of climate variability and change on household food sufficiency among small-scale farmers of Yatta district, Kenya *J. Environ.* **3(2)** 19-27

[5] Naylor R, Battisti D S, Vimon D J, Falcon W P and Burke M B 2007 *Assessing Risks of Climate Variability and Climate Change fo Indonesia Rice Agriculture - Proceeding of National Academy of Science of the United State of America PNAS* (Washington: Nasional Academy of Sciences) 104(19) 7752-57

[6] Ofuoku A U 2011 Rural farmers' perception of climate change in central agricultural zone of Delta state, Nigeria *Indonesian Journal of Agricultural Science* **12(2)** 63-8

[7] Surmaini, Elza, Runtunuwu E and Las E 2011 Agricultural Sector Efforts in Facing Climate Change (In Indonesian) *Jurnal Litbang Pertanian* **30(1)** 1-7

IOP Conf. Series: Earth and Environmental Science **129** (2018) 012034 doi:10.1088/1755-1315/129/1/012034

Assessment of water quality from water harvesting using small farm reservoir for irrigation

W S Dewi[1,4], Komariah[1], I Y Samsuri[2] and M Senge[3]

[1] Department of Soil Science, Agriculture Faculty of Universitas Sebelas Maret, Indonesia. Jl. Ir. Sutami No. 36, Kentingan, Surakarta, Central Java, Indonesia.
[2] Alumni of Department of Agrotechnology, Agriculture Faculty of Universitas Sebelas Maret, Indonesia.
[3] Faculty of Applied Biological Sciences, Gifu University, Yanagido 1-1, Gifu 501-1193, Japan
[4] Corresponding author: Email: wsdewi2000@gmail.com

Abstract. This study aims to assess the quality of rainfall-runoff water harvesting using small farm reservoir (SFR) for irrigation. Water quality assessment criteria based on RI Government Regulation number 82 the year 2001 on Water Quality Management and Pollution Control, and FAO Irrigation Water Quality Guidelines 1985. The experiment was conducted in the dry land of Wonosari Village, Gondangrejo District, Karanganyar Regency. SFR size was 10 m x 3 m x 2 m. Water quality measurements are done every week, ten times. Water samples were taken at 6 points, namely: distance of 2.5 m, 5 m, and 7.5 m from the inlet, at depth 25 cm and 175 cm from surface water. In each sampling point replicated three times. Water quality parameters include dissolved oxygen (DO), Turbidity (TSS), water pH, Nitrate (NO_3), and Phosphate. The results show that water harvesting that collected in SFR meets both standards quality used, so the water is feasible for agricultural irrigation. The average value of harvested water was DO 2.6 mg/l, TSS 62.7 mg/l, pH 6.6, P 5.3 mg/l and NO_3 0.16 mg/l. Rainfall-runoff water harvesting using SFR prospectus for increasing save water availability for irrigation.

1. Introduction

Water is a vital natural resource for living things to survive, including crops. Water availability, water scarcity, and water crisis are the widely discussed topics that recently concerned dispite climate change and population growth [1]. It dramatically affects the plants water availability, even water catchment area is threatened by climate change [2]. Increased atmospheric temperature globally has an essential impact on hydrological cycle changes [3], both in tropical and sub-tropical regions. The saturated water vapor pressure in the atmosphere is very sensitive to temperature; therefore climate change will affect the global water cycle disruption that will impact on precipitation and run-off [4]. By hydrological context, climate change accelerates its cycle on a global scale and causes uneven intensity and distribution of hydrological resources. The extreme rain phenomenon in some parts of the world are increasing, yet some areas drought out [5], and rainfall variability is also increasing [1]. The hydrological cycle changes lead to seasonal shifts, longer dry seasons and shorter rainy seasons with higher rain intensity, increased dry season and wet season anomalies, less soil moisture, resulting in farmland prone to drought. It is one of food security threats.

Agriculture is highly dependent on water resources [6]. The agricultural sector is the most significant water user in some countries [7]. Globally, agricultural activities require approximately

Content from this work may be used under the terms of the Creative Commons Attribution 3.0 licence. Any further distribution of this work must maintain attribution to the author(s) and the title of the work, journal citation and DOI.
Published under licence by IOP Publishing Ltd

3,100 billion m^3, or around 70% to 90% of the total water on the surface and groundwater [8][10], and will increase to 4,500 billion m^3 by 2030 if not efficiently to use. It is a major limiting factor for plant growth in tropical and sub-tropical environments [9]. The seasonal rainfall quantity is not the only major limitation for agricultural production, but uncertain rain phenomena are also an important issue [9]. Climate change guides water availability problems. Water scarcity is a threat to meeting future food demands unless significant improvements in agricultural water productivity can be achieved. Sustainable water management strategies contribute to transforming degraded soils into healthy agroecosystems, and increasing resilience to climate change and variability [6]. Therefore, availability of sustainable and more productive water management technology for agriculture is a very urgent need.

Water harvesting (rainfall-runoff water harvesting) using small farm reservoir (SFR) is one of the strategies for sustainable water management [6]. That technique can reduce the risk of floods and droughts, expand the chance to grow in a dry season, maintain and enhance ecosystem resilience to climate variability and change [6]. Water harvesting techniques using SFR are also applied in Sub-Saharan to address dry spell [9]. Dry spells are a slightly extended dry period, ranging from a week to several weeks, between occurrences of rainy season [9]. Water harvesting is very important for rainfed farming systems, especially for supplement irrigation. Supplement irrigation - irrigation only during periods of growth and critical development - plus rainwater harvesting can increase yields by 2-3 times more than conventional rainwater farming [6].

Increased water scarcity will trace the water harvesting and recycling needs for agricultural irrigation and the low-quality water application [10]. Rainwater is relatively free of dirt unless it carries pollutants from the atmosphere. Pollution may occur at harvest time or during storage. Rainfall harvest from the agricultural catchment area using SFR will mix with run-off, potentially degrading water quality, as it carries various pollutants such as sediment, heavy metals, pesticide residues, organic and inorganic fertilizer residues, plant litter, bird droppings, garbage, insects, even disease-causing microbes. Thermotolerant coliform and E. *coli* are common microbial contaminants present in rainwater harvesting.

Is water harvesting using SFR in Gondangrejo Sub-district, Sragen Regency, Central Java, Indonesia, eligible for agricultural irrigation? Government Regulation No. 82/2001 on Water Quality Management and Water Pollution Control, Article 8, states that irrigation water belongs to classes II, III, and IV with various indicators of physical, chemical, biological and radioactive parameters [11]. In Government Regulation No. 20 of 2006 on Irrigation [12] does not mention the water irrigation quality , while to achieve food security, it is necessary to measure the water harvesting quality for irrigation. Some international standard operate water quality criteria for irrigation, such as Scofield, Oklahoma Water Quality Standard, Colorado Irrigation Water Quality Criteria, FAO, and Guidance for Irrigation Water Quality and Water Management in The United Kingdom of Saudi Arabia [13]. Rainwater harvesting quality using rooftop has been widely published but not much publication about the water harvesting quality using SFR. This paper presents the feasibility of rainfall-runoff water harvesting quality using SFR for irrigation.

2. Material and Methods

The research was conducted in rain-fed rice field located in Wonosari Village, Gondangrejo District, Karanganyar Regency, from April to June 2013, using SFR with size 10 x 3 x 2 m^3. The SFR site selection based on a place that can accommodate all runoff streams from different directions. The collected water in the SFR is a mixture of rainfall and runoff. Water samples were taken 6 points of SFR, covering 2.5 m, 5 m, and 7.5 m from the inlet, and each at a depth of 25 cm and 175 m from the surface. Water samples were collected every one week, for ten weeks by utilizing techniques which based on the Indonesian National Standard Collection of Public Works on Water Quality [14]. Water quality parameters analyzed include dissolved oxygen (DO), turbidity (TSS), pH, Carbon, Nitrate (NO_3-N), Phosphate, and Potassium based on Land Research Center method [15]. Assessment of water harvesting quality for irrigation based on Government Regulation of the No. 82/2001 on Water

IOP Conf. Series: Earth and Environmental Science **129** (2018) 012034 doi:10.1088/1755-1315/129/1/012034

Quality Management and Water Pollution Control, and the Food and Agriculture Organization (FAO) Irrigation Water Quality Guidelines 1985.

3. Result and Discussion

3.1. Description of research location

The research area at Gondangrejo Sub District, Karanganyar was located at 110°48'16,3 "- 110°53'4,36" E and 07°27'17,8 "- 07°32'42,16" S with height 100-200 m asl. It has a total area around 5677 ha with rainfed lowland area approximately 1250 ha. The average rainfall in Gondangrejo based on the last ten years data was below 3000 mm/year, with eight wet months and four dry months (Figure 1). During the wet months (October-April) the rainfall was relatively high, around 213.1 - 337.5 mm, whereas in dry months the rainfall ranges from 14.1 - 53.8 mm. This condition affects the planting pattern in Gondangrejo only two times rice planting. High rainfall needs to be harvested using SFR, which can eventually be used as irrigation water to meet the needs of plant water during a dry spell or the dry season. The soil type present in this region was Vertisol, with sandy clay texture up to clay.

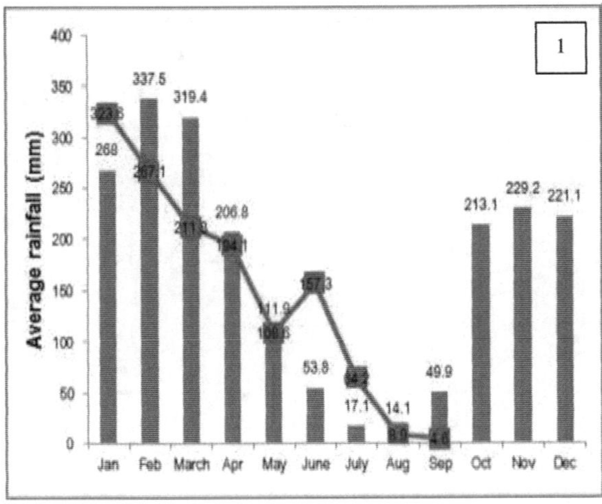

Figure 1. Histogram of average monthly rainfall (mm) in Gondangrejo, the year 2000-2010 (Source: BMKG Central Java). (Note: ▬ = average 10-year rainfall, ▬ = Average rainfall during the research).

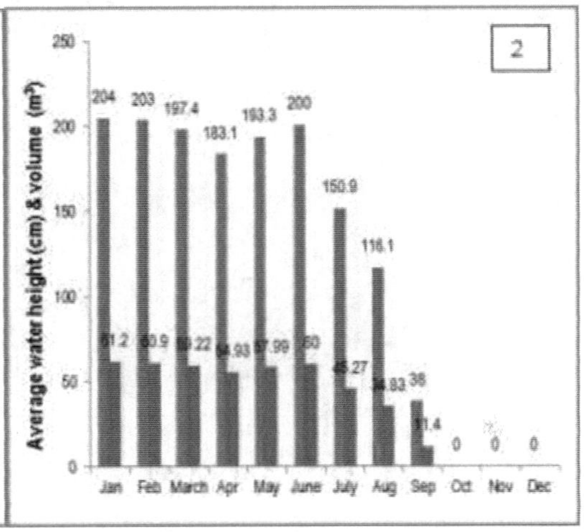

Figure 2. Histogram of average water height (cm) and water volume (m^3) in SFR during the study. (Note: ▬ = average water height (cm), ▬ = water volume (m^3).

During the research showed that six wet months (January to June), and three dry months (July to Sept). Average rainfall at that time categorized to the wet month (the rainfall between 100-200 mm). The average water density in SFR during April to June was 183.1 cm, 193.3 cm, and 200 cm, respectively (Figure 2). In July, water density in SFR started to fall down, and farmers began for using water from the SFR and lead to reducing the water level in the SFR. In September, the SFR became dry and could not harvest again.

3.2. Water Harvesting Quality in SFR

The average DO, TSS, pH, and nitrate at each point of all sampling points met the quality standards of agricultural irrigation water quality (class III and IV) according to RI Government Regulation no. 82 of 2001 and according to the Food and Agriculture Organization (FAO) Irrigation Water Quality Guidelines 1985 (Table 1).

Table 1. Some parameters of water harvesting quality at SFR (Dissolved Oxygen, Turbidity, pH, Phosphate, and Nitrate)

No.	Parameter	Sampling point	Content (mg/l)	Government Regulation. No. 82/2001 Class III	Government Regulation. No. 82/2001 Class IV	FAO 1985
1.	Dissolved Oxygen (DO), mg/l	I A	3.08	3	0	-
		I B	2.30			
		II A	3.10			
		II B	2.28	(minimum threshold)		
		III A	2.93			
		III B	2.02			
2.	Turbidity/ TSS,	I A	58.24	400	400	-
		I B	65.85			
		II A	57.01			
		II B	65.91			
		III A	60.06			
		III B	69.29			
3.	pH	I A	6.61	6 - 9	5 - 9	6.5 – 8.4
		I B	6.50			
		II A	6.60			
		II B	6.57			
		III A	6.60			
		III B	6.64			
4.	Fosfat, mg/l	I A	6.09*	1	5	-
		I B	4.99			
		II A	5.5*			
		II B	5.36*			
		III A	5			
		III B	5.11			
5.	Nitrat (NO_3), mg/l	I A	0.15	20	20	5 - 30
		I B	0.16			
		II A	0.15			
		II B	0.18			
		III A	0.16			
		III B	0.13			

Note :
I A: Water sample in 25 cm depth from water level, at inlet nearest point (2.5 m)
I B: Water sample in 175 cm depth from water level, at inlet closest point (2.5 m)
II A: Water sample in 25 cm depth from water level, at midpoint (5 m)
II B: Water sample in 175 cm depth from water level, at the midpoint (5 m)
III A: Water sample in 25 cm depth from water level, at outlet nearest point (7.5 m)
III B: Water sample in 175 cm depth from water level, at outlet closest point (7.5 m)
Class III: water quality criteria for freshwater fish cultivation, livestock, water to irrigate crops.

Class IV: water quality criteria for irrigating crops.
*): the parameters of water samples exceed the threshold of water quality standards for irrigation established by Government Regulation. No. 82/2001.

The phosphate concentration at point IA, IIA, IIB, and IIIB exceeded the predefined standard threshold (> 5 mg/ml), but surrounding soil at SFR was Vertisol which P deficient, therefore a high phosphate content can have a good effect on agricultural production. High levels of phosphate in water harvesting at the SFR affected by the location of the catchment area in the agricultural regions where the average farmer uses a lot of fertilizers, pesticides, and insecticides for the cultivation crop. The agricultural activities impacts will result in the runoff with a high content of nitrate and phosphate sediments [16]. Nutrients or other contaminants transported to SFR along with runoff and rainwater. Phosphate content in SFR that exceeds the needs of organisms has potential lead to eutrophication in a short time.

3.3 Water Quality Dynamics in SFR

To understand the water harvesting quality in SFR safe or not used for irrigation, it was necessary to see the quality as long as it is in SFR, before being used for irrigation. In this experiment, water quality assessment in SFR was conducted weekly for ten weeks, from April to June (Figure 1).

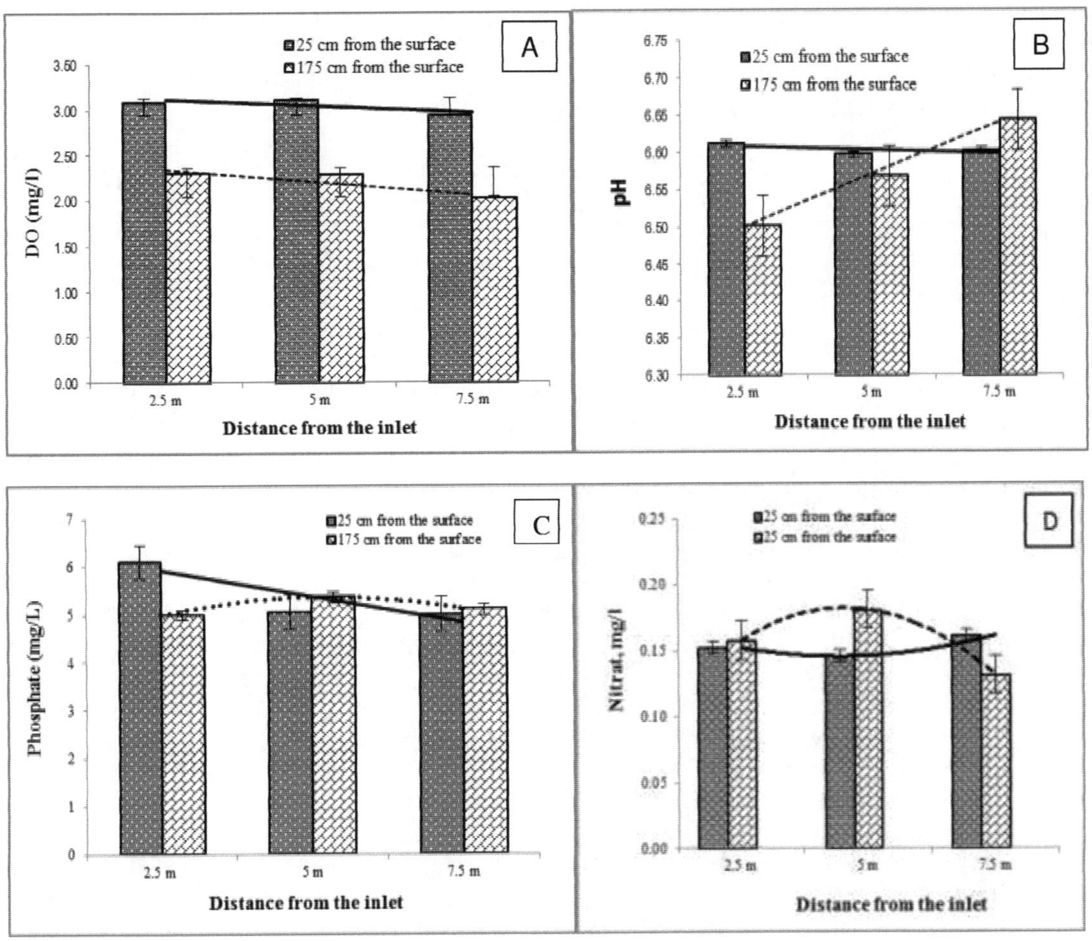

Figure 3. Dynamics of several water quality parameters during harvest in SFR: Dissolve Oxygen (DO) (A), pH (B), Phosphate (C), and Nitrate (D)

During the experiment, the harvesting water stayed for ten weeks in SFR. The average DO near the surface (25 cm from the surface) was 3.02 mg/l, higher and significantly different ($p < 0.05$) with DO at a depth of 175 cm from the surface of 2.02 mg/l (Figure 3A). Dynamics of DO changes were occurred because of diffusion process between water and free air, and photosynthesis process by algae. The farther from the water surface, there was a decrease in dissolved oxygen levels (Fig. 3A). Its related to declining of photosynthesis process and oxygen which widely used for respiration and oxidation of organic and inorganic materials by microorganism [17]. The unorganized organic matter, when discharged into water bodies, will be decomposed by bacteria that need oxygen to decaying material. Because the oxygen solubility in the water was relatively low, high level of organic waste affects reducing DO faster.

There was a significant difference ($p<0.05$) between DOs nearest and farthest from the inlet. The DO decreased with farther distance from the inlet. The average DO closest (2.5 m from inlet) was 2.69 mg/l, while the most distant (7.5 m from the inlet) was 2.48 mg/l (Fig. 3A). Decreasing DO due to the buildup of sediments and organic matter at the outlet area was carried by the runoff stream. Microorganism decomposed organic matter in sediment using oxygen was contained in the upper water layer. The high natural material was trapped in the SFR sediment which affects the increasing decomposition process by aquatic microorganisms and lead decreasing oxygen content in SFR. Reducing DO in sediment was affected by natural respiration and biochemical processes by bacteria against the organic matter input which ambushed in the sediments. Besides, oxidation processes such as nitrification that convert ammonia to nitrate also contribute to the decrease oxygen levels in sediments. The most significant oxygen consumption in the waters was in deposits and water columns near the residues, where there was an organic matter accumulation and active bacterial and benthic metabolism [18].

The average pH of SFR is 6.59. The pH near the water surface tends to be constant, both far to near from the outlet, which was about 6.60 (Figure 3B). While in the deeper water layer (175 cm from the surface), pH increases significantly ($p<0.05$) from the inlet to outlet, i.e., from 6.50 to 6.64. The increased pH was positively correlated with the amount of rainfall during the study ($r = 0.43^{**}$). Rainwater that falls on the catchment affect run-off flows that carry dissolved sediments, both organic and inorganic. The soil type in the water catchment area was Vertisol which has a near-neutral pH so that the deposits taken along with the run-off flow were likely to carry relatively high Ca-containing materials thereby increasing the pH. The Vertisols have neutral to slightly alkaline pH [19]. Besides, organic sediments on the bottom of the SFR from far to near the outlet tend to increase. Natural substances decomposition may increase or decrease the pH depending on the organic material type. Respiration process affects pH dynamic, the more carbon dioxide was produced by respiration process reducing pH, but if higher photosynthesis activity leads increasing pH [20].

Phosphate levels was decreased significantly ($p < 0.05$) from the surface to deeper water layer in SFR, i.e., from 5.38 to 5.15 mg/l. Likewise, there was a significant decrease ($p < 0.05$) of P concentration from the nearest inlet to the outlet, from an average of 5.54 to 5.06 mg/l (Figure 3C). A decrease in P content related to low P content in sediment carried along with runoff flow. Paddy fields have low P content [21]. Water catchment areas were also far from residential areas, so run-offs carry little domestic waste. Household waste, like detergent, affect soluble phosphates in SFR [22]. The soluble P can also be derived from the decomposition of organic matter in a water body. It still found in the deeper layers of water surface. The SFR bottom layer was contained a rich nutrients, both derived from sediments and decomposition of organic compounds [23].

The nitrate concentration at all sample points was not significantly different ($p> 0.05$). Average nitrate levels in SFR ranged from 0.13 to 0.34 mg/l. The nitrate concentration at 175 cm depth was slightly higher compared to the nitrate level at 25 cm depth from the surface (Figure 3D). The soluble nitrate in SFR can be derived from the N fertilizer contaminated with runoff, as well as the organic compounds decomposition which was accumulated in SFR. The higher nitrate concentrations close to the bottom layer related with nitrates in the surface layer were more used or consumed by phytoplankton [24]. Slightly more elevated nitrate concentrations near the bottom also were affected by sediment. Nitrate in deposit resulted from the biodegradation of natural materials into ammonia and oxidized to nitrate. Based on nitrate concentration levels, its oligotrophic waters SFR categorizes of which means poor nutrients because the nitrate content in the SFR was between 0-1 mg/l [18].

The climate change guided precipitation patterns transformation, consequently leaded to frequent short drought (dry spell) in Karanganyar Regency, Indonesia [25], wherein the research area that was implemented. Water harvesting was a good strategy for adaptation to climate change, especially in the arid region. SFR was one of the techniques for harvesting water that can be done by farmers around the cultivation area, therefore it can be utilized in a more accessible way for agricultural irrigation. Rainfall-runoff water harvesting was using SFR prospectus for increasing save water availability for irrigation.

4. Conclusions

The harvested water in SFR was feasible for agricultural irrigation because it meets the water quality standard for agricultural irrigation based on class III, and IV according to RI Government Regulation No. 82 of 2001 and according to the Food and Agriculture Organization (FAO) Irrigation Water Quality Guidelines 1985. The average value of harvested water in SFR was DO 2.6 mg/l, TSS 62.7 mg/l, pH 6.6, P 5.3 mg/l and NO_3 0.16 mg/l. Harvested water in SFR was useful for irrigation and supplements irrigation during dry spelling. Water harvesting technology which using SFR was one of the most important climate change adaptation strategies, especially for dryland areas.

Acknowledgments

The authors would like to thank the Japan Society for the Promotion of Science (JSPS) that funded this research in the fiscal year 2015.

References

[1] Cooper P J M, Dimes J, Rao K P C, Shapiro B, Shiferaw B and Twomlow S 2008 Coping better with current climatic variability in the rain-fed farming systems of sub-Saharan Africa: An essential first step in adapting to future climate change? *Agriculture, Ecosystems and Environment* **126**: 24–35

[2] Immerzeel W W, van Beek L P H and Bierkens M F P 2010 Climate Change Will Affect the Asian Water Towers *Science* **328**: 1382-85

[3] Barnett T P, Adam J C and Lettenmaier D P 2005 Reviews Potential impacts of a warming climate on water availability in snow-dominated regions *Nature* **438**: 303-09.

[4] Milly P C D, Dunne K A and Vecchia A V 2005 Global pattern of trends in streamflow and water availability in a changing climate *Nature* **438**: 347-350.

[5] Rana A, Foster K, Bosshard T, Olsson J and Bengtsson L 2014 Impact of climate change on rainfall over Mumbai using distribution-based scaling of global climate model projections. *Journal of Hydrology: Regional Studies* **1**: 107–28.

[6] "Agricultural water management water and development strategy implementation brief", usaid.gov, 2017. [Online], Available: https://www.usaid.gov/sites/default/files/documents/1865/Ag%20Water%20Management_Implementation%20Brief_Jan%202015.pdf. [Accessed: 16-Oct-2017].

[7] Amarasinghe U A and Smakhtin V 2014 *Global Water Demand Projections: Past, Present and Future* [Sri Lanka: International Water Management Institute].

[8] Wisser D, Frolking S, Douglas E M, Fekete B M, Vörösmarty C J and Schumann A H 2008 Global irrigation water demand: variability and uncertainties arising from agricultural and climate data sets *Geophysical Research Letters* **35**: 1-5.

[9] Fox P, Rockstrom J and Barron J 2005 Risk analysis and economic viability of water harvesting for supplemental irrigation in semi-arid Burkina Faso and Kenya *Agricultural Systems* **83**:231–50.

[10] Jensen P K, Matsuno Y, Hoek W V and Cairncross S 2001 Limitations of irrigation water

quality guidelines from a multiple use perspective *Irrigation and Drainage Systems* **15**: 117–28.

[11] The Republic of Indonesia, Government Regulation 2001 *Government Regulation No. 82 the year 2001 on the Management of Water Quality and Control of Water Pollution* (Jakarta: Government of Indonesia)

[12] The Republic of Indonesia, Government Regulation 2006 *Government Regulation No. 20 the year 2006 on Irrigation* (Jakarta: Government of Indonesia)

[13] Yusuf I A 2014 Review of water quality criteria for irrigation *Jurnal Irigasi* **9(1)**: 1-15.

[14] The National Standardization Agency of Indonesia 1990 *SNI 06-2412-1991 on sampling methods of water quality* (Jakarta: BSN).

[15] Soil Research Institute 2005 *The chemical analysis of soil, plants, water, and fertilizer* (Bogor: Agency for Agricultural Research and Development, Ministry of Agriculture).

[16] Casali J R, Gimenez J, Alvarez M J D V, Lersundi M, Goni M A, Campo Y, Chahor R, Gastesi J and Lopez 2010 Sediment production and water quality of watersheds with contrasting land use in Navarre (Spain) *Agricultural Water Management* **97**: 1683-94.

[17] Dugan P R 1972 *Biochemical Ecology of Water Pollution* (New York: Plenum Press) p 159

[18] Wetzel JT 1975 *Limnology* (Philadelphia: Sounders).

[19] Getahun M, Selassie Y G 2017 *Characterization, Classification and Mapping of Soils of Agricultural Landscape in Tana Basin, Amhara National Regional State, Ethiopia*. *In*: Stave K, Goshu G and Aynalem S (eds) Social and Ecological System Dynamics. AESS Interdisciplinary Environmental Studies and Sciences Series (New York: Springer)

[20] Kordi K M G H 2000 *Cultivation of Crab and Milkfish in Pond Polyculture System* (Jakarta: Dahara Prize Publisher)

[21] Dewi W S 2015 Improvement soil biology characteristics of paddy field by system of rice intensification *Journal of Soil Science and Agroclimatology* **12 (2)**: 67-75.

[22] Effendi H 2003 *Water Quality Study for the Management of Resources and Water Environment*. (Bogor: Canisius Publisher).

[23] Edward, and Tarigan M S 2003 The season effects on fluctuations of phosphate and nitrate levels in the Banda Sea *Journal of Makara Science* **7 (2)**: 73-82.

[24] Sarangi R K 2011 Remote-sensing-based estimation of surface nitrate and its variability in the southern peninsular Indian waters *International Journal of Oceanography* **2011**:1-16.

[25] Rachmadiyanto A N 2014 Crop water productivity (CWP) with the small farm reservoir (SFR) as a supplemental irrigation for cash crop on rainfed area in Karanganyar Regency, Indonesia *Global Advanced Research Journal of Agricultural Science* **3(6)**:158-64.

Environmental heat stress enhances crystallization in urine

H Setyawan, Q C Pratiwi, I Sjarifah, T B Atmojo and Khotijah

Occupational Health and Safety Department, Faculty of Medicine, Universitas Sebelas Maret, Jl. Ir. Sutami No. 36A Kentingan, Surakarta, Central Java 57126 Indonesia

Abstract. Over the past several decades, agriculture and plantations have been used as the main livelihood of most of the Karanganyar residents. However, these two sources of living are now replaced by industrial areas that employ thousands of people in that district. The development of this industry triggers multiple environmental impacts, including ecosystem and temperature changes. In consequence, there is an increase in air temperature that can cause a variety of diseases, especially in the workplace. According to the International Labour Organization (ILO) data in 2013, one worker dies every 15 second due to a work accident and 160 workers are suffering from the occupational disease. In Indonesia, the incidence of crystallization in urine is actually still unknown, but it is estimated that there are 170,000 cases annually. A high temperature or called heat stress is one among many factors causing this disease to appear. The workers in the textile industry, especially in the Finishing Department Kusumahadi Co. Ltd that exposed heat stress from the finishing machines and inadequate ventilation. This hot working climate causes the human body to adapt in the form of body cooling mechanism or called sweating This adaptation can cause an increase in sweat production and decrease the production of urine. If it is not followed by consuming the recommended amount of water intake, it can result in the precipitation of body salts that, in a long time, will cause crystallization in urine. The research used the analytic observational designs for a cross-sectional study. There were 34 samples collected from 57 finishing workers. The data were analyzed using Spearman correlation test. The results showed that heat stress (p=0,015) and water intake (p=0,034) has a significant correlation with crystallization in urine.

1. Introduction

Karangnyar is a regency in the Indonesian province of Central Java. Considered as one of the fastest growing industrial areas, the development of the industry in this region triggers multiple environmental impacts, including ecosystem and temperature changes. As a result, there is an increase in temperature from year to year. Over the past several decades, agriculture and plantations have been used as the main livelihood of most of the Karanganyar residents. However, these sources of living are now replaced by industrial areas that employ thousands of people in that district. Of course, these developments have both positive and negative impacts on humans and the environment. The negative impact that can be clearly seen is the change in weather and temperature in this region. The old Karanganyar city, known for its cold temperatures, is now becoming a hot environment. On the other hand, there are several positive impacts such as an increase in the income of the community, the opening of job vacancy and the increasing of development in Karangnyar. One of the popular industries that recruit lots of workers in Karanganyar is a textile industry. The textile industries in Indonesia develop significantly and play an important role in the economic growth in the society. According to Statistic Center Bureau, the textile industry hires 1.47 million people in 2011. It increases by 4.78% of the total labor from the previous year (around 1.4 million people). According to

Content from this work may be used under the terms of the Creative Commons Attribution 3.0 licence. Any further distribution of this work must maintain attribution to the author(s) and the title of the work, journal citation and DOI.
Published under licence by IOP Publishing Ltd

the International Labour Organization (ILO) data in 2013, one worker dies every 15 second due to a work accident and 160 workers are suffering from occupational disease [1].

In the textile industry, the prevention of disease at work is very important as the company's effort to improve the work productivity, because if there are many workers suffer from disease complaints, the greater money and time will be wasted for the process and cost of treatment [2]. The production process on textile industry generates heat as the result of a series of production processes that comes from the equipment, ventilation, or human factor. Hot working climate causes the human body to adapt in the form of body cooling mechanism or called sweating [3]. Urine crystallization happened as a result of human excess dehydration. The relatively high of fluid discharge will affect the balance and concentration of body fluids, so the body fluid will decrease (dehydrated), followed by concentration and formation of urine crystals [4]. Currently, urinary infection becomes a hot issue as it affects millions of people in the world. In many countries especially America, Europe, and Asia, the data of total disease caused by urinary infection increase two times in the last three decades [5].

This study was conducted at Kusumahadi Santosa Corporate, Karanganyar. This company is engaged in the export of textile apparel and its production includes weaving, finishing, printing, and dying. Finishing department is where the fabric undergoes the refinement process from grey to ceramic color. In its every process, heavy machinery are used and heat dominates the room. Poor layout and ventilation cause air circulation to be hampered and then the working place becomes hotter. Based on the heat pressure measurement, it was found that the average heat pressure on finishing room was 30.65°C and it exceeds the maximum standard defined by the Ministry of Manpower. On the other hand, most workers complained about the heat in their working environment. They were often thirsty and having a high intensity of sweat and a small amount of urine. This study was aimed at determining the relationship between the heat pressure and water consumption with the urine crystallization in the workers from the finishing department at Kusumahadi Santosa Co. Ltd.

2. Subject and Method

This study was an analytic observational study with a cross-sectional approach. A simple random sampling technique was employed to collect the sample. This study was conducted in the finishing department at Kusumahadi Santosa Co. Ltd. There were totally 57 workers as the population of this study, whereas the samples used in this study were 34 respondents. The heat pressure was measured using a Quesstemp Heat Stress Monitor. The total water consumption was the average of water consumption in a day and a night by workers from the finishing department at Kusumahadi Santosa Co. Ltd. Urine crystallization indicated the existence of crystal in the worker's urine who were exposed to the heat pressure and less water consumption. The laboratory urine examination showed that there was a positive result with urine crystal in the form of calcium oxalate crystal, uric acid crystal, and amorphous phosphate crystal. The data were then analyzed using SPSS version 16.0 with correlation bivariate test of Spearman rank.

3. Result

This study was conducted in Karanganyar, Indonesia. The respondents in this study were the workers from the finishing department at PT. Kusumahadi Santosa. Table 1 showed the result of a measurement of four spots at the finishing department of PT. Kusumahadi Textile. The data showed that there were three spots that exceeded the maximum level of The Minister of Work and Transmigration Regulation Number 13 in 2011 (about 28°C). Table 2 showed the characteristics of the risk factor of worker's urine crystallization. Most of them showed that the respondents had a high risk of urine crystallization including age, work periods, heat pressure, and water consumption. Table 3 showed that heat pressure and water consumption had a close relationship with urine crystallization. Coefficient correlation level of heat pressure (r) was 0.415 as moderate correlation level. As there was a positive relation (+), it can be known that the higher the respondents received the heat pressure, the higher urine crystallization that they formed. On correlation test between water consumption, an obtained correlation coefficient level (r) of -0.352 showed that the relationship between each variable was low and indicated the opposite correlation between each variable (-). Therefore, it can be known that the higher the respondents consumed the water, the lower urine crystallization that they formed.

Table 1. Results of pressure heat measurement at 4 points location of finishing section.

Measurement Location	Wet ball in temperature index(°C)	Explanation
Point A	31.25	Exceeded the threshold value
Point B	31.66	Exceeded the threshold value
Point B	31.87	Exceeded the threshold value
Point B	27.85	Below the threshold value

Table 2. Distribution of respondents with crystallization in urine.

Variable	Category	Frequency	Percentage
Age	30 – 39	16	47.1
	≥ 40	18	52.9
Work period	5-9	15	44.1
	≥ 10	19	55.9
Body mass index	Thin	3	8.9
	Normal	25	73.5
	Fat	6	17.6
Heat pressure	≥ threshold value	27	79.4
	< threshold value	7	20.6
Water consumption	≤ 3 liter/day	30	88.2
	> 3 liter/day	4	11.8
Urine crystallization	Positive	17	50
	Negative	17	50

Table 3. Bivariate analysis of crystallization in urine using Spearman rank test

Risk factor for Crystallization in Urine	P value	Correlation coefficient (r)
Age	0.507	0.118
Work period	0.315	-0.178
Body mass index	0.760	0.054
Heat pressure	0.015	0.415
Water consumption	0.034	-0.352

4. Discussion

Age is one of the internal factors causing of urolithiasis 1 of 2 people per 1000 population [6]. The age of labors in this research starts from 30 years old. According to the results of this research, the percentage of the respondents whose their age ≥ 40 years are about 52.9%. The research sample is chosen from the 30-year-old worker because in that age, the body decreases its ability to protect itself from heat pressure. The ability to tolerate heat will decrease, as the workers get older. The condition is very real when the age reaches 40 years. In this age, the body's ability to cool down through the evaporation of sweat becomes slower [7]. It showed that the respondent of this research, in the finishing department of PT. Kusumahadi Santosa is observed from the age, the experienced reduction in heat stress protection, and the slower evaporation of sweat. It is the same with the research from Morocco showing that urolithiasis happens as the age increase to above 48 years old because the body resistance to adapt to the hot environment will slow down and decrease at that age. Thus, the prevalence of urine crystal incidence will increase because the body's ability of elderly people in returning body temperature to normal will be slower. [8]. The statistical test result on labor age variable shows that there is no significant relationship between age and the occurrence of urine crystallization, seen from the number of p-value = 0.507. In this study, it is known that age does not contribute to urine crystallization.

Respondents in this study are workers with the service period of 5 years. The period of work of respondents that mostly appear is $\geq$ 10 years, with a percentage of 55.1%. Longer working lives are likely to result in the accumulation of the effects of exposure to the hazard. Heat exposure on workers can generally be overcome naturally by the body through the process of acclimatization of the body to heat temperature. However, this form of body adjustment can also affect the normal functioning of the body, so that the body can slowly accumulate the negative effects of heat exposure such as an increase in expenditure [9]. Urinary tract stones tend to recur, with the average recurrence occurs 50% in 5 years and 70% in 10 years. So the identification of the first cause of the stone is important for the prevention of further kidney damage[10]. This shows that the respondents in this study had a 70% increased risk of urinary tract stones if they had already worked for 10 years. Based on the statistical test, the relationship between working period and urine crystallization shows that there is no significant relationship that can be seen from the value of p = 0.315. This means that the increase of urine crystallization can happen to the worker after working on the heat pressure with the working period of more than 15 years.

It is very difficult to determine that one of the causes of urinary crystal formation is a body mass index since diet can also affect the formation of urine crystals in the body[11]. In this study, 25 workers (73.5%) belong to in the normal nutritional status category, whereas three people (8.8%) are in the thin category, and 6 people (17.65%) are in the fat category. The production of heat in the body is related to the body weight and body mass. Overweight BMI may increase the risk of heat-related disorders (heat-related disorder) because the fat layer that they have is thicker than people with a normal BMI.

The symptoms may include light head, pounding, and dehydration. People are said to experience excess weight based on certain criteria that are determined by calculating Body Mass Index (BMI)[12]. Body mass index of most respondents in this study was in a normal status; hence, there were no risks of the formation of urine crystallization. Based on the statistical test results, there is no significant relationship between body mass index with urine crystallization. This can be seen with the value of p = 0.760 ($p > 0.05$). Therefore, it can be concluded that body mass index in this study does not contribute to the urine crystallization.

The pressure of heat in the workplace can be caused by various sources, including from the heat of the sun, inadequate ventilation and the use of heat removing apparatus. The research in China showed that workers exposed to the heat have a 4-fold risk than workers who are not exposed to the heat[13]. Based on the statistical test results, there is a significant relationship between heat pressure and urine crystallization. It can be seen from the value p = 0.015, with correlation strength (r) = 0.415. The interpretation of the test results showed that the higher the exposure to heat received by the workers, the higher the risk of urine crystallization. This is reinforced by the research conducted by Mulyani stating that workers exposed to high temperatures have a risk of urinary tract stones up to nine times greater than the other workers who are not exposed to heat[14]. As the workforce is under the influence of a hot working environment, the sweating rate becomes maximal. With this condition, the body will experience a lot of loss of mineral salts, causing the body to dehydrate[15]. The hot environment affects the amount of fluid and electrolyte of the body that is one of the factors of the stone formation. Physiologically, the mechanism of dehydration begins with the expenditure of sweat to cool the body. The relatively high fluid release will affect the fluid balance in the body. It is followed by urine concentration so that urine supersaturation will occur. Urine supersaturation is an important cause since there is an increase in the concentration of salts that triggers the formation of calcium crystals, uric acid, or phosphate[16]. It showed that the respondents in this study, who worked under the hot work environment, had increased the expenditure of sweat, dehydration, reduction of urine volume and marked with positive laboratory test results.

The average drinking water consumption of respondents was mostly ≤ 3 liters. Moreover, based on statistical test results, there is a significant relationship between drinking water consumption and urine crystallization. It can be seen from the value p = 0.034 with the value of correlation coefficient (r) of -0.352, which means the direction of the opposite correlation between variables. It can be concluded that consuming >3 liters of water can lower the formation of urine crystallization. Kidney stones can be formed due to the body's response to the environmental and/or metabolic risk factors [17] It is supported by the research conducted in French showing that the respondents who drink water more than 3 liters per day can reduce the incidence rate of crystals formed in the urethra by 55% [18]. The low urine volume and significant environmental factors, as well as the low fluid intake or direct fluid loss, will increase the risk of stones by increasing urine saturation from rock-forming salts. In addition, drinking water consumption is essential in the process of preventing urinary tract stones [19]. If a person lacks in drinking water, the supersaturation of urinary stone forming material can occur [20]. For people with a chronic dehydration, the pH of urine likely tends to fall. Meanwhile, the density of the urine rises. The supersaturation of uric acid rises and causes the attachment of uric acid crystals. A lack of water consumption results in a calcium deposition in the renal pelvis due to the incoming fluid imbalances. It results in a decrease in the amount of urine volume. The consumption of adequate drinking water can prevent urine from becoming concentrated or colorful [20].

5. Conclusion

There is a significant relationship between the heat pressure and drinking water consumption with urine crystallization. To prevent the crystallization of urine in the workers of finishing department, the company should take control in reducing the heat pressure exposed to labor by building the natural ventilation and adding the exhaust fan so that the indoor heat can flow outside the room. In addition, the availability of drinking water in the company is also very important to keep the labor hydrated and prevent the formation of urine crystallization. Moreover, a worker who has urine crystals in the urinary tract should immediately be examined at the nearest hospital for the immediate treatment in order to prevent the formation of kidney stones

Acknowledgement

The authors would like to thank the management and the workers of Kusumahadi Sentosa Corporate who have participated in this study.

References

[1] "Safety and health", http://www.ilo.org, 2016 [Online] Available: http://www.ilo.org/global/topics/safety-and-health-at-work/lang--en/index.htm. [Accessed: 08-08-2017]

[2] Setyawan H 2017 Risk Factors for carpal tunnel syndrome among food-packing workers in Karanganyar *Kesmas Natl Public Heal J.* **11(3)** pp 123–6

[3] Alzeer A H and Otair H A K Al 2014 Sweat chloride concentration in patients with heatstroke *J Taibah Univ Med Sci.* **9(1)** pp 50–3

[4] Abou-Elela A 2017 Epidemiology, pathophysiology, and management of uric acid urolithiasis: A narrative review *J. Adv Res.* **8(5)** pp 513-7

[5] Durner L, Bourdoumis A and Buchholz N 2016 Comptes rendus chimie metabolic syndrome and urolithiasis la lithiase urinaire dans le syndrome metabolique *Comptes rendus - Chim.* **19(11–12)** pp 1451–5

[6] Bycroft J 2016 The management of urolithiasis *Surgery* **34(7)** pp 352–60

[7] Raquel M, Vega A, Angel L, Torres P De, Carrasco J, José M, et al 2016 Prevalence of

IOP Conf. Series: Earth and Environmental Science **129** (2018) 012035 doi:10.1088/1755-1315/129/1/012035

urolithiasis in the 40 to 65-year-old Spanish population : The PreLiRenE study ☋ *Medica Clin.* **146(12)** pp 525–31

[8] Bouatia M, Benramdane L, Idrissi MOB and Draoui M 2015 An epidemiological study on the composition of urinary stones in Morocco in relation to age and sex *African J. Urol.* **21(3)** pp 194–7

[9] Shen D and Zhu N 2015 Influence of the temperature and relative humidity on human heat acclimatization during training in extremely hot environments *Build Environ.* **94** pp 1–11

[10] Ratu G, Badji A and Hardjoeno 2006 The analysis of urethral stone profile at the clinical pathology laboratory *Indones J. Clin Pathol Med Lab.* **12(3)** pp 114–7

[11] Bu B and Oguz U 2015 Influence of body mass index on pediatric urolithiasis *J. Pediatr Urol.* 11:35 e-1-350e-6

[12] Antonelli JA, Maalouf NM, Pearle MS and Lotan Y 2014 Platinum priority – stone Disease use of the national health and nutrition examination survey to calculate the impact of obesity and diabetes on cost and prevalence of urolithiasis in 2030 *Eur Urol.* **66(4)** pp 724–9

[13] Luo H, Turner LR, Hurst C, Mai H, Zhang Y and Tong S 2014 Science of the total environment exposure to ambient heat and urolithiasis among outdoor workers in Guangzhou, China *Sci Total Environ.* **472** pp 1130–6

[14] Mulyani R 2004 *Association Between Heat Exposure and Urolithiasis in Workers* Thesis Dissertation (Jakarta: Program Pendidikan Dokter Spesialis Kedokteran Okupasi)

[15] Nurullita U 2007 The effect of workload and physical environmental factors to the blood pressure and pulse of workers in arc furnace and rolling mill PT Inti General Yaja Steel Semarang *J. Kesehat Masy Indones.* **4(2)** 72–7

[16] Sja'bani M and S Bakri P R 2009 *Urinary Tract Stone In The Book Of Science Of Internal Medicine* 5th ed (Jakarta: Interna Publishing)

[17] Weinberg A E, Patel C J, Chertow G M and Leppert J T 2014 Diabetic severity and risk of kidney stone disease *Eur Urol.* **65(1)** 242–7

[18] Y Lotan, I Buendia Jiménez, I Lenoir-Wijnkoop, M Daudon, L Molinier, I Tack and M J C Nuijten Increased Water Intake as a Prevention Strategy for Recurrent Urolithiasis: Major Impact of Compliance on Cost-Effectiveness *The Journal of Urology* **189(3)** pp 935-4

[19] Jabbar F, Asif M, Dutani H, Hussain A, Malik A, Amjad M, et al 2015 Assessment of the role of general, biochemical and family history characteristics in kidney stone formation *Saudi J Biol Sci.* **22(1)** pp 65–8

[20] Tarplin S, Monga M, Stern K L, Mccauley L R, Sarkissian C and Nguyen M M 2016 Predictors of Reporting Success With Fluid Intake Among Kidney Stone Patients *Urology* **88** pp 49–56

IOP Conf. Series: Earth and Environmental Science **129** (2018) 012036 doi:10.1088/1755-1315/129/1/012036

Climate-sensitive urban design through Envi-Met simulation: case study in Kemayoran, Jakarta

K D Kusumastuty[1,3], H W Poerbo[2] and M D Koerniawan [2]

[1] Magister Student of Urban Design, Department of Architecture, School of Architecture, Planning and Policy Development, Institute of Technology Bandung, Jl. Ganesha No.10 Bandung 40132, West Java, Indonesia

[2] Lecturer, Department of Architecture, School of Architecture, Planning and Policy Development, Institute of Technology Bandung, Jl. Ganesha No.10 Bandung 40132, West Java, Indonesia

[3] Corresponding author: karinadwik@gmail.com

Abstract. Indonesia as a tropical country which the character of its climate are hot and humid, the outdoor activity applications are often disrupted due to discomfort in thermal conditions. Massive construction of skyscrapers in urban areas are caused by the increase of human population leads to reduced green and infiltration areas that impact to environmental imbalances and triggering microclimate changes with rising air temperatures on the surface. The area that significantly experiences the rise of temperature in the Central Business District (CBD), which has need an analysis to create thermal comfort conditions to improve the ease of outdoor activities by an approach. This study aims to design the Kemayoran CBD through Climate Sensitive Urban Design especially in hot and humid tropical climate area and analyze thermal comfort level and optimal air conditioning in the outdoor area. This research used a quantitative method by generating the design using Climate Sensitive Urban Design principle through Envi-met 4.1 simulation program to find out the value of PMV, air temperature, wind speed and relative humidity conditions. The design area considers the configuration of buildings such as the distance between buildings, the average height, the orientation of the building, and the width of the road.

1. Introduction

Indonesia, a tropical country with hot-humid climate, the outdoor space utilization is often disrupted by inadequate thermal conditions. Rapid urban growth and less attention to providing public spaces have impacted on urban microclimate [1]. The escalating of the human population in urban areas motivates a skyscraper massive construction which leads to cut down green areas and recharge that provoking environmental imbalance and triggering microclimate changes by the advancement of surface air temperature. Its existence can affect the thermal comfort, that initiate inconvenience thermal conditions in the surrounding circumstances and stabilizes it through the day [2,3].

Nevertheless, the Climate Sensitive Urban Design concept implementation in Indonesia is still not optimally developed, yet some other tropical cities in Asia have issued design, planning, and management to improve thermal comfort. This research was conducted by designing a Central Business District area through Climate Sensitive Urban Design approach in tropical climate consideration of physical and non-physical parameters and analyzing the outdoor thermal condition by simulation.

Content from this work may be used under the terms of the Creative Commons Attribution 3.0 licence. Any further distribution of this work must maintain attribution to the author(s) and the title of the work, journal citation and DOI.
Published under licence by IOP Publishing Ltd

IOP Conf. Series: Earth and Environmental Science **129** (2018) 012036 doi:10.1088/1755-1315/129/1/012036

2. Outdoor Thermal Comfort in Tropics

Based on ASHRAE Standard 55, thermal comfort is a condition of satisfaction with the surrounding thermal conditions and is assessed by subjective evaluation. It is also affected by the climate zone which is different in any region. People who lived in hot climates are more resistant to high temperatures but sensitive to low or cold temperatures, and vice versa [4]

Outdoor thermal comfort is an essential factor to measure the perception level of urban microclimate quality and take into design consideration which could improve understanding of sustainable development [5]. The main problem related to thermal comfort in the tropics is the cold and hot air movement [6]. He explains that thermo-physiological differences between indoor and outdoor spaces which differ by clothing type, activity level, and solar radiation.

Table 1. Classification of thermal comfort sensation [7]

PMV	PET	Thermal Sensation	Level of Heat Stress
- 3.5	14	Very cold	Extreme cold stress
- 2.5	18	Cold	Strong cold stress
- 1.5	22	Cool	Moderate cold stress
- 0.5	26	Slightly cool	Slight cold stress
0		Neutral	No thermal stress
+ 0.5	30	Slightly warm	Slight heat stress
+ 1.5	34	Warm	Moderate heat stress
+ 2.5	38	Hot	Strong heat stress
+ 3.5	42	Very hot	Extreme heat stress

3. Climate-Sensitive Urban Design

Climate-Sensitive Urban Design is defined as a process that takes into the basic elements of the microclimate (sun, wind, and temperature) for designing purposes [8]. Its application is not only beneficial to a specific site but also it's surrounding through appropriate planning and design. As a concept in a tropical climate, the availability of access to public spaces and pedestrian pathways improved quality become an aspect of tropical climatic conditions approach [9]. In other words, inter-building spaces must be designed sensitive to climate and utilized good accessibility persons.

A design strategy to improve the physical environment quality is a specific step to embody a thermally comfortable development area. There are two parameters that considered in designing responsive areas for micro-climate change, physical and non-physical parameters [9]. Urban geometry such as building average height, a gap within, building H/W ratio, and buildings configuration can affect climate conditions in the area [10]. The effect of buildings on wind conditions and on roads, as well as wind and solar conditions on pedestrian comfort, were examined to find out the strong relationship between urban form and climate, where the road position and the building dimensions affect the city climatic conditions [11]. There are the most frequently used factors in various studies for non-physical parameters, such as air temperature (° C-Celsius, ° F-Fahrenheit, ° K-Kelvin), relative humidity (%), wind speed (m/s, km/h, knots, and mph), and solar radiation (MJ/m^2)

The designs that respond to microclimate conditions have been formulated by urban designer experts with specific principles for each climate area worldwide to enhance the passive energy utilization. This principle can be applied on an urban scale especially in the tropics and can be divided into urban and building environment.

Table 2. Design principle of climate sensitive urban design in tropical climate

Urban Environment	Building Environment
The widening of streets along the prevailing wind direction is considered of high effectiveness, especially for large sites facing narrow urban canyon with building setback on each side of the street [12]	Compactly integrated developments and podium structures with full of large ground coverage on extensive sites which impeding air movement should provide setback parallel to the prevailing wind and reduce site coverage of podium to allow more open space at grade [12]
Buildings should be parallel or up to 30 degrees to the prevailing wind direction to maximize the penetration of prevailing wind through the district [12,13]	Permeability and shading are the main requirements of the building. All vertical surfaces and openings should be protected but not blocking the movement of the wind. Arrangement of landscape and architectural elements should be combined to create a comfortable indoor and outdoor environment [12]
The position of the road on the slope 30° and 60° from the wind direction can enlarge the movement of the wind [13]	Creating free space between buildings to maximize indoor and outdoor ventilation [12,14]
Outdoor space should be shaded from buildings and the availability of vegetation must not block free passage of air [14]	A varying height profile with strategic disposition of low-rise and tall buildings in the dense urban context can help instigate wind flowing throughout the area [12,15]
Cultivate vertical landscaping with vines on the wall or pergola [16]	Using ground cover materials with light and reflective colors can help reduce the urban heat island effect [12]
Waterfront area should be given to appropriate scale, height and disposition of building blocks along the waterfront to avoid blockage of sea/land breezes and prevailing winds [12]	The most suitable orientation for Indonesian territory especially Jakarta which located at 6° south latitude is 10-30 ° from the southern direction which can be applied to areas with latitude 5°-9° [9]
	To circulate the wind flow at the pedestrian level, sufficiently dimensioned corridors between podium should be assured. [15]
	The ratio of the width and length of the building to the optimum humid tropics is 1: 3. [17]

4. Methodology

The quantitative method was exploited in this research, which emphasized the data collection and analysis of information in numbers, collected both individual and group measuring instruments values, and stressed in comparing between groups or linking the individual or group factors within experiments, correlation studies, and surveys [18]. This method was taken by analyzing a simulation result using Envi-Met program to find out the amount of air temperature, wind speed, relative humidity, and PMV in the particular area.

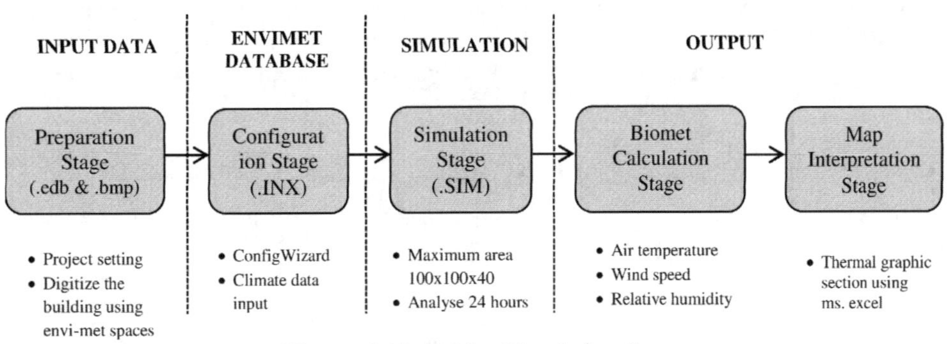

Figure 1. Envi-Met Simulation Stages

5. Study Area

The site was located in Gunung Sahari Utara, Sawah Besar District, Central Jakarta, in Pekan Raya Jakarta (PRJ) southern area 7.18 hectares of designed area and functioned as trading and services office zone with low Building Coverage Ratio (BCR). According to Panduan Rancang Kota Bandar Kemayoran 2014, the building intensity provisions has 71.892 m^2 of total area, 40% of BCR with a maximum FAR was 7, a maximum 48 floors of building height, and green plot ratio was 30%.

Table 3. Data Configuration of Envi-Met Simulation

Configuration Data	**February 15th**	**October 17th**
Boundary model	450m x 450m	450m x 450m
Grids	90x90	90x90
Start Simulation at Time (HH:MM:SS)	06:00:00	06:00:00
Total Simulation Time in Hours	24	24
Save Model State each? Min	30	30
Wind Speed in 10 m ab. Ground (m/s)	1,6	1,9
Wind Direction (0: N..90: E..180: S..270: W)	315° (+30°)	270° (+30°)
Roughness Length z0 at Reference Point	0,01	0,01
Initial Temperature Atmosphere (C)	24	29
Specific Humidity in 2500 m (g/kg)	7,0	7,0
Relative Humidity in 2 m	88	72
[BUILDING] Building Properties		
Inside Temperature [K]	293	293
Heat Transmission Walls [W/m^{2}K]	1,94	1,94
Heat Transmission Roofs [W/m^{2}K]	6	6
Albedo Walls	0,65	0,65
Albedo Roofs	0,75	0,75
[SOILDATA] Settings for Soil		
Initial Temperature Upper Layer (0-20cm) [K]	301	301
Initial Temperature Middle Layer (20-50cm) [K]	301	301
Initial Temperature Deep Layer (below 50cm) [K]	301	301
Relative Humidity Upper Layer (0-20cm)	60	60
Relative Humidity Middle Layer (20-50cm)	70	70
Relative Humidity Deep Layer (below 50cm)	70	70

The thermal comfort comparison was employed in outdoor space, the examination was done by developing two alternative designs using climate-sensitive design principle. The first alternative was a building configuration which was according to the building ideal shape in a tropical area with length and width ratios was 3: 1 [17] and the second one was a building arrangement which was based to the building typology, whether a residential, hotel, office or commercial building type.

IOP Conf. Series: Earth and Environmental Science **129** (2018) 012036 doi:10.1088/1755-1315/129/1/012036

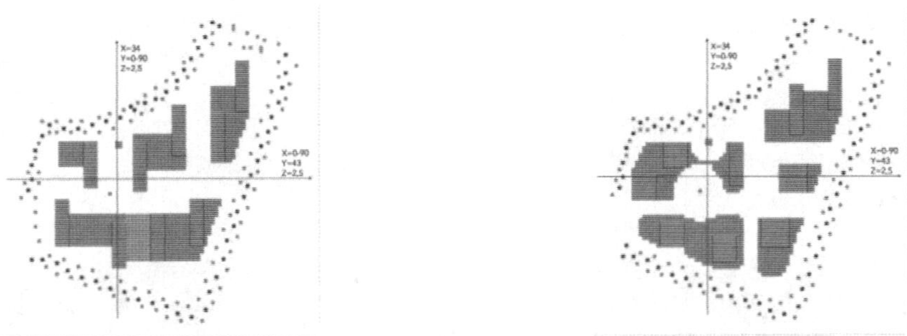

a. Design alternative 1 b. Design alternative 2

Figure 2. Site digitized through Envi-Met spaces features

6. Results and Discussion

6.1. Air Temperature

The air temperature at 5 pm in the afternoon was higher than 7 am in the morning and afternoon at 12 pm. The area of the plaza and public open space which were located at the border area at 2.5 meters of altitude to represent the pedestrian level that has shown the air temperature was quite similar as shown in the graph below which the line tends to be stable.

The air temperature result of both design alternatives at 07 am, on February 15th, 2016 as much as 34% of the area is mostly in 24.40°C - 24.45°C of air temperature, while the most area of 2nd alternative design was in 24.35°C - 24.40°C of the air temperature about 27%. On October 17th, 2016, as many as 41.17% of air temperature in 1st alternative design was in the range of 27.40 ° C - 27.60 ° C, while the 2nd design alternative the most air temperatures were in the same range as the 1st but as much as 43%. Based on the comparison, it can be seen that the alternative air temperature of the 1st alternative design was slightly lower than the 2nd.

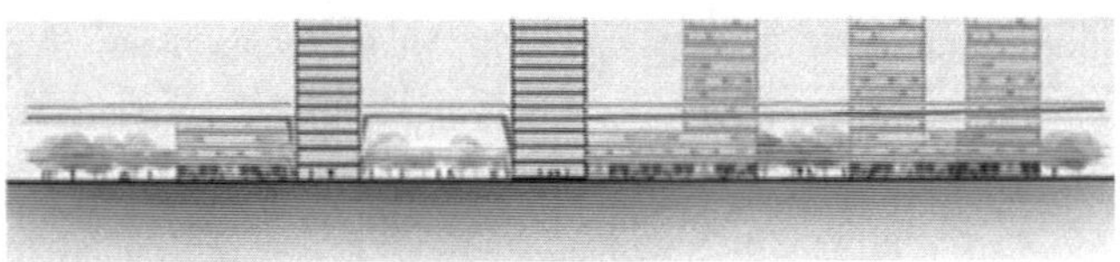

Figure 3. Graphic of air temperature design alternative 1

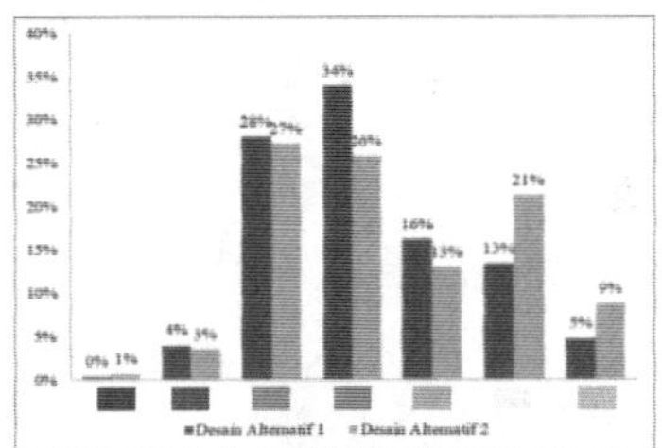

a. February 15th, 2016

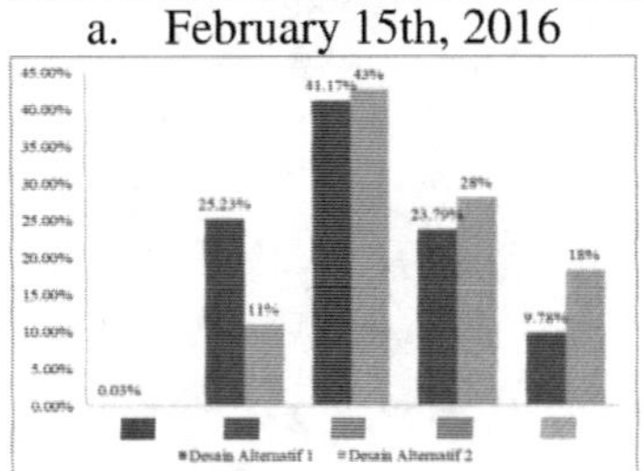

b. October 17^{th,} 2016

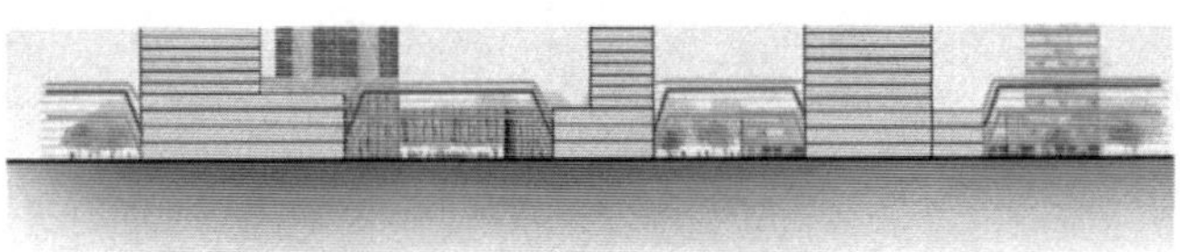

Figure 4. Graphic of air temperature design alternative 2

Figure 5. Simulation result of air temperature at 7 am

6.2. Wind Speed

Wind speed in the plaza area located was the middle of the area in the 1st alternative design that looks higher in the graph than other alternatives, can be observed in Figure 5 and 6 which shows the cut graph at all simulation times. Visible graphs increase in the plaza area which indicating 1 m/s of wind movement. On February 15th, the wind speed between 07 am, 12 pm and 5 pm was 0 - 2 m/s. And on October 17th, the wind speed in the central area of a wind condition design in the green space area was almost the same, with the wind power perceived by Beaufort scale largely quiet to a slight wind (0.28 – 3,32 m/s). Overall wind speed at 07 am in the morning also tends to be higher than the other two simulation time which was at 12 pm and 5 pm.

On October 17th, 2016, 18% of the wind speed in 1st alternative design was 1.25 m/s - 1.50 m/s, while in 2nd alternative design as much as 23% by the same wind speed range of 1.25 m/s - 1.50 m/s. Based on these comparisons, it appears that the wind speed on February 15th, design alternative 1 tends to be higher than the design alternative 2. While on October 17th, the 2nd has higher wind speed than design alternative 1.

Figure 6. Wind speed graphic of design alternative 1

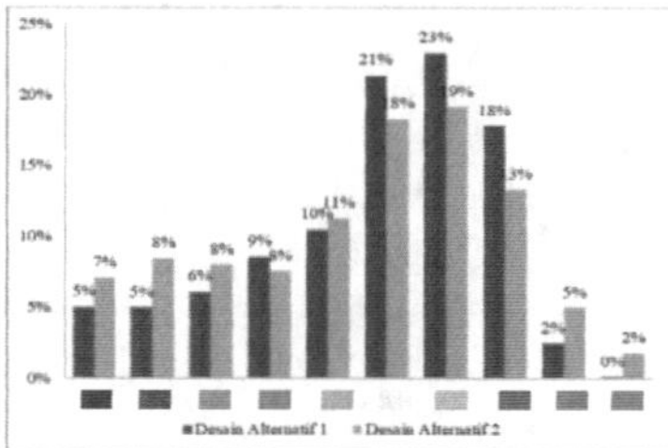

a. Design alternative 1

Figure 7. Wind speed graphic of design alternative 2

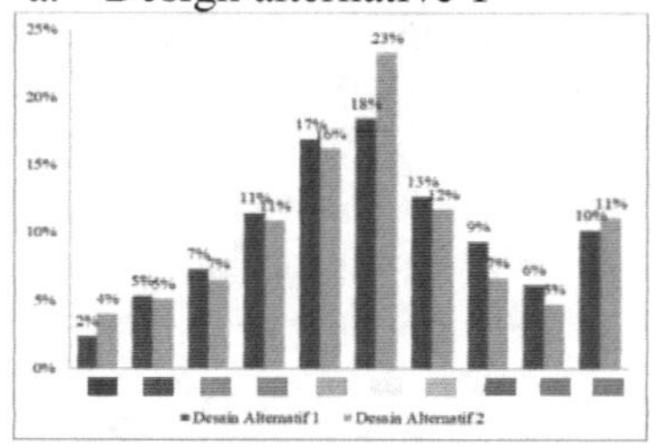

b. Design alternative 2

Figure 8. Simulation result of wind speed at 7 am

6.3. Relative Humidity

Relative humidity of the area at three specified times has a range of values between 50% - 80% of the highest relative humidity being at 07 am with 70% - 80% of humidity. The humidity were decreased by about 65% - 73% at 12 pm and at 5 pm the relative humidity became 54% - 63%.

The relative humidity comparison from both alternatives design can be seen that relative humidity at 07 am was higher than at 12 pm and 5 pm. The relative humidity value was lower in the afternoon and evening, just as wind speeds were getting lower at 12 pm and 5 pm. The relative humidity in 1st and 2nd alternative designs were measured at three times and it has a humidity between 54% - 77%.

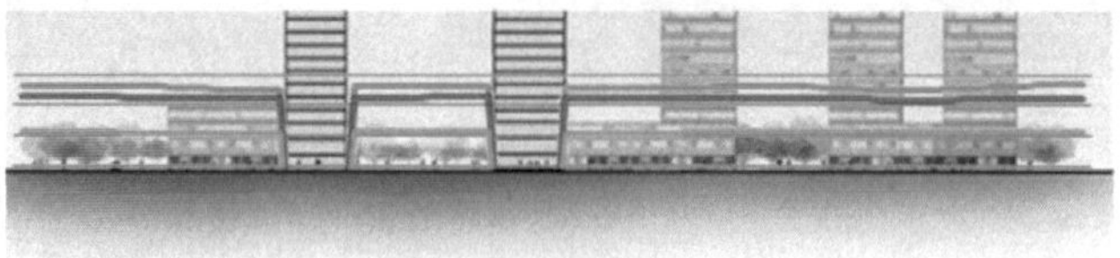

Figure 9. Relative humidity graphic of design alternative 1

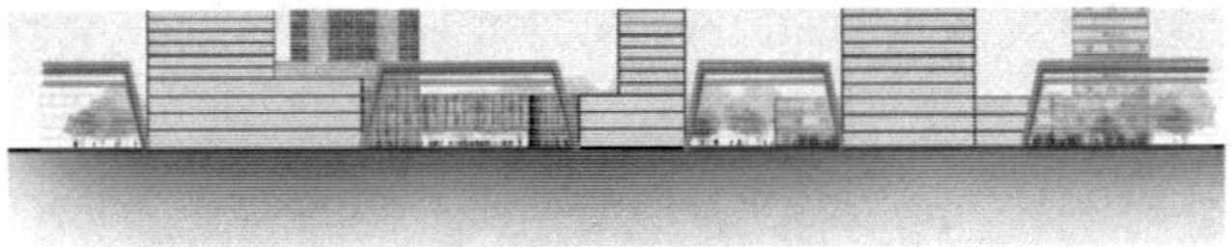

Figure 10. Relative humidity graphic of design alternative 2

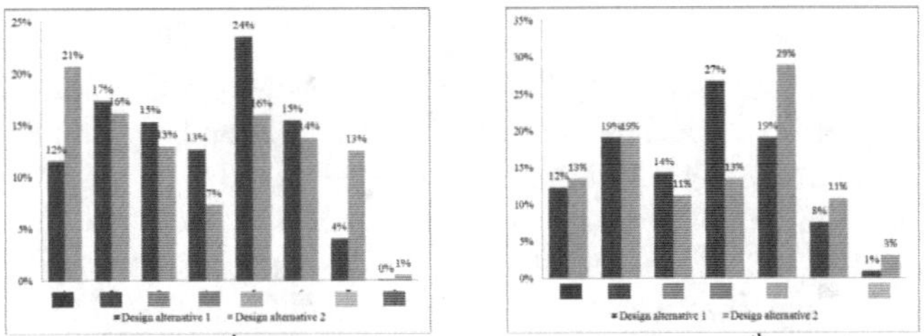

a. February 15th 2016 b. October 17th 2016

Figure 11. Simulation result of relative humidity at 7 am

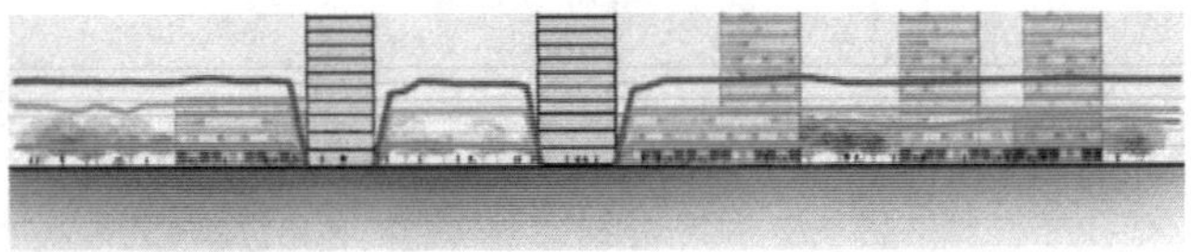

Figure 12. PMV graphic of design alternative 1

Figure 13. PMV graphic of design alternative 2

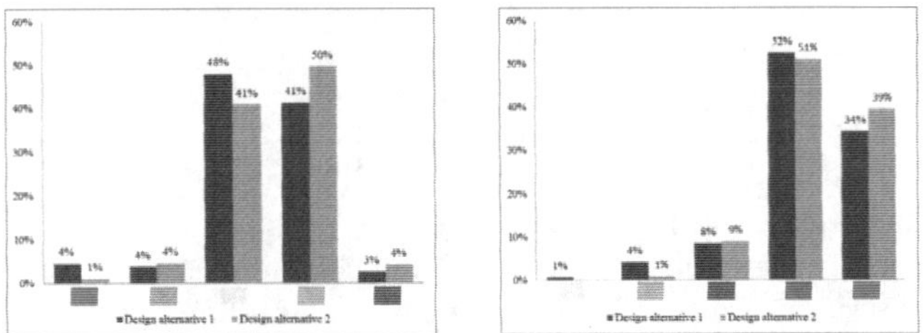

a. February 15th, 2016 b. October, 17th 2016

Figure 14. Simulation result of PMV at 12 pm

6.4. PMV

Based on the simulation results, the PMV condition at 07 am were in the range of +0,5 to + 2,5 by the thermal conditions from slightly warm to hot. While at 12 pm, the PMV rises to +3.5 were very hot conditions, and at 5 pm the condition is slightly more comfortable than at 12 pm specifically from +2.5 to +3.5.

The PMV comparison values at both alternatives design were analyzed by looking at the PMV value at 12 pm due to the value was highest compared to other times at 7 am and 5 pm, therefore, it can be used to see the thermal comfort level which felt by some people in hottest thermal condition.

Figure 13 shows that the PMV value at 1st alternative design tends to be lower than the 2nd. The values were indicated by the minimum and maximum value either on February 15th or October 17th.

6.5. Surface Area to Volume Ratio

Surface area to volume (S/V) ratio was one of the important factors that determined the heat transfer. The greater surface area affected the heat to received and released faster in the building, so the minimum value of S/V ratio imposes the minimum heat transfer [19].

After calculating the S/V value of each building block and the total areas, the value of 1st alternative design has a value of 0.123, higher than the 2nd design which has a value of 0,015. It means that buildings at 2nd alternative design received less heat from outside, therefore the thermal conditions in the building will be lower than the 1st design.

6.6. Solar Radiation Analysis

To find out the level of solar radiation in the area through 1 year, Sun Hours program was used to calculate the solar exposure which was indicated by a color grid that represents the intensity level. The darker the color of the area, the more exposed to solar radiation in 1 year.

The comparison results in Figure 15 show about 60% - 80% both of the area was exposed to solar radiation or more than 2373 hours per year, while the 2nd alternative design has a higher solar radiation intensity than the 1st, therefore it was cooler than the 2nd alternative.

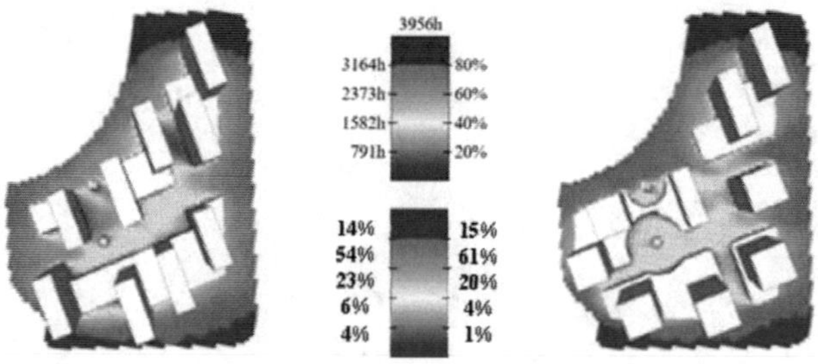

Figure 15. Solar radiation analysis

7. Conclusion

Based on the analysis, thermal conditions in Jakarta in the afternoon on 5 pm were the highest air temperature rather than 7 am and 12 pm. It was caused by the buildings material that stores heat from noon to evening and just released at night. Wind speed has an important role for improving thermal comfort in outdoor space in Jakarta compared to building shading based on solar radiation analysis as the solar radiation level result in outdoor were quite high about 60-80% for 1 year, through the site.

A building configuration with a shape ratio of 1: 3 according to the tropics was more optimally thermal outdoor space than the building configuration with typology which based on shading and air movement analysis.

Acknowledgement

The author would like to thank the master student of SAPPK ITB for basic guidance about using Envi-Met simulation and Institut Teknologi Bandung as fund donators in support this research.

References

[1] Salata F, Golasi L, Vollaro RdL and Vollaro AdL 2016 Urban microclimate and

outdoor thermal comfort: a proper procedure to fit ENVI-met simulation outputs to experimental data *Sustainable Cities and Society* **26** pp 318-43

[2] Koerniawan M D and Gao W 2015 Investigation and evaluation of thermal and walking comfort in hot-humid climates case study: the open spaces of Mega Kuningan-superblock in Jakarta *International Journal of Building, Urban, Interior, and Landscape Technology* **6** pp 53-72

[3] Koerniawan M D and Gao W 2015 Thermal comfort investigation in three hot-humid climate theme parks in Jakarta *American Journal of Environmental Sciences* **11(3)** pp 133-44

[4] Koch-Nielsen H 2002 *Stay cool: a design guide for the built environment in hot climates* (London: Earthscan) pp 32-65

[5] Coccolo S, Kampf J, Scartezzini JL and Pearlmutter D 2016 Outdoor human comfort and thermal stress: a comprehensive review of models and standards *Urban Climate* **18** pp 33-57

[6] Emmanuel R 2016 *Urban climate challenge in the tropics: rethinking planning and design opportunities* (London: Imperial College Press) ISBN 9781783268405

[7] Lin T and Matzarakis A 2008 Tourism climate and thermal comfort in Sun Moon Lake, Taiwan *International Journal of Biometeorology* **49** pp 236-42

[8] Tapias E and Schmitt G 2014 Climate-sensitive urban growth: outdoor thermal comfort as an indicator for the design of urban spaces *The Sustainable City IX: Urban Regeneration and Sustainability* **1** pp 623-34

[9] Emmanuel R 2005 *An urban approach to climate-sensitive design* (London and New York: Spon Press Taylor & Francis Group) pp 2-136

[10] Johansson E 2006 Influence of urban geometry on outdoor thermal comfort in a hot dry climate *Building and Environment* **41** pp 1326-38

[11] Bosselmann P EAKDRW 1995 Urban form and climate: case study, Toronto *Journal of the American Planning Association* **61** pp 226-39

[12] The Government of the Hong Kong Special Administrative Region, Planning Department 2015 *Hong Kong Planning Standards and Guidelines Chapter 11 Urban Design Guidelines* Hong Kong

[13] Givoni B 1994 Urban design for hot humid regions *Renewable Energy* **5** pp 1047-53

[14] Konya A 1980 *Design primer for hot climates* (London: The Architectural Press Ltd) pp 33-58

[15] Department of Urban Planning and Spatial Design 2013 *Adapt HCMC: Guidelines on Climate Change Adapted Urban Planning and Design for Ho Chi Minh City / Vietnam. Megacity Research Project* (Brandenburg: Brandenburg University of Technology Cottbus) Report ISBN 978-3-00-042085-6

[16] Sandifer S A 2009 Using the landscape for passive cooling and bioclimatic control: applications for higher density and larger scale *26th Conference on Passive and Low Energy Architecture*

[17] Egan MD 1975 *Concepts in thermal comfort* (New Jersey: Prentice-Hall)

[18] Creswell JW 2008 *Educational research: planning, conducting, and evaluating quantitative and qualitative research* (New Jersey: Pearson Education, Inc) pp 2-27

[19] Besh B 2002 Building form as an option for enhancing the indoor thermal conditions *6th Nordic Symposium* **18** pp 759-66

IOP Conf. Series: Earth and Environmental Science **129** (2018) 012037 doi:10.1088/1755-1315/129/1/012037

A preliminary study of mechanistic approach in pavement design to accommodate climate change effects

S R Harnaeni[1,3,4], F P Pramesti[2], A Budiarto[2] and A Setyawan[2]

[1]Doctoral Program in Civil Engineering, Faculty of Engineering, Universitas Sebelas Maret, Jl. Ir. Sutami 36A Kentingan, Surakarta Central Java 57126 Indonesia
[2]Civil Engineering Department, Universitas Sebelas Maret, Jl. Ir. Sutami 36A Kentingan, Surakarta Central Java 57126, Indonesia
[3]Civil Engineering Department, Universitas Muhammadiyah Surakarta, Jl. A. Yani Pabelan Kartasura Tromol Pos 1, Surakarta, Central Java 57102 Indonesia
[4]Corresponding author: srh289@ums.ac.id

Abstract. Road damage is caused by some factors, including climate changes, overload, and inappropriate procedure for material and development process. Meanwhile, climate change is a phenomenon which cannot be avoided. The effects observed include air temperature rise, sea level rise, rainfall changes, and the intensity of extreme weather phenomena. Previous studies had shown the impacts of climate changes on road damage. Therefore, several measures to anticipate the damage should be considered during the planning and construction in order to reduce the cost of road maintenance. There are three approaches generally applied in the design of flexible pavement thickness, namely mechanistic approach, mechanistic-empirical (ME) approach and empirical approach. The advantages of applying mechanistic approach or mechanistic-empirical (ME) approaches are its efficiency and reliability in the design of flexible pavement thickness as well as its capacity to accommodate climate changes in compared to empirical approach. However, generally, the design of flexible pavement thickness in Indonesia still applies empirical approach. This preliminary study aimed to emphasize the importance of the shifting towards a mechanistic approach in the design of flexible pavement thickness.

1. Introduction

Road repairs which are performed on regular basis due to the prevalence of short-term damage are high cost [1]. Road damage is caused by some factors, including climate changes, overload, and inappropriate procedure for material and construction process. Climate change is a phenomenon which cannot be avoided, caused by natural process or human activities. Carbon dioxide which is continuously produced and released into the air will damage the ozone layer in the atmosphere and will lead to the occurrence of global climate change. The effects of climate change include air temperature rise, sea level rise, rainfall changes, and higher intensity of extreme weather phenomena [2]. Several studies confirmed the link between climate changes and road damage [3,4,5,6,7,8,9,10,11,12].

To anticipate the occurrence of road damage, several measures should be devised during the road planning and construction, in order to save the cost of road maintenance. In addition, the recent road construction design must also involve the longer design life and higher capacity to hold load and weather effects. The attempt to design longer design life is in accordance with the concept of long life pavement that has been applied for several years in Europe and USA. The Forum of European

Content from this work may be used under the terms of the Creative Commons Attribution 3.0 licence. Any further distribution of this work must maintain attribution to the author(s) and the title of the work, journal citation and DOI.
Published under licence by IOP Publishing Ltd

National Highway Research Laboratories defines long life pavement as a type of pavement which will be no deterioration formed in the foundations or the road base layers since the correct surface maintenance is carried out [13]. The concept of long life pavement aims to obtain long life in order to get lower life-cycle costs. These life-cycle costs include the costs of construction, maintenance, rehabilitation, user delay, vehicles maintenance, and the cost spent to restore the impacts on the environment [14].

The design of flexible pavement thickness can apply three approaches, namely mechanistic approach, empirical approach and mechanistic-empirical (ME) approach [15,16,17,18]. According to the European Asphalt Pavement Association, the mechanistic approach have been widely applied in the design of flexible pavement thickness [19]. The advantages of this approach have been recognized, including its efficiency and reliability in the design of flexible pavement thickness, in addition to its capacity to accommodate climate change effects in compared to the empirical approach [20]. However, the design of flexible pavement thickness in Indonesia still applies the empirical approach, instead of the mechanistic approach. Therefore, this preliminary study aimed to reveal the importance of the shifting toward the utilization of mechanistic approach in the design of flexible pavement thickness.

2. Climate change

Climate is defined as the regular weather condition (temperature, rain, pressure wind, humidity) at a particular area. Weather is the day to day manifestation of this climate [8]. Climate is subject to change periodically due to both the natural and manmade factors, whether at the global, regional and local level. In all cases, changes in climate are generally unpredictable in which climatologists have to observe and take a period of time, approximately 50 years, in prior to establishing a trend for a particular jurisdiction. Nevertheless, since the second half of the last century, they have identified significant and extreme climatic events that tend to suggest the departures from events that were usually observed in the past. Given that these departures are attributed to an increase in human activities that lead to climate change, it is projected that even more profound climatic events will occur if human activities are continually uncontrolled [21].

A primary cause triggers the climate change is the greenhouse effect which is caused by greenhouse gases. These gases are generated from the natural and man-made sources, such as carbon dioxide, methane, and nitrous oxides, which accumulated in the atmosphere. Furthermore, over the past century, the anthropogenic sources of greenhouse gas emissions have substantially increased particularly since the pre-industrial era. Consequently, the impact of the greenhouse effect is greater than it should be and it contributes to the gradual heating of the Earth's surface [21]. Further investigation on anthropogenic induced climate change revealed that human activities, such as fossil fuel use and land use change, are likely to produce a positive radiative forcing, warm the earth's surface and trigger other changes in climate [22]. Lately, some of the changes have been identified while others have been estimated to occur in the future. Among those changes are the rise in global mean sea level, the decrease in the extent of ice and snow cover, and the rise of temperature in the lowest eight kilometers of the atmosphere during the past four decades [23,24].

Climate change also potentially leads to an increase in precipitation that will contribute to floods [25]. An increase in precipitation will also cause an increase in soil moisture resulting in slope instability and landslides [26].

The Transportation Research Board (TRB) issued Special Report 290 which stated that climate change will have significant impacts on how the transportation professionals will plan, design, construct, operate and maintain infrastructure in general. Subsequently, it lists the five major impacts of climate changes on transportation, namely, the increase of very hot days and heat waves, the increase of Arctic temperatures, the sea level rise, the increase in intense precipitation events, and the increase in hurricane intensity. Based on this list, potentially the greatest impacts of climate change on North America's transportation system are identified as flooding of coastal roads, railways, transit systems and runways as a result of global sea level rise, coupled with storm surges and exacerbated in some locations by land subsidence [27].

In association with the impact of climate change on transportation at the regional level, the United States' Gulf Coast identified the four key "climate drivers" in the Gulf region, namely, rising temperatures, changing precipitation patterns, rising sea levels and increasing storm intensity. The findings of the study showed that the highways, pipelines, ports, rail lines and airports at the study site will suffer severe damages when extreme climate changes at regional level occur [28].

Climate change is already having an influence on UK highways, for examples drier summers causing more incidences of subsidence in South East England and wetter winters creating a greater frequency of flooding. There are also less obvious effects on pavement deterioration with increased average temperatures and changes in rainfall patterns. Nevertheless, the impacts of climate change are not experienced uniformly across the UK. In the South East of England, high summer temperatures and drought are the major concerns, while in Scotland, intense rainfall and storms have more impacts [8].

3. The impacts of climate change on pavement damage

3.1. Environmental impacts on pavement performance

Environmental conditions (weather and climate) have a significant impact on pavement design and performance. Environmental conditions combined with factors such as traffic-related loads, construction methods, constituent layer materials and maintenance and rehabilitation regimens, are the key variables in the assessment of pavement performance [29]. The interaction between those variables and their effects on pavement performance are shown in Figure 1.

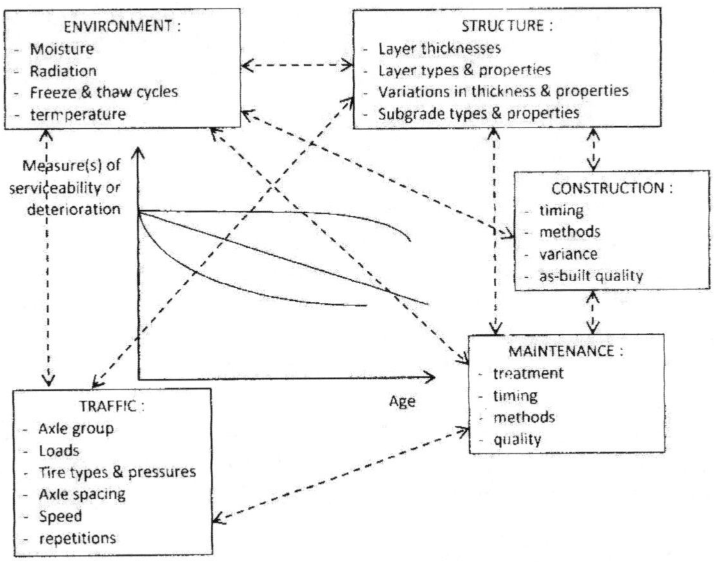

Figure 1. Factors affecting road performance [29].

3.2. Overview of the Impact of Climate Change on Pavement Damage

Climate change is a phenomenon which cannot be avoided. There are three main factors of climate change that influence the infrastructure of road transportation, namely: the increase of temperature, the increase of rainfall, and the rise of sea level [3].

3.2.1. The Increase in Temperature. The increase in temperature is potential to cause the road pavement construction damage. The temperature will cause the material of road pavement to expand at a very high temperature and to shrink at low temperature. This change leads to the susceptibility of the road from fatigue, which eventually causes the pavement layer to become peeled off, thinner, broken, cracked, bulging, perforated, and the aggregate material to become scrapped off easily. Further evidence from the study at 17 sites located in Southern Canada indicated that permanent rutting on asphalt roads, longitudinal cracking, and alligator issues would be exacerbated by climate change with transverse cracking becoming less of a problem [7].

3.2.2. The Increase of Rainfall. The trend of increased duration and intensity of rainfall will influence the stability of road construction. Several results from previous researches show that rainfall is potential to cause the road infrastructure damage. Soil erosion caused by rain run-off often happens, on the main roads and as a consequence, they cannot function appropriately. Furthermore, additional rainfall resulting in flooding and lane closures is also a potential engineering vulnerability [5].

3.2.3. The Rise of Sea Level. The rise of the seawater causes a tide. This tide can damage road which is near the coast and damage the existing road infrastructure. The road construction which is submerged in water will weaken due to the impact of water, which causes the decrease of bound between aggregate and asphalt and thus hinders the road construction process.

The report "The Changing Climate: Impact on the Department for Transport" stated the key implications of climate change for highway maintenance, which were identified as increased risk of flooding from rivers, sea, and inadequate drainage; deterioration and damage to highway structures from subsidence, heave and high temperatures; and increased road safety problems as a result of adverse driving conditions and deterioration of infrastructure [4]. Subsequently, the London Climate Change Partnership listed potential impacts caused by climate change, including carriageway rutting, embankment subsidence, deterioration of concrete, the problem with expansion joints and reduction in skid resistance [6].

Several recommendations related to climate change are provided to minimize pavement damage in the UK, that are: to undertake a risk assessment of the effects of climate change on structures, drainage and highway management; to prioritize development of guidance for highway authorities into managing the risks of climate change; and to encourage the asphalt industry to develop solutions to the problem caused by climate change [8]. All pavement types are susceptible to deterioration given a potential change in climate occurs [9][20]. From distresses at the surface to collapse of constituent layers, pavements are likely to undergo drastic deformation should they experience extremes in weather or climate. Flexible pavements under the same conditions are affected by bleeding, weathering, bumps, rutting, depressions, potholes, longitudinal and transverse cracking [11].

The research of road damage in Pacitan district, East Java also shows that climate changes (temperature, rainfall, humidity) have impacts on road damage. The data were taken by collecting data on temperature, rainfall, humidity, and road damage from 2002 to 2011 [12].

4. The design of flexible pavement thickness

The design of flexible pavement thickness can apply three approaches, namely the mechanistic approach, empirical approach and mechanistic-empirical (ME) approach [15][16][17][18]. An empirical approach is an approach based on the testing/trial or repeated experiences. Empirical equity is employed to connect the phenomenon being observed (pavement characteristics) and the output (pavement performance). Mechanistic approach is the approach which explains material phenomena (stress, strain) only refers to physical causes (traffic loading and environmental conditions). The relationship between phenomena and physical causes is explained by using mathematics model. Mechanistic-Empirical (ME) pavement design is a method that combines mechanistic models by observing the primary response of pavement in terms of stresses, strains, and displacement and empirical models, and then relating the generated response with pavement performance.

In Indonesia, the empirical method generally used to design flexible pavement thickness is SNI 1732-1989-F Method (Component Analysis Method). This method is based on AASHTO 1972. The modification has been done to adapt it to the condition of nature, environment, the basic nature of the soil, and the types of pavement layer in Indonesia. Then this method is revised as Design Manual of Flexible Pavement Thickness Pt T-01-2002-B follows AASHTO 1993. Only national and arterial roads which are designed based on mechanistic-empiric pavement design method, namely Design Manual of Road Pavement 02/M/BM/2013.

The California Department of Transportation (Caltrans) has changed the implementation of an empirical method to Mechanistic-Empirical (ME) Pavement Design Method. In 2005, Caltrans approved the adoption of Mechanistic-Empirical (ME) Pavement Design Method, an issue memorandum that calls for ME pavement design methodology to replace the existing empirical methods. There are many issues of importance to Caltrans and local governments that empirical methods cannot adequately consider [17].

According to European Asphalt Pavement Association, nowadays there are many types of research using the mechanistic approach in the design of flexible pavement thickness [19]. One of the advantages of mechanistic approach is more efficient and reliable in the design of flexible pavement thickness and relatively accommodates climate change is compared to empirical approach [20]. In mechanistic approach, modulus of elasticity of each layer will vary according to the climatic conditions. The subgrade and aggregate base modulus will change in accordance with moisture content and whether or not it is frozen or thawed [20]. Figure 2 shows how the layer modulus will change according to the climatic conditions.

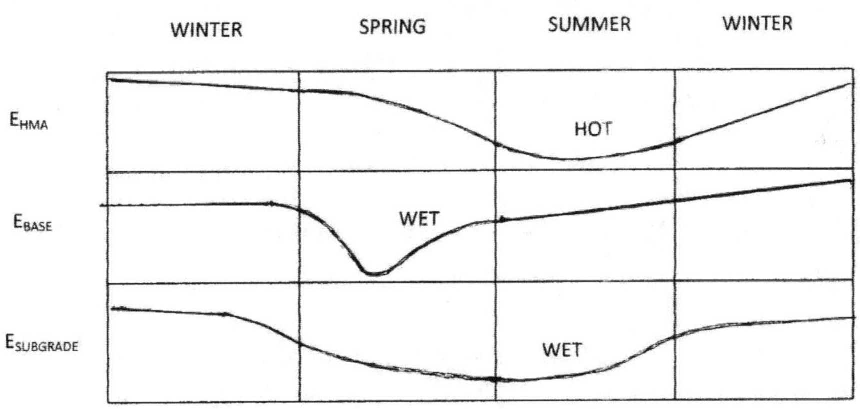

Figure 2. Modulus values changing based on climatic conditions [20].

Several researchers have applied Mechanistic-Empirical (ME) Pavement Design Method in their studies. Li *et al* (2011) used the concept of local calibration for MEPDG prediction models, a framework to incorporate the climate change effects into the mechanistic-empirical-based pavement design was developed by calibrating the coefficients performance models [30].

Federal Highway Administration reported the results of an evaluation of climate data from Modern-Era Retrospective Analysis for Research and Applications (MERRA) for further use in the Long-Term Pavement Performance (LTPP) Program and for other infrastructure applications. MERRA data were compared against the best available ground-based observations both statistically and in terms of effects on pavement performance as predicted using the Mechanistic-Empirical Pavement Design Guide (MEPDG). These analyses included a systematic quantitative evaluation of the sensitivity of MEPDG performance predictions to variations in fundamental climate parameters [31].

The most critical input parameters in terms of their effects on pavements damage and their influence on the determined number of corrective maintenance cycles to be performed during the design life of pavements using Empirical-Mechanistic Pavement Design Guide (MEPDG) method for flexible was identified by Getso (2016). The MEPDG method is used because it is the most comprehensive method that incorporates the effects of both truckloads and climate. The results showed that for flexible pavements, obtaining average monthly temperatures is critical because of the large effect temperatures have on the number of corrective maintenance cycles [32].

5. Conclusion

Nowadays, the current researchers on the design of flexible pavement thickness had already employed the mechanical approach. Previous studies also reported the advantages of mechanistic approach or mechanistic-empirical (ME) approach as well as its higher efficiency and reliability in the design of flexible pavement thickness. In addition, it relatively accommodates the impact of climate change in compared to the empirical approach. However, the design of flexible pavement thickness in Indonesia rarely applies this approach, instead, it is generally constructed based on the empirical approach. Therefore, a discourse on the importance of the shifting towards the using of mechanistic approach is required.

Acknowledgement
The authors gratefully acknowledge the funding support by Universitas Muhammadiyah Surakarta.

References

[1] Bina Marga 2015 *Strategic Plan of Direktorat Jenderal Bina Marga 2015-2019* (Jakarta: Bina Marga)
[2] Ridwan and Chazanah N 2013 Handling of global climate impacts on railway through mitigation and adaptation approaches *Journal of Theory and Application on Civil Engineering* **20(2)** 133-42
[3] Andrey J and Mills B N 2003 *Chapter 9-Climate Change and the Canadia Transportation System: Vulnerabilities and Adaptations. In: Weather and Transportation in Canada.* (Ontario: Department of Geography University of Waterloo)
[4] Department of Transport 2004 *The Changing Climate : Impact on the Department for Transport* (UK: Departement for Transport)
[5] GNB "Government of New Brunswick" 2005 "Transportation News Release: Roads closed due to flooding" http://www.gnb.ca/cnb/news/tran/2005e0556tr.htm [Accessed: 09-Sep-2017]
[6] London Climate Change Partnership 2006 *Adapting to Climate Change-Lessons for London* (London: London Climate Change Partnership)
[7] Mills B N, Tighe S, Andrey J, Smith J, Parm S, and Huen K 2007 *The Road Well-traveled: Implications of Climate Change for Pavement Infrastructure in Southern Canada report prepared for Government of Canada Climate Change Impacts and Adaptation Program* (Canada: University of Waterloo)
[8] Willway T, Baldachin L, Reeves S, and Harding M 2008 *The Effects of Climate Change on Highway Pavements and How to Minimise them : Technical report* (UK: Published Public Report 184)
[9] Meyer M, Amekudzi A, and O'Har J P 2010 Transportation asset management systems and climate change in transportation research record *J. Transp. Res. Board* **2160** 12-20
[10] Meyer M, and Wiegel B 2011 Climate change and transportation engineering, preparing for a sustainable future *J. Transp. Eng.* **137** 393-403
[11] Baladi G Y 1990 *Highway Pavement (NHI Course No. 13114* (McLean VA: FHWA)
[12] Rendrarini D 2014 *The Effect of Climate Change to Road Damage in Pacitan District*

(Surakarta: UNS)
[13] FEHRL 2004 *A Guide to the Use of Long-Life Fully-Flexible Pavements* (Belgium: FEHRL)
[14] Bina Marga 2013 *Design Manual of Road Pavement No. 02/M/BM/2013* (Jakarta: Bina Marga)
[15] Read D and Whiteoak J 2003 *The Shell Bitumen Handbook* (Thomas Telford Publishing: London)
[16] Sukirman S 2010 *Structural Thickness of Flexible Pavement Planning* (Bandung: Nova)
[17] Harvey I, John and Basher 2011 California's transition to mechanistic-empirical pavement design *Technol. Transf. Progr* **3(1)** 1–12
[18] Hardiyatmo H C 2015 *Pavement Planning and Soil Investigation, Second edition* (Yogyakarta: UGM Press)
[19] EAPA 2007 *Sustainable Roads - Long-Life Asphalt Pavements Version for "Bankers"* (Belgium: EAPA)
[20] Newcomb D, and Timm D E 2001 Mechanistic pavement design *Hot Mix Asphalt Technology*. September/October 2004 49-51
[21] IPCC 1990 *Climate Change, the IPCC Scientific Assessment* Edited by J T Houghton, G J Jenkins and J J Ephraums (Cambridge: Cambridge University Press)
[22] IPCC 1995 *The Science of Climate Change, Contribution of Working Group I to the Second Assessment Report of the Intergovernmental Panel on Climate Change* Edited by J T Houghton, L G Meira Filho, B A Callander, N Harris, A Kattenberg and K Maskell (Cambridge: Cambridge University Press)
[23] IPCC 2001 *Climate Change 2001 : The Scientific Basis. Contribution of Working Group I to the Third Assessment Report of the Intergovernmental Panel on Climate Change* Edited by Houghton J T, Y Ding, D J Griggs, M Noguer, P J van der Linden, X Dai, K Maskell (Cambridge: Cambridge University Press)
[24] IPCC 2007 *Summary for Policymakers. In : Climate Change 2007 : The Physical Science Basis. Contribution of Working Group I to the Fourth Assessment Report of the Intergovernmental Panel on Climate Change* Edited by Solomon S, D Qin, M Manning, Z Chen, M Marqu (Cambridge: Cambridge University Press)
[25] Clean Air Partnership 2006 "A Scan of Climate Change Impacts on Toronto" http://www.cleanairpartnership.org/pdf/climate_change_scan.pdf [Accessed: 09-Sep-2017]
[26] AIT "Alberta Infrastructure and Transportation" 2007 *The Vulnerability and Adaptation of the Transportation Sector to Climate Change* (Alberta: AIT)
[27] Transportation Research Board 2008 *Potential Impacts of Climate Change on US Transportation. Transportation Research Board Special Report 290* (Washington: TRB)
[28] U.S. Climate Change Science Program 2008 *Impacts of Climate Change and Variability on Transportation Systems and Infrastructure: Gulf Coast Study, Phase I. Department of Transportation* (Washington: Department of Transportation)
[29] Haas R, Falls L C, MacLeod D, and Tighe S 2004 *Climate Impacts and Adaptations on Roads in Northern Canada, Cold Climates Conference* (Regina: Cold Climates Conference)
[30] Li B Q, Mills L, and Mcneil S 2011 *The Implications of Climate Change on Pavement Performance and Design* (Delaware: UD-UTC)
[31] Federal Highway Administration 2015 *Evaluation of LTPP Climatic Data for Use in Mechanistic-Empirical Pavement Design Guide Calibration and Other Pavement Analysis* (McLean VA: FHA)
[32] Getso T 2016 *Mechanistic-Empirical Pavement Design Guide (MEPDG) Method Implemented to Estimate Damage in Flexible and Rigid Pavement* (New York: City University of New York)

IOP Conf. Series: Earth and Environmental Science **129** (2018) 012038 doi:10.1088/1755-1315/129/1/012038

The precautionary principle in fisheries management under climate change: How the international legal framework formulate it?

E Latifah[1,3] and M N Imanullah[2]

[1] International Law Department, Faculty of Law, Universitas Sebelas Maret, Jl. Ir. Sutami 36A, Surakarta, Central Java 57126 Indonesia.
[2] Civil Law Department, Faculty of Law, Universitas Sebelas Maret, Jl. Ir. Sutami 36A, Surakarta, Central Java 57126 Indonesia.
[3] Corresponding author: emmy.latifah@yahoo.com

Abstract. One of the objectives of fisheries management is to reach long-term sustainable benefits of the fish stocks while reducing the risk of severe or irreversible damage to the marine ecosystem. Achieving this objective needs, the good scientific knowledge and understanding on fisheries management including scientific data and information on the fish stock, fishing catch, distribution, migration, the proportion of mature fish, the mortality rate, reproduction as well as the knowledge on the impact of fishing on dependent and associated species and other species belonging to the same ecosystem, and further the impact of climate change and climate variability on the fish stocks and marine ecosystem. Lack of this scientific knowledge may lead to high levels of uncertainty. The precautionary principle is one of the basic environmental principles needed in overcoming this problem. An essence of this principle is that, in facing the serious risk as a result of the limited scientific knowledge or the absence of complete evidence of harm, it should not prevent the precautionary measures in minimizing risks and protecting the fish stocks and ecosystem. This study aims to examine how the precautionary principle in fisheries management be formulated into the international legal framework, especially under the climate change framework.

1. Introduction

Since the second half of the 1940s, the problem of over-fishing has been on the international agenda, yet international fisheries law well into the 20th century focused on "congestion, conservation, or fishing rights". Conservation, in terms of the body of law, is referred to "the aggregate of measures rendering possible the optimum sustainable yield from those resources so as to secure a maximum supply of food" and did not encompass the environmental consequences of fishing activities. During the latter part of the 20th century, it became abundantly clear that international fisheries policy and law had not been particularly successful at the conservation of fish stocks and evidence emerged that fishing activities were negatively impacting the environment. Moreover, climate change also has accelerated environmental degradation, especially the marine environment. Therefore, in the context of international environmental instruments, during the 1990s, it started to address the issues related to fisheries management. As a result, the international law of fisheries, if not the practice of over-fishing, has changed considerably. The introduction of the precautionary principle and, related thereto, ecosystem-based management approaches are amongst the most important substantive changes. In

Content from this work may be used under the terms of the Creative Commons Attribution 3.0 licence. Any further distribution of this work must maintain attribution to the author(s) and the title of the work, journal citation and DOI.
Published under licence by IOP Publishing Ltd

institutional terms, the most important development is probably the introduction of a general international law to regulate high seas fisheries and thereby curtail the freedom of fishing.

The Precautionary Principle has been widely incorporated, in various forms, in international environmental agreements and declarations and further developed in a number of national laws. An element common to the various formulations of the precautionary principle is the recognition that lack of certainty regarding the threat of environmental harm that should not be used as an excuse for not taking action to avert that threat.

This article would like to describe how the international legal framework formulates the precautionary principle, especially in fisheries management.

2. Literature review

2.1 The concept of precautionary principle

There is no clear and uniform understanding of the meaning of the precautionary principle among states and other members of the international community. At the most general level, it means that state agrees to act carefully and with foresight when making decisions, which concern on the activities that may have an adverse impact on the environment. A more focused interpretation provides that the principle requires activities and substances which may be harmful to the environment to be regulated, and possibly prohibited, even if no conclusive or overwhelming evidence is available as to the harm or likely harm that they may cause to the environment [1].

A more fundamental change would be adopted by an interpretation of the precautionary principle, one increasingly wide held, which would shift the burden of proof. According to traditional approaches, the burden of proof currently lies with the person opposing an activity to prove that it does or is likely to cause an environmental damage. A new approach, supported by the precautionary principle, would tend to shift the burden of proof and require the person who wishes to carry out an activity to prove that it will not cause harm to the environment. This interpretation would require polluters and polluting states to assert that their activities and the discharge of certain substances would not adversely or significantly affect the environment before they were granted the right to release the potentially polluting substances or carry out the proposed activity.

2.2 The concept of fisheries management

There is no generally accepted definition of fisheries management. One of the sources that provide the definition is FAO Technical Guideline for Responsible Fisheries, which state that [2]:

> "...the integrated process of information gathering, analysis, planning, consultation, decision-making, allocation of resources and formulation and implementation, with enforcement as necessary, of regulations or rules which govern fisheries activities in order to ensure the continued productivity of the resources and the accomplishment of other fisheries objectives."

From that definition, it can be seen that fisheries management embraces a complex and wide-range set of tasks, which collectively have the achievement of sustained optimal benefits from the resources as the underlying goal.

A good management requires data and information as a basis for all stages process begun from the formulation of the policy, development management plans, evaluation progress until updating policy and planning to provide continuous improvement. Based on FAO Technical Guidelines for Responsible Fisheries, the information that feeds into a fishery management plan should include: (1) the area of operation of fishery and its jurisdiction; (2) the various stakeholders; (3) the gear and vessel types to be employed in fishery; (4) the history, management, and socio-economic importance of fishery; (5) if possible, the distribution area of the most commercial species in the catch; (6) the relevant information about the life histories of species; (7) the effects of fishery on the recruitment, abundance, spatial distribution and age or size structure of the target species, as far as possible; (8) any available monitoring data; and any management procedures already in place, with descriptions and a performance evaluation [2].

3. Result and discussion

In the fisheries management context, the precautionary principle has been recognized in several international instruments.

3.1 The 1982 United Nation Convention on the law of the sea

The 1982 United Nation Convention on the Law of the Sea (UNCLOS) has implied the precautionary approach in fisheries conservation. Article 61.2 asserted that [3]:

> "...the Coastal State, *taking into account the best scientific evidence* to it, shall ensure through proper conservation and management measures that the maintenance of the living resources in the exclusive economic zone is not endangered by over-exploitation..."

Although this article lacks any express reference to the precautionary principle, it already implied a precautionary approach to fisheries conservation [3]. This article obligates to the Coastal State to take into account 'the best scientific evidence available' in determining fisheries conservation and management measures. This formulation would raise the question 'whether, in the absence of convicting scientific evidence, measures should be designed to ensure continued exploitation or to ensure conservation [4]. It is then argued that the UNCLOS provisions, by making conservation as the primary obligation, place a presumption in favor of conservation on the potential exploiter. The scientific evidence must be adduced to show that projected harvesting meets the requirement to maintain or restore populations at a level that can produce a maximum sustainable yield (MSY), rather than the other way around. In other words, if an adequate scientific evidence is not available, the primary conservative obligations of UNCLOS prevail [4].

Maximum sustainable yield (MSY) is also not defined by UNCLOS but is generally defined as the largest annual catch or yield of a fishery that can be taken continuously from the stock, based on the renewability of the resource, however, the concept is widely criticized because of the difficulties in determining MSY in practice, due to the natural variability of stocks and other uncertainties. Besides, it is seen as largely inadequate to the task of managing an already fully exploited or even declining resources and ignores the effects of fishing on non-target species [5].

MSY is a biological concept defined as the largest annual catch that can be taken continuously from the stock. It thus marks the upper limit beyond which harvesting levels are no longer sustainable. If this strict biological limit is to be 'qualified' by environmental and economic factors, this can only be in terms of lower catch levels than the concept of MSY would actually permit. Higher catch levels are *per se* contrary to the concept and cannot pass for a qualification. Besides, catch levels beyond MSY would naturally prevent the maintenance or restoration of populations 'at levels which can produce the maximum sustainable yield' (Article 62(3)). By definition, a stock or population that is exploited beyond the MSY level cannot continue to produce the same catch levels. Levels of harvesting beyond MSY are therefore contrary to the primary obligation contained in Article 61(3) [6]. Moreover, continuous catch levels beyond MSY would inevitably lead to over-exploitation and eventually contradict the general obligation under Article 61(2) UNCLOS, once populations become endangered.

According to Article 61(1) UNCLOS, the coastal state 'shall determine the total allowable catch of the living resources in its exclusive economic zone'. While, Burke suggests that the purport of this provision is to enable only the coastal state, to the exclusion of other entities, to determine the allowable catch in its EEZ [7]. The language is clearly mandatory. Besides, as Article 56(1)(a) UNCLOS attributes to the coastal state exclusive sovereign rights for the conservation and management of natural resources in its EEZ, Burke's interpretation would render Article 61(1) meaningless.

Burke further argues against a legal obligation to determine the total allowable catch (TAC) that especially developing countries may not be able to establish the requisite scientific basis [7]. However, Article 61(1) UNCLOS is only concerned with the basic duty to limit resource exploitation, not with pertinent data requirements. The latter issue is left to Article 61(2), calling on the coastal state to 'take into account the best scientific evidence available to it' in taking conservation and management measures, which seems easy but it is a small burden.

The wording of Article 61(1) UNCLOS appears to suggest that a total allowance catch (TAC) must be established for every fish stock within the EEZ [34]. However, the use of the term 'resources', rather than 'stocks' or 'species', may imply that the obligation applies only to such stocks or species that are affected by exploitation [8]. On the other hand, the determination of total allowance catch (TAC) is a requisite for the identification of the potential surplus that exceeds the coastal state's own harvesting capacity and must be made available to foreign fishing vessels [9]. When it is alleged that a coastal state has arbitrarily refused to determine the TAC and its harvesting capacity at the request of another state with respect to stocks which other states are interested in fishing, the dispute is subject to conciliation (Article 297(3)(b)(ii) UNCLOS). This implies that a coastal state would have to determine the TAC for a stock that is of interest to other states, even though it is not harvested by the coastal state itself. Yet the conciliation provision supports the view that the obligation to establish a TAC does not apply to all living resources in the EEZ, as it appears 'highly unlikely that a dispute would arise over a failure of the coastal state to determine an allowable catch for a species or population that is only of theoretical interest for harvesting' [8]. This interpretation is also supported by practical considerations.

Article 61(2) UNCLOS requires the coastal state to take 'into account the best scientific evidence available to it' in determining conservation and management measures, albeit not to base its action solely on such evidence. Limited data are sufficient, as long as they are the best available to the coastal state. Thus, there is no expression of a positive duty on coastal states to undertake scientific research. However, the primary obligation to conserve the living resources in the EEZ 'reasonably imposes the burden of acquiring data that make this obligation achievable' within the limits of the coastal state's financial resources. In any case, the 'available' data are not only the data generated by the coastal state but includes the data from other sources, such as other states involved in the fishery and international organizations, that can reasonably be obtained. Article 61(5) UNCLOS places a positive duty on all states participating in a given fishery 'where appropriate' to exchange a range of scientific information and the data relevant to the conservation of fish stocks on a regular basis through competent international organizations.

3.2 The 1995 United Nations Fish Stocks Agreement

The 1995 United Nations Fish Stock Agreement (hereinafter called as 'Agreement') marks a significant shift of emphasis and approach and remains probably the most important of many international fisheries agreements [10]. The Agreement establishes obligations for signatory States that affects both management within national waters of straddling or highly migratory stocks and management of high seas stocks by the international and regional fishing organizations. Within these constraints, The Agreement provides a legal basis for the application of several of the most important provisions of the United Nations Food and Agriculture Organization (FAO) Code of Conduct for Responsible Fisheries 1995.

Environmental considerations are strongly highlighted in the preamble of The Agreement, and its effects throughout the operative provisions. The Agreement is the first international agreement requiring a precautionary approach to fisheries management. Article 6 of The Agreement requires that to preserve the marine environment as well as protect marine living resources, the precautionary approach should be applied in conservation, management, and exploitation measures. It includes requirements that States apply a prescribed methodology for precautionary measures (set out in Annex II), implements improved techniques for dealing with risk and uncertainty, takes into account both ecological and socio-economic uncertainties, and develops research and monitoring programs and plans aimed at conserving non-target and dependent species. Annex II sets out guidelines for precautionary measures based on the establishment of reference points, and actions to be taken when such points are approaches and exceeded. Reference to maximum sustainable yield (MSY) is retained in Annex II guidelines, but as a limit reference point, constraining harvest, rather than a target for management.

3.3 The 1995 FAO code of conduct for responsible fisheries

The FAO Code of Conduct for Responsible Fisheries (hereinafter called as 'the Code') includes an exhortation to apply the precautionary approach (this Code use the term 'precautionary approach' rather than 'precautionary principle) widely in the conservation, management, and the utilization of living aquatic resources, directed as States, sub-regional, and regional fisheries management organizations and arrangements. While the code of conduct is voluntary, it is widely supported by most of the States. However, although most of the States have signed up to the Code, progress at implementation has been slow [11].

The technical guidance for implementation of the precautionary approach has been developed by the FAO. These guidelines represent probably the most detailed treatments of the operational meaning of precautionary in a natural resources management or conservation arena and offer valuable lessons for other sectors. The FAO guidance first characterizes the general concept of the precautionary approach, setting out that the precautionary approach requires, inter alia:

- Avoidance of irreversible changes;
- Prior identification of undesirable outcomes;
- Initiation of corrective measures without delay;
- Priority is given to conserving the productive capacity of the resource;
- Harvesting and processing capacity commensurate with estimated sustainable levels of the resources;
- That all fishing activities have prior management authorization and are subject to periodic review;
- Legal and institutional frameworks for fisheries management, with management plans implementing the above for each fishery;
- Appropriate placement of the burden of proof through meeting these requirements.

Detail guidance is then developed for the implementation of the precautionary approach in relation to fisheries management, research, technology development/transfer, and species introductions, including management planning and design, monitoring, stock assessment methods, review and evaluation of new technologies, and cooperation and information system on invasive species. The FAO continues to actively develop the precautionary approach, developing guidance across a range fisheries [12,13,14]. The precautionary approach has also been endorsed by and incorporated into ongoing work under FAO auspices on developing guidance for the ecosystem approach to fisheries [15].

4. Conclusion

Precautionary principles have been applied to international conventions, particularly those related to fisheries management. However, this principle still needs operational rules so that it can be applied to the national law and become effective in order to achieve fisheries management objectives.

Acknowledgment

The author would like to thank the Directorate General of Higher Education, Ministry of Research, Technology and Higher Education, the Republic of Indonesia for funding on Basic Research Schema in 2017. Additionally, the authors also would like to say thank the editor, who has helped in technical matters so that this article become eligible to be published.

Reference

[1] Philippe S 2003 *Principle of International Environmental Law* (Cambridge: Cambridge University Press)
[2] "FAO Technical Guideline for Responsible Fisheries" *https://documents-dds-ny.un.org* 2012 [Online] Available: https://documents-dds-ny.un.org/doc/UNDOC/GEN/N95/274/67/PDF/N9527467.pdf?OpenElement [Accessed: 07-08-2017]
[3] "The 1982 United Nations Convention on the Law of the Sea" *http://www.un.org* 1982 [Online] Available: http://www.un.org/depts/los/convention_agreements/texts/unclos/unclos_e.pdf [Accessed: 06-08-2017]
[4] Freestone D 1999 *International Fisheries Law Since Rio: The Continued Rise of the Precautionary Principle; International Law and Sustainable Development: Past Achievements and Future Challenges* ed Boyle A and Freestone D (New York: Oxford University Press) pp 135-64
[5] Barnes R 2006 *The Convention on the Law of the Sea: An Effective Framework for Domestic Fisheries Conservation?; The Law of the Sea: Progress and Prospects* ed Freestone D, Barnes R and Ong D M (Oxford: Oxford University Press) p 233
[6] Kaye S M 2001 *International Fisheries Management*; *International Environmental Law and Policy Series* (The Hague: Kluwer Law International)
[7] Burke W T 1994 *The New International Law of Fisheries: UNCLOS 1982 and Beyond (*Oxford: Clarendon Press)
[8] Wolff N 2002 *Fisheries and the Environment: Public International and European Community Law Aspects* Schriften des Europa-Instituts der Universität des Saarlandes – Rechtswissenschaft Vol 40 (Baden-Baden: Nomos Verlagsgesellschaft)
[9] Christie D R 1999 *The Conservation and Management of Stocks Located Solely within the Exclusive Economic Zone* ed E Hey Developments in International Fisheries Law (The Hague: Kluwer Law International) pp 395-419
[10] "The 1995 United Nations Fish Stocks Agreement for the Implementation of the Provisions of the UNCLOS relating to the Conservation and Management of Straddling Fish Stocks and Highly Migratory Fish Stocks" *https://documents-dds-ny.un.org* 1995 [Online] Available: https://documents-dds-ny.un.org/doc/UNDOC/GEN/N95/274/67/PDF/N9527467.pdf?OpenElement [Accessed: 06-08-2017]
[11] FAO 2005 *Progress in the Implementation of the Code of Conduct for Responsible Fisheries and Related Plans of Action* (Rome: UN Food and Agriculture Organisation)
[12] FAO 2001 *Progress in Implementation of The Code of Conduct for Responsible Fisheries and Related International Plans of Action; Report to the FAO Committee on Fisheries 24th Session 26 Feb - 2 Mar 2001* (Rome: UN Food and Agriculture Organisation)
[13] Caddy J F 1998 *A Short Review of Precautionary Reference Points and Some of Their Use in Data-poor Situations* (Rome: FAO).
[14] Caddy J F and Mahon R 1995 *Reference Points for Fisheries Management*; *FAO Fisheries Technical Paper 347* (Rome: UN Food and Agriculture Organization)
[15] FAO 2003 *The Ecosystem Approach to Fisheries*; *FAO Technical Guidelines for Responsible Fisheries No. 4 Supplement 2* (Rome: UN Food and Agriculture Organization)

IOP Conf. Series: Earth and Environmental Science **129** (2018) 012039 doi:10.1088/1755-1315/129/1/012039

The future of climate change policy by provincial government in Indonesia: A study on the vision and mission of elected governors in 2017 election

T Kurniawan

Department of Public Administration, Universitas Indonesia, FISIP UI Campus, Building M, 2nd Floor, Depok 16424, Indonesia

Corresponding e-mail: teguh.kurniawan@ui.ac.id

Abstract. The IPCC in 2015 has recognized the critical role of local governments in scaling up the adaptation of communities to climate change. The role will be executed properly if the leaders have an awareness of climate change as a strategic agenda in his administration. In the context of Indonesia, the vision and mission of a regional head can be a clue as to how he supports efforts in tackling climate change. The five-year regional development plan is an elaboration of the development agendas offered by a regional head in his vision and mission. Therefore, this paper aims to analyze the future of climate change policy in some provincial governments based on the vision and mission of the elected governors. To achieve the goal, the author employs a qualitative approach to analyze the contents of the vision and mission and their preference to climate change. The results of the study indicate that the elected governors are still not very aware of climate change and did not specify it in their vision and mission. Therefore, in the preparation of five-year regional development plan, it is necessary for some stakeholders to address and specifically mention the issues of climate change.

1. Introduction

The Intergovernmental Panel on Climate Change (IPCC) in 2015 has recognized the critical role of local government in scaling up the adaptation of communities to climate change [1]. According to the IPCC, local governments and the private sector have different functions that vary regionally. They are increasingly recognized as critical to progress in adaptation, given their roles in scaling up the adaptation of communities, households, and civil society and in managing risk information and financing.

To perform the role in handling the climate change adequately, the leadership in local government plays a very decisive role. Peter John, as quoted by Isaac-Henry [2], suggests that the politics of decentralization, networks, participation, partnership, bureaucratic reform, rapid policy change, and central intervention need powerful, but creative figures to give a direction to policymaking. Therefore, leadership is a crucial factor in determining the direction and policy of a regional government.

In the context of Indonesia, a regional government consists of provinces and regencies or cities. A province is headed by a governor, and a regency is headed by a regent while a city is headed by a mayor. All of them are elected directly by the people through general elections. In determining the direction of development, the governors, regents, and mayors have a very significant role. The five-year regional development plan in the region is largely determined by their vision and mission submitted during the election.

Content from this work may be used under the terms of the Creative Commons Attribution 3.0 licence. Any further distribution of this work must maintain attribution to the author(s) and the title of the work, journal citation and DOI.
Published under licence by IOP Publishing Ltd

IOP Conf. Series: Earth and Environmental Science **129** (2018) 012039 doi:10.1088/1755-1315/129/1/012039

Since the role of regional heads in determining the five-year regional development plan is very important, the successful implementation of local government's role in making policies to address climate change will be largely determined by the alignment of the vision and mission of the Regional Head. Thus, this paper tries to examine the future of climate change policy by the provincial government of Indonesia based on the vision and mission of the regional head. Of the thirty-four provinces in Indonesia, seven provinces have held gubernatorial elections in early 2017. Accordingly, the vision and mission discussed in this paper are the vision and mission of elected governors in the gubernatorial election in early 2017.

2. Methods of research

To achieve the expected objectives, this paper employs a qualitative approach to literature studies on some documents and data sources, namely the vision and mission of the elected governors. The data are then analyzed qualitatively to explore the content of the vision and mission that addresses climate change and relevant environmental issues in general. Climate change occurs due to the increasing number of greenhouse gases emitted from burning coal, oil and gas. This causes several environmental impacts such as deforestation, increasing livestock farming, fertilizers containing nitrogen, and fluorinated gases. Therefore, the content of the vision and mission are also tried to be associated with various causes, in the sense that there are possible linkages between the programs proposed in the vision and mission and the various causes.

There are seven governors elected in the governor election held in early 2017, namely Irwandi Yusuf (Governor of Aceh), Erzaldi Rosman Djohan (Governor of Bangka Belitung), Wahidin Halim (Governor of Banten), Anies Rasyid Baswedan (Governor of Jakarta Capital Special Region), Rusli Habibie (Governor of Gorontalo), Ali Baal Masdar (Governor of West Sulawesi), and Dominggus Mandacan (Governor of West Papua). Of the seven elected governors, from online search results, the author obtained only the vision and mission documents in the gubernatorial election for the six elected governors. The author did not get the vision and mission documents from Dominggus Mandacan. Therefore, this paper only discusses the vision and mission of the six elected governors. This paper only uses the vision and mission documents as the main data without conducting in-depth interviews with the elected governors so that it makes a major limitation in this paper.

3. Results and discussion

The result will be discussed in two sections. In the first section, the author will explain the relationship between the vision and mission of the elected governors with regional development planning in Indonesia. This explanation is expected to provide an overview on the urgency of vision and mission on policies taken by a regional government. In the second section, the author will describe and analyze the content of the vision and mission of the elected governors and relate it to the future of climate change policy.

3.1. Vision and mission in the context of development planning in Indonesia

Based on the provisions of Article 5 paragraph 2 of Law No. 25 of 2004 on the National Development Planning System, the five-year regional development plan is an elaboration of the vision, mission, and program of the regional head [3]. The vision and mission of the elected regional head will determine the regional development strategy, general policy, priority programs, and the direction of regional financial policy. Therefore, the climate change policy of the regional government in Indonesia will also be determined by the vision and mission of the regional head when they follow the election process of the regional head.

3.2. Vision and mission of the six elected governors

The following sub-section will give a brief description of the vision and mission of elected governors related to climate change and analysis of the future of climate change policy.

3.2.1. Irwandi Yusuf [4]. Of the nine missions offered by Irwandi Yusuf, the seventh mission is related to climate change. The seventh mission is to provide clean and renewable energy sources to meet the electricity needs of the people and industry, as Aceh's commitment to the development of low emissions. To run his mission, Irwandi Yusuf offers fifteen priority programs. The two programs are related to the seventh mission, namely Aceh energy (the fourth program) and Aceh green (the thirteenth program). Aceh energy is a program in the framework of the fulfillment of electrical energy for the people of Aceh and industries derived from renewable clean energy. Aceh Green is a program to reaffirm the development of a sustainable environment since Aceh is sensitive to the risk of natural disasters. To implement the Aceh energy program, there are six subprograms that will be conducted, including: (1) the preparation of comprehensive and integrated Aceh energy master plan; (2) the implementation of public-private partnership for the utilization of medium and large-scale energy resources, especially geothermal and hydropower; (3) the assurance of the commitment of the State Electricity Company (PLN) to improve the transmission system; (4) the acceleration of realization of Geothermal Seulawah power plant and hydropower plant (mini and micro hydro) with medium capacity; (5) the granting of licensing permits for private parties interested in investing in clean and renewable energy sectors; and (6) strengthening the existence of the Aceh Oil and Gas Management Agency (BPMA) as the embodiment of the mandate set out in the Law on Aceh Government (UUPA). Meanwhile, to implement the Aceh Green program, there are five subprograms that will be conducted, namely: (1) designing Aceh green growth plan; (2) taking strategic steps to mitigate and adapt to climate change; (3) implementing disaster risk management and mitigation strategies; (4) developing critically endangered species recovery plans ; and (5) building manufacturing facilities for the supply of wood polymer composite in order to anticipate the increase of world's wood demand so that it can prevent forest degradation. Thus, based on the vision and mission of the elected governor of Aceh, it can be seen that climate change has been taken into account so that it will become a policy that can be realized. In the programs and subprograms offered, it can be seen that there are alignments to activities that contribute to the reduction of climate change, dealing with clean and renewable energy, as well as deforestation. One of the subprograms also specifically mentions about the mitigation and adaptation to climate change. Therefore, having a closer look at the vision and mission of Irwandi Yusuf, the climate change policy will get an adequate portion in the implementation of Aceh government.

3.2.2. Erzaldi Rosman Djohan [5]. Of the eleven missions offered by Erzaldi Rosman Djohan, the two missions including the second and the eighth mission are likely linked to climate change. The second mission is about the mining field, that is tin mining on an ongoing basis by weighing the environment as the main factor to maintain the balance of nature. The eighth mission is about energy field, namely the development of power plant as well as renewable energy so that Bangka Belitung Province can increase local revenue and energy independence through the development of factories to manage raw materials in the fields of mining, agriculture, plantation, fishery, and others. To run the mission, Erzaldi Rosman Djohan offers eight priority programs in which the two programs are related to the second and eighth mission, namely the seventh program (infrastructure and energy) and the eighth program (mining). To carry out infrastructure and energy programs, there are six subprograms to be undertaken: (1) the routine development and maintenance of infrastructure, interconnection roads between districts that are the responsibility of the governor; (2) the construction of Batu Beriga port and bridge between Bangka-Belitung; (3) the construction of a floating Mosque as the new icon of Bangka Belitung; (4) the construction of GMT museum in Bangka and Belitung; (5) the improvement of the status of the Depati Amir airport (Central Bangka) and HAS Hanandjoedin (Belitung) as an international airport; and (6) the initiation of intensive coordination between Bangka Belitung Province, Ministry of Energy and Mineral Resources, PLN, and private investors to build Steam Power Plant (PLTU) so that PLTU in Bangka Belitung can be realized to build energy independence. Meanwhile, to implement the mining program, there are two subprograms that will be done, namely: (1) the zonation or management-specific of areas that are shared proportionally and sustainably; and (2) the payment and management of corporate social responsibility of tin in an integrated and fair way for all tin companies. Based on the vision and mission of the governor of Bangka Belitung elected, it can be seen that climate change is not a concern so that the climate change policy will be a difficult thing to materialize. Although the two missions have the relevance to climate change in the proposed program and subprogram, there are no programs or subprograms that are adequate to carry out both missions.

3.2.3. Wahidin Halim [6]. There are no specific missions related to environmental issues or climate change although in the identification of common problems in Banten there are two environmental problems stated, namely the tenth problem (waste and the environment) and the twentieth problem (spatial and land). About waste and environmental problems, Wahidin Halim's vision and mission documents mention the increasing production of garbage, environmental degradation of forests, mangroves, and river pollution. Meanwhile, dealing with spatial and land problems, vision and mission documents from Wahidin Halim mention the decrease of green open space stating that there is still not an effective spatial arrangement in the region. Of the five priority programs proposed for carrying out its vision and mission, in the fourth program on "improving access and equity of quality health services" there are thirteen subprograms with four subprograms related to environmental issues, raised in the fifth sub-program: the normalization of 32 rivers, the sixth subprogram: the normalization of 41 lakes to restore the function, the eleventh subprogram: the environmental control and waste management of the province scale to maintain the environmental carrying capacity and the implementation of sustainable development, and the twelfth subprogram: the evaluation of the planning of the utilization of provincial spatial control. Thus, based on the vision and mission of the governor of Banten elected, it can be seen that climate change is not a concern so that the climate change policy will be a difficult thing to materialize. Environmental issues, in general, are not accommodated in specific missions or programs. In addition, the environmental issues are also not adequately addressed.

3.2.4. Anies Rasyid Baswedan [7]. Of the three missions offered by Anies Rasyid Baswedan, the second mission is related to climate change. The second mission is to build a sustainable city environment in Jakarta with a plan that takes into consideration the environmental and social carrying capacity. To run this mission, Anies Rasyid Baswedan offers six priority programs in which one program is related to the second mission, stated in the sixth program: Jakarta's chronic problem management, i.e., water problems (flood and clean water supply), mobility (congestion), and garbage management. The development program is conducted by involving the three pillars of development including human, environment, and welfare. The environmental approach is to build the earth as a common property, equal access and equal responsibility for all citizens. In the second pillar (development of urban environment), there are two main programs, namely the handling of flood and clean water problems and the management and reduction of waste. In addition, there are five subprograms in the flood prevention program, namely: (1) the revitalization of dikes and water pumps; (2) the provision of a direct cash assistance (BLT) for flood victims; (3) the establishment of a water distribution system and green environment; (4) the implementation of zero run-off policy downstream, which in essence is that all water is maximized to be absorbed again into the soil rather than diverted to the channel; and (5) more effective communication in land acquisition matters. Meanwhile, there are three sub-programs in water management, namely: (1) planning and constructing absorption wells and/or water treatment facilities to reduce surface pools and increase groundwater supply (water conservation); (2) cooperating with the surrounding areas of Jakarta for reforestation of upstream catchment areas (forests and mountains); and (3) treating wastewater to be reused or absorbed into the soil. As for the waste management and reduction program, there are five subprograms, namely: (1) self-care waste handling training at household level; (2) incentives for community-based recycling programs; (3) socialization of the use of organic materials for household or industry; (4) incentive programs for the packaging industry and producers using organic materials or recyclable materials; and (5) incentive programs for the production and distribution of innovative composting tools and the recycling industry. In the pillar of welfare development, in one of its programs is the third program (the prevention of mobility and congestion) consisting of six subprograms of eleven subprograms that have a relevance to climate change, namely subprograms on carpooling, public transportation, and mass rapid transportation. Thus, based on the vision and mission of the elected governor of the Special Capital Region of Jakarta, it can be known that climate change has received attention so that climate change policy will become a thing that can be realized. However, the programs or sub-programs offered do not specifically mention the problems of climate change. Moreover, since Jakarta is a big city having the problem of traffic congestion as a result of the massive use of private transportation, the congestion reduction efforts through the use of public transport should become the part of the environmental pillars linked to climate change rather than in the welfare pillars. Therefore, it is still seen that the problem of climate change has not received an adequate portion although there is already a growing attention to the problems of development prioritizing the environmental aspects.

3.2.5. Rusli Habibie [8]. Of the five missions offered by Rusli Habibie, the first mission is related to climate change. The first mission is to realize the management of tourism and natural resources that are environmentally sustainable. This mission is oriented to the management of natural resources, especially tourism, agriculture, fisheries, and marine. This is aimed at maintaining energy and water security while maintaining the preservation of the environment and forest areas. To carry out the mission, Rusli Habibie sets eight policy directions in which one of them is related to the first mission, the eighth policy direction: a more sustainable environment. This program is oriented to the management of various natural resource potentials, especially tourism, agriculture, marine, mining, which is more environmentally friendly while maintaining the interests of the community and, above all, prioritizing environmental sustainability. Based on the vision and mission of the elected governor of Gorontalo, it can be seen that climate change is not a concern so that climate change policy will become a difficult thing to materialize. Although there are missions that have a likelihood of linkage with climate change in the direction of the policy, there is no adequate direction for carrying out the mission.

3.2.6. Ali Baal Masdar [9]. Of the five missions offered by Ali Baal Masdar, the fifth mission has the possibility of linkage with climate change. The fifth mission is to mainstream the environment for sustainable development. The following are nine proposed programs to carry out this mission:(1) facilitating the arrangement of green open space in six districts as the lungs of the city; (2) building the capacity of the apparatus related to development and the environment; (3) establishing an environmental committee; (4) West Sulawesi go clean and green!; (5) developing the information system and report of environmental management plan; (6) facilitating environmentally sound school movements; (7) developing workshops of family of love environment; (8) establishing and mobilizing 16,225 environmental cadres; and (9) establishing environmental posts, as one form of the development of integrated service post function. Based on the vision and mission of the governor of West Sulawesi elected, it can be seen that climate change is not a concern so that climate change policy will be a difficult thing to materialize. Although there is one mission that has potential relevance to climate change in the program proposal, there is no adequate program to carry out the mission.

4. Conclusions

Based on the discussions conducted, it can be concluded that the elected governors are still not very aware of the climate change and did not specify climate change in their vision and mission. There is only one governor (Irwandi Yusuf) who explicitly mentions programs and subprograms on climate change. The other governors did not specify the climate change in his vision and mission. There are also governors who have an awareness of environmental issues but do not directly affirm their allegiance to climate change. Moreover, there are also governors who have environmental awareness in their mission but it is not reflected in their programs and subprograms.

Concerning such conditions, in the process of discussing the five-year regional development as the elaboration of the vision and mission of elected governors in some provinces, it is necessary for some stakeholders to accommodate and specifically mention the issues of climate change. The involvement of the stakeholders is also needed to safeguard the five-year regional development that accommodates climate change

Acknowledgements
The author would like to thank the research cluster of Democracy and Local Governance (DeLoGo) at the Faculty of Administrative Science of the Universitas Indonesia who has provided a financial support for the publication of this paper.

References

[1] IPCC 2015 *Climate Change 2014: Synthesis Report* (Geneva: International Panel on Climate Change)
[2] Isaac-Henry K 2003 Leadership, Politics and Accountability: G.W. Jones' Contribution to Local Government *Publ. Pol. Admin.* **18** p 89
[3] Law No. 25 of 2004 on National Development Planning System
[4] Yusuf I and Iriansyah N 2016 *Vision and Mission* pp 1-10
[5] Djohan E R 2016 *Vision-Mission* pp 1-19
[6] Halim W and Hazrumy H 2016 *Vision, Mission & Program* pp 1-20
[7] BaswedanA R and Uno S S 2016 *Vision, Mission and Work Program* pp 1-23
[8] Habibie R and Rahim I 2016 *Vision and Mission* pp 1-11
[9] Masdar A B and Anwar E A 2016 *Vision, Mission, and Promise Work* pp 1-13

IOP Conf. Series: Earth and Environmental Science **129** (2018) 012040 doi:10.1088/1755-1315/129/1/012040

An adaptation strategy of sandland peasants in Yogyakarta toward climate change

E Rusdiyana and Suminah

Faculty of Agriculture, Universitas Sebelas Maret, Jl.Ir.Sutami No 36.A, Kentingan, Jebres, Surakarta, 57126, Indonesia

Abstract. This study aims to explore and describe the adaptation strategies of sandland peasants toward climate change. Qualitative research method was employed and the data were collected through observation. In addition, the recording of the data, interview and the validity of data were determined by triangulation of sources. The results of the research showed that the adaptation strategies of sandland peasants toward climate change were; (1) the adjustment of crop varieties, (2) the utilization of productive crops as wind breaking, and (3) the irrigation system using "sumur panthek".

1. Introduction

Sandland is a marginal area having low productivity because of a low water savings factor, infiltration and high evaporation, fertilization and organic material and also low water efficiency [1]. However, it has a great potency to support development in the agribusiness sector with its strength as a large area, flat, less flood, great sunshine quantities, and also the shallowness of sand water depth. This sand land potency is substantial when correlated to the potency of Indonesia as a maritime nation with the longest coastal line in the world [2].

Coastal sand land in *Kulon Progo* Regency, *Yogyakarta* Province is a coastal sand land developed for farming activity with 1.753 hectares, spreading in four sub-districts: Panjatan, Galur, Wates, and Temon. Coastal sand land peasants try to cultivate food plants, horticulture, plantation, and livestock suitable for the land characteristics. Chili is a primary farming plant on coastal sand land that is cultivated by the peasants, intercropped with watermelon and horticulture commodity. The objective of this research is to describe the adaptation strategies of coastal sand land peasants in facing the climate change.

2. Material and method

This study was a descriptive research with a qualitative approach. The primary and secondary data were garnered by an interview, observation, and document and archive study. The location of this research was selected purposively, based on the criteria of coastal sand land peasants in Kulon Progo, Yogyakarta. Data validation was carried out using data source triangulation. The informant and information exploration in this research are presented in table 1.

The data were analyzed using qualitative analysis in the first research and during the research process. The data collected were then processed systematically, starting from the interview, observation, editing, classification, reduction, data presentation, to drawing conclusions. This research employed the following interactive model of data analysis.

Content from this work may be used under the terms of the Creative Commons Attribution 3.0 licence. Any further distribution of this work must maintain attribution to the author(s) and the title of the work, journal citation and DOI.
Published under licence by IOP Publishing Ltd

Table 1. Informant on research

No	Informant	Information
1	Coastal Sandland Peasant	The history of coastal sand land, productive behavior, attitude join group activity
2	Agriculture Department of Kulon Progo	Data of Chili production in Kulon Progo, the existence of coastal sand land agriculture in government's view, agriculture department role
3	Chairman of Peasant (farmer) Group	Peasant behavior, agriculture system of coastal sand land, chairman of peasant group role
4	Seller	Quality of farming product from coastal sand land
5	Agricultural Extension Agent	The role of Agricultural Extension Agent

Source: Field Data

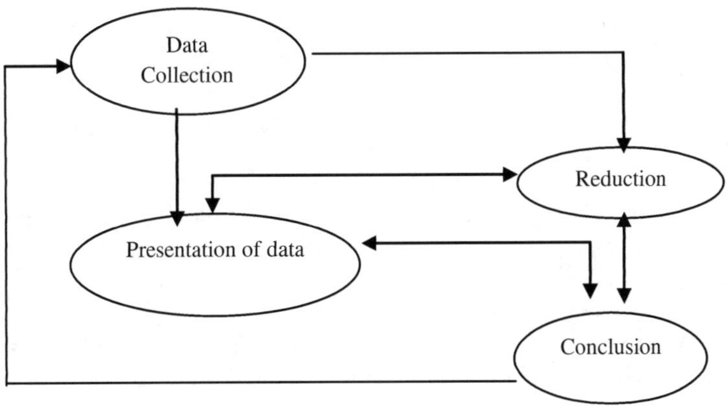

Figure 1. Interactive model of data analysis

3. Results and discussion

3.1 The history of coastal sand land agriculture in Yogyakarta

Kulon Progo Regency, Yogyakarta has approximately 116-120 hectares of coastal sandland in each village spreading from Wates (Karangwuni), Panjatan (Garongan, Bugel, Pleret), Galur (Karangsewu) to Trisik (Banaran). Coastal sand land in Panjatan district, Kulon Progo regency spreads in three villages: Bugel, Garongan, and Pleret. The coastal sand land has been used by the society as productive farmland since 1980.

In 1980, the people in southern coastal areas of Kulon Progo Regency, particularly those living in Panjatan Sub District began to utilize the existence of beach sand areas as one of productive land for farming. The farmers' accidentalness of finding chili plants that can flourish on beach sand land inspired the peasants to plant chili by following the chili planting pattern in a rice field. In fact, chili plants can flourish with fertilization and irrigation like in the rice field. Then, from this chili commodity, the peasants in Panjatan develop melon and watermelon plants with a very good result as well. Considering the experience, many peasants living in the area near Panjatan Sub District imitate this innovation. Afterwards, this innovation is disseminated along the south coastal areas of Kulon Progo Regency with red chili as a typical commodity.

The width of coastal sand land utilized by the peasants is ± 1500 hectare along the coastal area of Kulon Progo in 2017 with the average chili productivity of 15-20 ton/hectare, and ± 150 hectares in a rice field with the productivity of 5-8 ton/hectare. The result of chili marketing through auction market is, on average, 2.5-4 ton red chili per day in every auction group. Until today, the auction market has accommodated around ± 70% of peasants' chili marketing, while the rest of 30% are sold through the local seller.

Figure 2. Peasant activity in coastal sand land of Kulon Progo, Yogyakarta

3.2 Sandland peasants' strategy of adapting to climate change

3.2.1. Adjustment of Plant Variety. From their 30-year experiences, the sand land peasants conclude that Lado, Kyo and Helix varieties of chili can grow well in the sand land of Kulon Progo. This variety choice is made based on the chili cultivated widely by peasants so far. Actually, there is no provision related to this chili variety; however, because chili selling is conducted collectively in the group in auction market (chili will be mixed by variety), the chili variety not cultivated widely by peasants will be sold collectively in a small quantity so that it cannot be auctioned. The peasants receive some offers of other varieties, but based on the demplot tryout or planting by the peasants themselves, other varieties cannot grow as well as those three varieties. This chili variety adaptation, in addition, related to compatibility and land adaptation, is also adjusted with the auction market need as the marketing medium in the group.

In this chili cultivation, to result in a good chili, the peasants should fulfill the need for chili plant in order to get maximum production. The measure is manifested in choosing adaptive variety, using fertilizer, irrigating, maintaining, and harvesting. Rusdiyana [3] mentioned the gain of sand land peasants' productive behavior in Kulon Progo as shown in Table 2.

From Table 2, it can be found that overall, the percentage of peasants' productive behavior is 81.55% with the total mean score of 23.65. This figure shows that chili peasants' productive behavior belongs to the high category; in other words, the peasants perform a productive behavior to achieve good chili harvest. Since in chili agribusiness, in profitable time, the broker requires a good quality of chili, the good quality and optimum production are achieved through good productive behavior.

In addition to having a stipulation in planning the chili variety, the farmer group of coastal sand land also has a consensus dealing with the time of planting, corresponding to a planting calendar based on the peasants' local wisdom calculation. Chili planting is conducted simultaneously in the ninth season (*mangsa kesanga*) usually occurring in March. The planting is conducted simultaneously based on the approved time. If the peasants plant the chili before the date, they will be imposed with a sanction in the form of the removal of plant existing on the land. Every year during climate condition change, the peasants hold a "pranata mangsa" meeting to determine the date of planting season beginning that should be approved by sand land peasants along the coastal line of Kulon Progo.

Asosiasi Pasar Tani (ASPARTAN) (Farmer Market Association), as a means of managing chili auction market in Kulon Progo, has had a time calculation to determine the opening of the auction market in each area and it has been socialized well in each of farmer groups. It is ASPARTAN that serves as a sand land peasant consensus institution in the coastal line of Kulon Progo. When there is a change of time-related to chili harvest different from the prior condition (postponed/longer), the auction market opening schedule will be reviewed to determine the time when the auction is reopened.

Table 2. Gain of chili peasants' production behavior

No	Indicator	Score Interval	Gain Score	Gain Level (%)
Fertilization				
1	Using compost for fertilizing chili	0-3	2.90	97
Land cultivation				
2	Cultivating the land before planting chili	0-4	3.73	93
Chili planting				
3	Establishing contact with farming laborer	0-2	1.18	59
4	Planting chili corresponding to planting pattern approved by the group	0-4	3.61	90
Weeding, Spraying, Watering and Harvesting				
5	Establishing contact with farming laborer to weed grass	0-3	1.14	38
6	Weeding	0-4	3.36	84
7	Spraying to anticipate disease's attack against chili plant	0-3	2.29	76
8	Watering chili everyday	0-3	2.87	96
9	Establishing contact with chili harvesting laborer	0-3	2.57	86
	Total	0-29	23.65	81.55

Source: Rusdiyana [3]

(a) (b)

Figure 3. Farmer group meeting in the attempt of socializing planting season (a) and opening group auction market (b)

(a) (b)

Figure 4. Chili marketing through group auction market (a) and chili sorting process in auction market (b)

3.2.2. The utilization of productive plant as a windbreaker and intercropping plant. The characteristic of the coastal area is, among others, high wind exposure, either onshore wind or sea breeze. This challenge becomes heavier since the wind often brings sand or water material with it, leading to the plant collapse and the failed conception. The adaptive strategy that the coastal sand land peasants take is to utilize hard plants as a windbreaker aiming to reduce the speed of wind that will pass through the plant cultivated. The plants chosen for this function has strong roots and stems, and thick leaves. The hard plants utilized by peasants include bushes and casuarina tree in the forefront. Then, close to the land, there are kelor and coconut trees. A kelor plant row is covered with creeping plants, like squash. On the other hand, peasants plant intercropping plants like cassava, tomato, eggplant, and etc. It means that peasants have a strategy of adapting to wind and weather challenge through wind breaking system with the production plants. The product of wind breaking plant becomes a vegetable source for self-consumption and by-product sold to the middleman (as additional income).

(a) (b)

Figure 5. Wind Breaking Plant (Coconut, *Kelor*, *Pare*) in Peasant Land (a) and Dry Coconut Leaves and Bushes (b)

3.2.3. Irrigation with sumur panthek (artesian well) system. Hot sunlight exposure and irrigation water source existence become the challenge to coastal sand land farming. In the beginning of coastal sand land farming in the 1980s, the peasants created an ordinary well by constructing a tube well. Water was drawn from the well and poured onto chili plants by using a pail or watering can. Such technique needs strong laborers and too long time recalling that the manual water transportation speed with human energy is often limited by the width of the land to be watered. In 2000, the peasants found an irrigation system called as a "*sumur panthek*" (artesian well) system. The well is constructed with a drilling system using a pipe and hose to flow the water into farmland. This drilled well modification has an advantage in term of watering system speed and watering affordability because the long hose facilitates the farmers to afford the wide land. Watering becomes important in the high temperature of sand land, especially in the dry season. The adaptation strategy using the artesian well benefits the peasants as they can water the plant twice a day to meet its irrigation need.

The adaptation strategy of the coastal sand land peasants in Kulon Progo can provide sand land farming sustainability for about 37 years. Local wisdom patterns in adapting to nature are even institutionalized strongly through a group consensus and agreement binding its members. Through the local wisdom of adapting to the environment, peasants organize themselves in a consensus to ensure the environment-friendly sustainable farming. It is in line with Umstot [4], Ife and Tesoriero [5] and Putnam [6] finding that a group has the power to control its members' behavior. The strongest way is the norm, that is, the behavioral rule that should be obeyed and implemented by individual members. As the member of farmer (peasant) group, peasants are expected to attend some activities supporting the sustainability of coastal sandland farming activity.

In a theory suggested by Ife and Tesoriero [5], a consensus is not only equality but also pertains to the consensus established in a group. Moreover, the consensus is a commitment in which group or society is committed to trying to look for a solution or a series of action acceptable to and owned by everyone and decision made and approved by those within it for mutual interest. The consensus-based community development is better than the conflict-based one. As suggested by Putnam [6] the establishment of structure and process in the society is very desirable to perpetuate the life existing in the society. It is intended to realize a sustainable farming system as expected by Roling and Wagemakers [7] and Maunder [8].

To achieve a sustainable farming, according to Roling and Wagemakers [7], a process or procedure is required. The following are the procedures of realizing sustainable farming: (a) improving the economy and environment efficiency, (b) integrating new technology and strengthening it (regenerative technology), and (c) ordering the sustainable farming community life requiring self-regulatory in managing the ecological system and plant diversity. In addition, the society should be independent (self-reliance) in its planning attempt. The coastal sand land peasants have performed four matters suggested by Roling and Wagemakers [7] as indicated by the presence of local wisdom that has been institutionalized in the group rule.

Social Action theory, according to Maunder [8], has been adjusted to the present paradigm, community's active participation, in which society is not in given position. This theory assumes that in rural society, the role of legitimacy institution is still very substantial. Nevertheless, the image of a legitimizer should refer to the village elders who not only always lead a customary rite but also can be opinion leader or agent of change in the village. This implementation is very strategic to be performed by the chiefs of farmer groups serving to bridge innovation and to lead their members to keep being subjected to the specified consensus.

4. Conclusion

The results of the research showed that the adaptation strategies of sandland peasants in Yogyakarta toward climate change were:

- The adjustment of crop varieties,
- The utilization of productive crops as wind breaking, and
- The irrigation system using the artesian good system.

References

[1] Kertonegoro B 2010 *Sand Land Area in Yogyakarta: Potential and Utilization for Sustainable Agriculture* Proceedings of the National Seminar on the Utilization of Local Resources for Sustainable Agriculture Development October 2, 2010. (Yogyakarta: Universitas Wangsa Manggala)

[2] Rusdiyana E 2012 *Consensus as the Basis for Community Organizing (Study of Auction Market: A Power of Farmer of Panjatan Coastal, Kulon Progo Regency) Proceeding 4th International Graduate Students Conference on Indonesia* (Yogyakarta: Universitas Gadjah Mada)

[3] Rusdiyana E 2014 *Behavior of Chili Farmer in Auction Market in Panjatan District, Kulon Progo Regency* Thesis (Yogyakarta: Universitas Gadjah Mada)

[4] Umstot D D 1988 *Understanding Organizational Behavior* (St. Paul: West Publishing Company)

[5] Ife J and Tesoriero F 2008 *Community Development* Translated by Manulang, Yakin and Nursyhid (Yogyakarta: Pustaka Pelajar)

[6] Putnam R 1993 *Making Democracy Work: Civic Tradition in Modern Italy* (New Jersey: Princetone University Press)

[7] Roling N G and Wagemakers M A E 1998 *Facilitating Sustainable Agriculture "Participatory Learning and Adaptive Management in Times of Environment Uncertainty* (Cambridge: Cambridge University Press)

[8] Maunder A H 1972 *Agriculture Extension "A Reference Manual"* (Rome: FAO - The United Nation)

A roadmap to effective urban climate change adaptation

R Setiadi

Department of Urban and Regional Planning, Diponegoro University, Jl. Prof. H. Soedarto, Tembalang, Kota Semarang, Central Java 50275 Indonesia

Abstract. This paper outlines a roadmap to effective urban climate change adaptation built from our practical understanding of the evidence and effects of climate change and the preparation of climate change adaptation strategies and plans. This roadmap aims to drive research in achieving fruitful knowledge and solution-based achievable recommendations in adapting to climate change in urban areas with effective and systematic manner. This paper underscores the importance of the interplay between local government initiatives and a national government for effective adaptation to climate change and takes into account the policy process and politics. This paper argues that effective urban climate change adaptation has a contribution to build urban resilience and helps the achievement of national government goals and targets in climate change adaptation.

1. Introduction

A framework is something larger than a theory that serves a particular function at least to describe a particular subject in a systematic manner and ideally to provide key elements that allow thorough analysis to the subject [1]. This indicates that a framework is required because there is complex and unstructured yet past and existing knowledge on the subject of interest. Climate change in general and climate change adaptation in specific may fall to this subject of interest. Hulme [2] states that climate change as ideas has traveled far away from its original root in the atmospheric study since the 1950s and it has crossed various disciplines, from economics to politics and from social to psychology.

The development of a framework will make it easier for a researcher, in particular, to position his or her research in such complex and massive jungle of climate change knowledge. It helps a scientist to shape his or her focus on this contested and rapidly changing terrain. A good framework may assist a group of researchers or scientists to communicate each other, through hopefully constructive debates on each key element of the framework – although sometimes it could end up with heightened unproductive one. However, the proposed framework in this paper is more an invitation to constructive debate and hopefully end up with a collaborative research program to improve or even modify the framework.

2. Literature Review

2.1. Why effective urban climate change adaptation

As indicated in the earlier sub-section, climate change adaptation science has developed rapidly. Climate change adaptation as a notion is substantially more diverse and complex than its definitions. Adaptation to climate change has different meaning depending on the framing. Fünfgeld and McEnvoy [3] and Preston et al [4] for example have shown this case and it shapes different adaptation practices in over the globe. Füssel [5] has successfully unpacked complex scope of climate change adaptation in several categorizations, such as in terms of types of adaptation (autonomous vs.

Content from this work may be used under the terms of the Creative Commons Attribution 3.0 licence. Any further distribution of this work must maintain attribution to the author(s) and the title of the work, journal citation and DOI.
Published under licence by IOP Publishing Ltd

planned), agents (individual vs. collective), adaptation horizon (short vs. long-term), and etc. Although, an empirical survey on adaptation practices [6] shows that most of the adaptation is short-term oriented, small-scale, and follow business as usual rather than reduce the effect of climate uncertainty in the future.

Effective climate change adaptation is a phrase that may put all of these notions in the same basket. Effective climate change adaptation contradicts to 'mal-adaptation', adaptation practices that only increase human and natural systems at the same level or higher risk due to climate change and variation, in which Eriksen and Brown [7] and Eriksen et al [8] call as 'sustainable adaptation'. Effective climate change adaptation drives adaptation practices not only away from mal-adaptation but also achieve their additional goals harmoniously. Effective climate change adaptation has become one of the important themes in climate change research. It has become a focus of interest of many climate change researchers and scientists. A Google searching for phrase "effective climate change adaptation" offers more than 38,000 results, and with careful selection of scholarly sounds result in no less than 1200 documents, which address this phrase is found. Füssel [5] argues that the definitions of adaptation to climate change is not well developed and is open to various scopes and activities. However, the latest report of the fifth working group of the IPCC defines adaptation to climate change as 'the process of adjustment to actual or expected climate and its effects' [9].

Without a doubt, cities and urban areas play an important role not only in climate change mitigation but also in adaptation [10]; [11] because cities contribute to major global greenhouse emissions and are home for more than 60% of the global population today and 80% in 2050. Moreover, IPCC [9] also highlights the importance of cities in a single specific chapter of their latest report laying the foundations for strategies for effective urban climate change adaptation.

2.2. About the framework: some criteria

In addition to the literature review, this framework is built from author's practical experience in the preparation of action plans and urban climate change adaptation strategies and author's participation in a number of international collaborative programs. To begin with, this section will outline some criteria for the development of this framework.

There are a number of criteria in developing this proposed framework. First of all, this framework should be broad enough and flexible to cover multiple disciplines or subjects. It also capable to explain different dimensions, aspects or themes that are related to urban contexts such as human settlements, foods, energy, mobility, water, disasters (e.g. urban floods and drought), health, and etc. Secondly, the framework should cover various aspects that produce uncertainty in the future. At least, two crucial sources of uncertainty should be involved in the framework, namely: politics and policy-science relationships. Thirdly, the framework should consider various adaptation agents. It means that the framework provides space for analysis of adaptation initiatives delivered by governments, community groups, private sectors and other non-government actors or partnership between these actors including their adaptation performance. The performance of each actor may differ from one another following their organizational characteristics or institutional arrangements across the governance level. Fourthly, the framework should put attention to factors influencing the effectiveness of adaptation programs, projects or actions, such as mainstreaming, funding, cost-effective or affordable adaptation solutions, and etc. Finally, the framework should be able to link all these criteria for both explanatory and evaluative purposes, particularly to understand the ultimate goal of urban climate change adaptation. It is also essential to place the framework in broader and up-dated urban development discourses.

3. Discussion

A framework for research on effective urban climate change adaptation (Effective-UCCA) is illustrated in Figure 1. Besides the importance of local initiatives for effective adaptation to climate change, this research roadmap understands that adaptation to climate change also requires the support of the national government. Moreover, adaptation to climate change should also be framed in the context of governance or multilevel governance. Multilevel governance is defined as a government that relies on more than one layer of government [12]. Multilevel governance is a model of governance that is inclusive; relying on either the role of government actors and non-governmental organizations [13]. Betsill and Bulkeley [14] categorize that the characteristics of multilevel governance in climate change has become one of the three important issues in the study of cities and climate change. There are two indications that reflect the existence of multilevel governance [15]. First, structures or institutions and actors or agents involved in the development of policy are increasingly more complex. Second, the success in addressing complex problems usually does not only involve a single institution or agency, but also institutions and agencies, which are at different levels of government. Multilevel governance is regarded as a model of governance that makes the management of climate change in cities becomes effective [16].

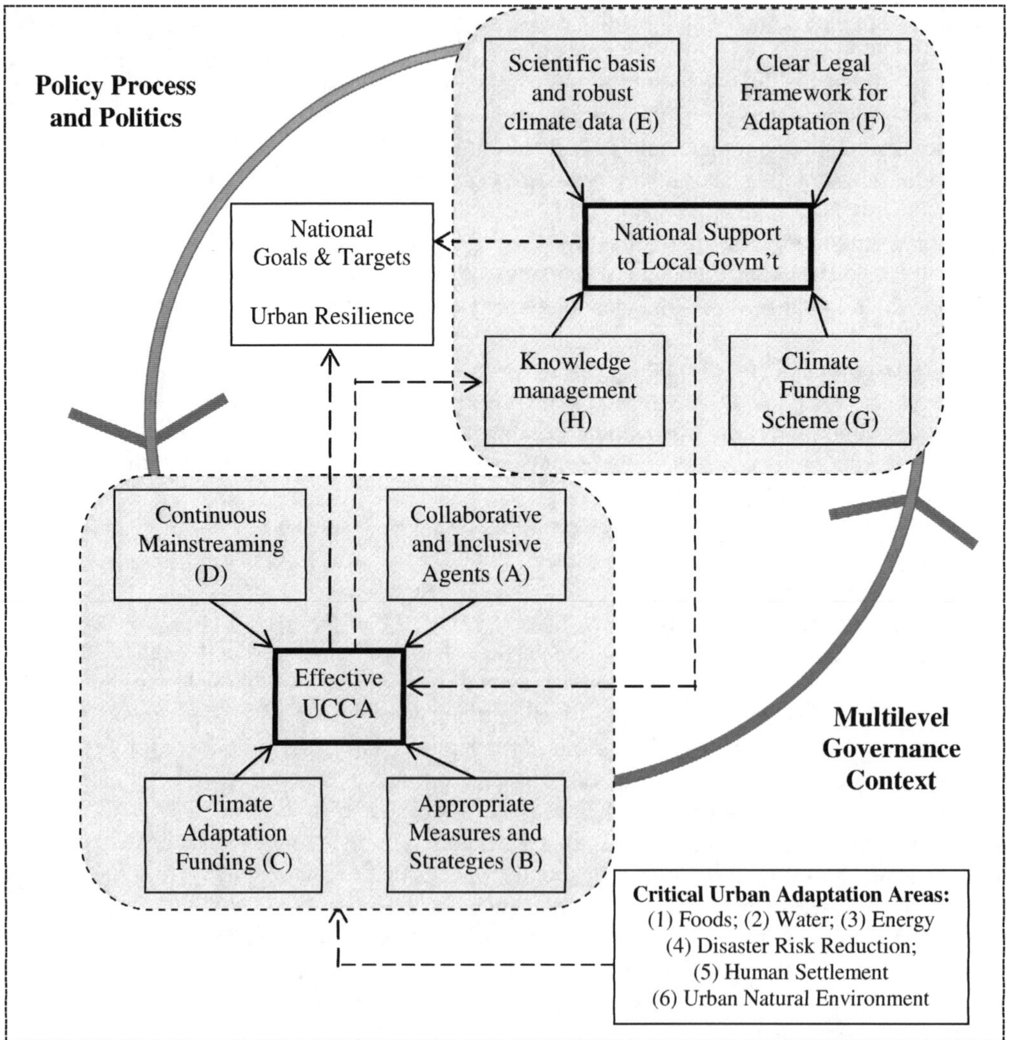

Figure 1. A roadmap to effective urban climate change adaptation.

Therefore, cooperation among stakeholders that is inclusive and collaborative is extremely needed and becomes the first theme (A) in this roadmap. Some emerging research questions from this point of view are:

- A.1. How does collaborative climate change adaptation can achieve effective results? and in what ways?
- A.2. What are benefits generated from and challenges in implementing inclusive and collaborative climate change adaptation?
- A.3. What are effective strategies to encourage private sector participation in climate change adaptation?
- A.4. What sort of models for effective community participation in climate adaptation?

Effective adaptation to climate change is also determined by the management and appropriate adaptation strategies. There are various approaches to adaptation to climate change, which can be broadly divided into physical adaptations or infrastructural adaptation and non-physical adaptations that are more related to e.g. strengthening the capacity of the actors and the strengthening of community's social and economic access to adaptation [5]. Some studies also identified the use of technology for adaptation to climate change, particularly in the food and agriculture sectors. Other studies introduce ecosystem-based adaptation that is promoted by IUCN and a number of other scientists [17]; [18]. Among various adaptation activities, Biagini et al [6] evaluated that most of the adaptation actions are currently focused on social reinforcement and capacity building. Meanwhile, Satterthwaite and Dodman [19] state that physical adaptations are also required by developing and underdeveloped countries because at this time they actually have experienced 'infrastructure deficit'. Therefore climate change is creating new problems in responding to the deficit. Therefore, some research questions that arise from this second theme (B) are:

- B.1. What are efforts or strategies for urban change adaptation in urban areas? And is there any best approach to implement them?
- B.2. How do different approaches to climate change adaptation bring benefits to the city and its inhabitants?

Recent works of Setiadi and Nalau [20] for example addresses B2 question type by assessing whether urban regeneration through public vertical housing development is effective to improve urban health resilience and to deal with the changing climate.

Financing aspects of climate change adaptation become the third essential theme of this roadmap, given the absence of appropriate and adequate financing prevents adaptation actions and implementations. Some studies [21] underline the importance of and encourage the commitment to financing adaptation to climate change, particularly in developing countries. Without any adequate financial support, adaptation to climate change will likely be spontaneous, unplanned, relatively small in scale, and so the impact generated from them is also relatively limited. This pattern of adaptation does not much to change and to provide greater benefits to society. A number of important questions that need to be resolved on this theme (C) include:

- C.1. How much cost required to produce adequate adaptation actions? And what are economic benefits resulting from adaptation activities?
- C.2. How do climate change adaptation activities be implemented at low costs but provide effective results and greater impacts to society?
- C.3. How are the best way to integrate climate change adaptation activities and other activities that have economic or market values?

Mainstreaming adaptation to climate change to be one of research themes that is not less important in climate change adaptation. Adaptation to climate change will not run effectively when climate change is not on the agenda of all development stakeholders. In this mainstreaming, supporting conditions are necessary and required. It is a unique process as it also involves timing and the availability of the key actors and local champions, and political alignments. The theories of the policy process are important to understanding this; they can explain why does climate change policy, including mainstreaming process work very well or does not work at all in the city government. A study conducted by Setiadi [22] for example describes the process of adaptation policies to climate change in urban areas by applying one policy theories known as the advocacy coalition framework

[23]. Yet, many other theories that can be used to explain the development of urban climate change policies, such as the institutional analysis and development, policy diffusion, and multiple streams. Back in the context of mainstreaming for effective adaptation to climate change (D), there are a number of fundamental questions such as:

- D.1. What are key drivers for effective mainstreaming of climate change adaptation? And how to establish that key driver at the municipal level?
- D.2. How to encourage adaptation to climate change at the municipal level?
- D.3. What are capacities required by local actors to get the highest effective mainstreaming of climate change?

This research and practical roadmap can be applied in the areas of climate change adaptation at the local level and in particular in urban areas. In other words, the question type A to D above can be tested to explain and evaluate critical areas of urban studies. These critical areas or fields include (1) food, (2) water, (3) energy, (4) disaster risk reduction, (5) settlements and the built environment, and (6) the urban natural environment.

As described in Figure 1, the support of the national government is another element that makes adaptation to climate change at the city level effective. The national government also needs the support of international community. The elements of central government support consist of four other sub-elements, namely: support to basic scientific and reliable climate data, a clear legal framework for climate actions and adaptation programs, adaptation financing schemes, and knowledge management. Associated with the support for the scientific basis and climate data, there are several underlying research questions (E), such as:

- E.1. What are key drivers, barriers, and challenges in providing scientific evidence and continuous climate data?
- E.2. What are roles played by scientific evidence in urban climate change policy development? Are they really matters?

Some recent works such as Morgan [24] and Tangney [25] has addressed E2 questions. While Morgan [24] examines the case of science-policy relationship in water resource management in Southeast Queensland, Tangney [25] compares complex relationships between climate change adaptation policy and evidence in UK and Australia. Additionally, Setiadi and Lo [26] also explore the role of policy research in local climate change policies in the Indonesian context.

Various existing regulations (ranging from legislation, presidential decree, ministerial regulation, to the local regulations) greatly affect the effectiveness of climate change adaptation. The regulation can encourage or inhibit climate change adaptation. It is associated with a pattern of local government culture, especially in Indonesia, which still puts themselves as the executor rather than as an innovator or creator of policy breakthroughs. These regulations sometimes are unclear and overlapping, which raises doubts and reluctance of local governments to run climate change adaptation programs. Some of the key questions on this theme (F) include:

- F.1. What are regulations that still unclear, potential to generate contradictions, and inhibit adaptation to climate change?
- F.2. What are breakthroughs that should be done to overcome this?

The availability of financing from national schemes has become one of the incentives to city governments to run programs for adaptation to climate change. In 2008 the Indonesian government through BAPPENAS, for example, established the Indonesia Climate Change Trust Fund (ICCTF). Despite attracting international funds to support climate change programs in developing countries, the establishment of the ICCTF represented an attempt to respond to demands for greater coordination and harmonization of climate change funding. Yet, most Indonesian local governments are unaware of the existence of and the funding opportunities provided by the ICCTF. The ICCTF has funded a limited number of climate change projects and all of them were proposed by and went to ministries and institutions at the national level [27], rather than government agencies or non-state actors at the provincial or city level.

- G.1. To what extent of climate change adaptation funding is available?

- G.2. What are roles played by such funding scheme that is publicly opened to climate change adaptation initiatives? and what are effects following this open adaptation funding on the ground?
- G.3. What are barriers and challenges faced by local governments and local community actors in accessing available funding sources?

Efforts in building and maintaining urban resilience to climate change are strongly influenced by the ability of cities to manage knowledge [28], sometimes referred to knowledge management. The knowledge that is built from local actions or practice is important and valuable to be properly managed. The knowledge that can be managed includes methods, solutions, experience, systems, etc. Additionally, providing the media in the form of a forum, or a platform to facilitate the transmission of knowledge so that they can be exchanged to the entire network of the city is also essential.

In essence, the learning process becomes a very important component to accumulate knowledge and experience, as well as to find out what works and does not, including the factors that influence them. The practices of adaptation to climate change that has proven successful in a city or a community need to be lifted and to be disseminated in the other cities and communities involved. Thus, urban climate change adaptation activities are not always departed from the beginning or zero. Under this theme, a number of fundamental research questions include:

- H.1. What are the extent and forms of support made by the national government in encouraging peer-learning among cities in climate change adaptation?
- H.2. What are roles played by the process of peer-learning between cities in the context of urban climate change adaptation?
- H.3. What is the most effective mechanism to improve peer-learning among city governments?

4. Conclusion

This paper has described the rationale, development criteria, and the element of urban climate change adaptation framework. The framework aims to organize various themes of research and then it is expected to result in a roadmap toward research programs and practical actions focusing on urban climate change adaptation. As this framework is at its infancy stage, the framework welcomes further constructive suggestion and means to invite further research collaboration in this field.

References

[1] Sabatier P A 1999 The Need for Better Theories *Theories of The Policy Process: Theoretical Lenses on Public Policy* ed P. A. Sabatier (Westview Press: Boulder) pp 3-17.

[2] Hulme M 2009 *Why We Disagree About Climate Change: Understanding, Controversy, Inaction and Opportunity* (New York: Cambridge University Press)

[3] Funfgeld H and McEvoy D 2014 *Environ & Plann. C: Gov. & Policy.* **32(4)** 603-22.

[4] Preston B L, Mustelin J and Maloney M C 2013 *Mitigation and Adaptation Strategy for Global Change.* **20(3)** 467-497

[5] Füssel H M 2007 *Sustainability Sci.* **2** 265–75 doi: DOI 10.1007/s11625-007-0032-y

[6] Biagini B, Bierbaum R, Stults M, Dobardzic S and McNeeley S M 2014 *Global Environ. Change.* **25** 97-108.

[7] Eriksen S and Brown K 2011 *Climate and Development.* **3** 3-6.

[8] Eriksen S, Aldunce P, Bahinipati C S, Martin R D, Molefe J I, Nhemachena C and Ulsrud K 2011 *Climate & Dev.* **3** 7-20.

[9] IPCC 2014 Summary for Policymakers *Climate Change 2014: Impacts, Adaptation, and Vulnerability. Part A: Global and Sectoral Aspects. Contribution of WG II to the Fifth Assessment Report of the Intergovernmental Panel on Climate Change* eds C. B. Field et al. (Cambridge University Press, Cambridge, UK and New York, NY) pp 1-32

[10] OECD 2010 Urbanisation, Economic Growth and Climate Change *Cities and Climate Change.* pp 37-79.

[11] UN-Habitat 2011 *Global Report on Human Settlements 2011: Cities and Climate Change-Policy Directions* (London: Earthscan)
[12] Bevir M 2008 *Key Concepts in Governance* (London: SAGE)
[13] Bache I and Flinders M 2004 *Multilevel Governance* (Oxford: Oxford University Press)
[14] Betsill M M and Bulkeley H 2007 *Local Environ.* **12(5)** 447-56.
[15] Crespy C, Heraudi J A and Perry B 2007 *Reg. Stud.* **41(8)** 1069-84.
[16] Corfee-Morlot J, Cochran I, Hallegate S and Teasdale P J 2011 *Clim. Change.* **104** 169-97.
[17] Munang R, Thiaw I, Alverson K, Mumba M, Liu J and Rivington M 2013 *Current Opinion in Environmental Sustainability.* **5(1)** 67–71. doi: http://doi.org/10.1016/j.cosust.2012.12.001
[18] Vignola R, Locatelli B, Martinez C and Imbach P 2009 *Mitig Adapt Strateg Glob Change.* **14(691)** doi: 10.1007/s11027-009-9193-6
[19] Satterthwaite D and Dodman D 2009 The costs of adapting infrastructure to climate change *Assessing the costs of adaptation to climate change: A review of the UNFCCC and other recent estimates* eds M. Parry et al. (London: IIED and Grantham Institute for Climate Change, Imperial College London)
[20] Setiadi R and Nalau J 2015 *Asian Cities Climate Resilience Working Paper Series* **23** (London: IIED)
[21] Bouwer L M and Aerts J C J H 2006 *Disaster* **30(1)** 49-63 doi: 10.1111/j.1467-9523.2006.00306.x
[22] Setiadi R 2015 The Emergence of Local Climate Change Adaptation Policy: An Advocacy Coalition in Indonesian Cities 1993-2013 [Dissertation] (Griffith University: Gold Coast)
[23] Sabatier P A and Jenkins-Smith H C1999 The advocacy coalition framework: an assessment *Theories of the Policy Process: Theoretical Lenses on Public Policy* ed P A Sabatier (Boulder: Westview Press) pp 117-166
[24] Morgan E 2014 *International Journal Sustainability Policy & Practice.* **9(2)** 37-54.
[25] Tangney P 2017 *Climate Adaptation Policy and Evidence: Understanding the Tensions between Politics and Expertise in Public Policy (*London: Earthscan)
[26] Setiadi R and Lo A Y 2017 *Urban Policy and Research.* doi: https://doi.org/10.1080/08111146.2017.1377607
[27] Gruning C, Menzel C, Shuford L S and O'Brien V S 2012 National climate finance institutions case study: The Indonesia Climate Change Trust Fund (ICCTF): (Frankfurt: Frankfurt School of Finance & Management). 1-22.
[28] Kernaghan S and Silva J D 2014 *Urban Clim.* **7** 47-63.

Relationship of various factors affecting the sustainable private forest management at Pajangan District, Special Regions Yogyakarta, Indonesia

B Widayanto[1,4], R Karsidi[2], Kusnandar[3] and J Sutrisno[3]

[1] Doctoral Program in Community Development and Empowerment, Graduate School, Universitas Sebelas Maret (UNS), Jl. Ir. Sutami No. 36A, Kentingan, Surakarta, Central Java 57126, Indonesia
[2] Department of Education, Universitas Sebelas Maret, Jl. Ir. Sutami No. 36A, Kentingan, Surakarta, Central Java 57126, Indonesia
[3] Department of Agribusiness, Universitas Sebelas Maret, Jl. Ir. Sutami No. 36A, Kentingan, Surakarta, Central Java 57126, Indonesia
[4] Corresponding author: budiwidayanto@ymail.com

Abstract. Forests have a role and function in providing good atmosphere with stable oxygen content and affecting global climate stability. Good forest management will provide stable climatic conditions in global climate change. A good forest is managed to provide a sustainable environment condition. This study aims to analyze the relationship of various factors affecting the sustainability of private forests management. This research is a quantitative research with survey method and determination of sampling are was by purposive sampling. Sampling method using multiple stage cluster sampling with 60 samples. From the results it was found that the successful sustainable private forest management influenced by various factors, such as group dynamics, stakeholder support, community institutions, and farmer participation. The continuity of private forest management is determined by the fulfillment of economic, social and environmental dimensions. The most interesting finding is that the group dynamics conditions are very good, whereas the sense of togetherness among community is very strong under limited resources managing private forests. The sense of togetherness resulted creativity to diversify business and thus reduced the pressure in exploiting the forest. Some people think that managing the people's forest as a culture so that its existence can be more sustainable.

1. Introduction

Forest is an ecosystem unit that exists from a natural process and its existence greatly affects the interaction between human's social, economic life, and their natural environment. A study of forest which is according to the Sustainable Development Goal's (SDG) at objective 15 deals with life on land. It discusses in detail life on land by emphasizing on protection, restoration, sustainable improvement of the land ecosystem utilization, sustainable forest management, reduction of barren or monarable/arid lands (i.e. combating desertification), prevention and recovery of land degradation, and putting an end to the loss of biodiversity.

According to Bauer *et al.* [1] the attainment of the SDG's target are, *inter alia*, a) to ensure conservation, restoration, and land utilization that are compatible with a sustainable ecosystem of land freshwater and the hinterland; b) to promote the implementation of sustainable management of all types of forest, to put a stop to deforestation, recover damaged forest, and step up afforestation and

Content from this work may be used under the terms of the Creative Commons Attribution 3.0 licence. Any further distribution of this work must maintain attribution to the author(s) and the title of the work, journal citation and DOI.
Published under licence by IOP Publishing Ltd

IOP Conf. Series: Earth and Environmental Science **129** (2018) 012042 doi:10.1088/1755-1315/129/1/012042

reforestation globally; c) more specifically, the idea of the management of long-lasting forest seems to focus on logging, while forest conversion into farmland is the main impetus to global deforestation.

Forest has a strategic function in determining global climate because its vegetation can serve to change CO_2 into organic matter and O_2 (O_2 availability in the atmosphere), so that it can reduce global warning. Indonesian tropical forests have a strategic position geopolitically in responding to global climate change, reduction of carbon emission, conservation of reserves and increase in carbon intake/absorption. Tropical forests have a role in the process of the equilibrium of carbon cycle in the earth's atmosphere to maintain global climate [2]. Forest has its importance in the aspect of social welfare individually and in the interest of a larger area. Law No. 41 of 1999 concerning forestry Article 18 paragraph 2 stipulates that a province or regency must reserve/set aside land utilization in the form of forest preserve of 30% in the *RTRW* (spatial and regional plan). The land area of *DIY* (The Special Region of Yogyakarta) is 318,580 hectares, so 95,574 hectares is needed. The data on the area of state-owned forest in *DIY* is 16,819.52 hectares (i.e. 5.28% of land area). This condition indicates that in order to fulfil the stipulation of Law No. 41 of 1999, there is a need to increase the acreage of forest beyond the state-owned forest with the alternative of extending and improving the quality of the people's private and the state-owned forest. According to the data in the district and provincial forestry office in 2004, the area of state-owned forest and the peoples' (private) forest was 56,211.34 (17.64% of *DIY* area), thus there was still a shortage of forest area as much as 38,388 hectares (12.36%). This necessitates a policy in the utilization of the people's forest which can meet the interest of the community and larger region.

The objective of this research is to carry out a study on the relationship of various factors that affect the sustainable private forest management. The factors under study were community development, community institutions, stakeholder support, group dynamics, farmer participation affecting the sustainable private forest management.

2. Literature review

Pajangan is private (privately owned) forest area of 950 hectares. Its people are joined together in a group which called Wono Lestari *UMHR* (private forest management unit). An institutional approach through UMHR is adopted to avoid the management and utilization of private forest by individual families instead of a collective action, so the private forest employment is operated without the collectivism/ togetherness principle in dealing with the environment. The individual families application often motivate excessive logging, which is resulting in an environmental imbalance. On the other hand, the community's capacity for establishing the private forest ecosystem is the their culture product which should be respected and appreciated by all concerned parties [3].

Wono Lestari *UMHR* has a system called *SVLK* (Timber Legality Verification System), which is a requirement to fulfil the timbers/ products legality that is devised according to a parties concerned consensus who set up standards, criteria, indicators, verifiers, verification method, and evaluation norms (Forestry Ministerial Regulation No. P.38/Menhut-11/2009 Article 1 para. 10). This certification has an implication for acceptance and recognition by the international market on the local timbers product and it bolsters the long-lasting forest management. *UMHR* also has a certiification called *PHBML* (Sustainable Community-based Forest Management), which is a forest and forest products management that is praticed by the traditional ways of community and individuals in a community unit, community-based operation unit, as well as by individuals on a small and medium scale run in a sustainable way [4].

Wono Lestari is a forum for the private forest farming community in Bantul Regency to step up the community's welfare by preserving and maximizing the forest potential. The vision is to become an organization that oriented to the private forest conservation, and the misson is to foster Wono Lestari as a strong organization, improve the members's skills and knowledge, and boost the forest products selling price [5]. As a system, the private forest should be managed by paying due attention to the various functions importance that ensure the harmony of the sustainable private forest system.

The people who live in the vicinity of the forest can actually become the pillars for the creation of sustainable forest management. Their behavior is a crucial component in managing and conserving forest [6]. In addition, initiative should be encouraged from outside and gradually the rural community can learn from their own experience [7].

Private forest is owned by the farmers so all utilizations are largely determined by the farmers' will and capability. To encourage the farmers to be willing to conduct a sustainable forest management, the most logical approach is a group approach where they are organized. Friedman [8] affirms that an individual's capability to organize themselves in a group is regarded as the most effective form of community level empowerment (collective self-empowerment).

A dialogical encounter that fosters and strengthens an awareness and solidarity will be occurred by employing a group. Empowerment is aimed at enhancing power out of a disadvantage. Community empowerment should be able to influence community institutions that play a role in shaping the people's behaviour in terms of the sociocultural aspect history. Besides, it should also be able to influence the stakeholders in efforts to promote a sustainable management. The behavioral change strategy says, the main strength in changing an individual's behavior is through the process of participation in a group [9].

The empowerment has the independence and participation element, and in this context is oriented to strengthen community institutions; therefore its community members is an initial step towards the all the members participation [10]. In other words, it is a participation road, particularly in the decision making process in order to foster their independence. It is the assets and capabilities expansion of poor people to participate in, negotiate with, influence, control, and hold accountable institutions that affect their lives [11].

The results of the World Bank's study on the agriculture development project show that the project success happens when the attention is given to institutional development and community participation [12]. Lineberry [13] suggests that the main problem in community enpowerment is determined by the institutional factor that exists in the community. Its afforestation and forest rehabilitation is intended to involve all the potentialities in society, especially universities, community organizations, and non-government organizations for cultivating sustainable forest development for an indefinite period [14]. The private forestry operations development needs to be managed and designed with a sustainable forest management system. For its purpose, its management should be participatory, providing opportunities to the people who in and arround the forest to actively take part for sustainable forest advancement by starting from decision making, planning implementation of activities, monitoring and evaluation as well as the products utilization. Private forest management should cover the economic, social, and environmental aspects [4].

3. Method

Research was conducted in Pajangan, Special Region of Yogyakarta with the consideration that the private forest management there has developed with the ownership of Sustainable Community-Based Forest Management (*PHBML*) and Timber Legality Verification System (*SVLK*). This research was a quantitative research using the survey method. It is the best available method in social sciences in gathering original data in order to describe a very large population observed directly [15]. By this method, the showed sampling was ensured that the samples' characteristics can reflect a large population. In the implementation, great care is needed to determining the same standard questionnaire construction for all respondents. The survey method was adopted to create generalization from the observation of samples number.

3.1. Population and sample

A population is all the individuals, conditions, or symptoms that serve as research objects. Considering the population distribution, the technique which was used is to multiple stage cluster sampling in accordance with groups based on area, or sample taking in stages.

IOP Conf. Series: Earth and Environmental Science **129** (2018) 012042 doi:10.1088/1755-1315/129/1/012042

The taken steps with regard to the research samples were as follows: 1) Selection of subdistrict villages: Sendangsari and Triwidadi were selected for the simple reason that they pioneered the existence of Wono Lestari private forest management unit (*UMHR*) and sustainable community-based forest management (*PHBML*); 2) Selection of 3 forest farmer groups (*KTH*) on the basis of their intensity in the private forest sustainable management of in each village, and with the officers consideration in charge, 6 *KTH* were selected; 3) Random respondents sampling from each *KTH*: 10 respondents were selected for their active participation in private forest management (as attested by the data from the groups). As a whole, from 2 villages (6 *KTH*) 10 respondents were selected so that there were 60 respondents.

In this research, the data were obtained by using a research instrument with scales measurement on attitudes by *Likert scale*. States that the basic principle of Likert scale is determining an individual position in a continuum of attitudes to the object of attitudes, starting from very negative to very positive [16]. The model comprises 6 variables consisting of 1 exogeneous variable (community empowerment/*X1-PM*) and 5 endogeneous variables (community institutions/*Y1-KM*), (stakeholder support/*Y2-DS*), (group dynamics/*Y3-DK*), (farmer participation/*Y4-PP*) (sustainable private forest management/*Y5-PHRB*). The details of the community institutions variables (*Y1-KM*) consists of 3 subvariables with 22 indicators; the stakeholder support variable (*Y2-DS*) consists of 4 subvariables with 16 indicators; the community empowerment variables (*X1-PM*) consists of 3 subvariables with 25 indicators; the group dynamics variable (*Y3-DK*) with 9 variables with 37 indicators; the farmer participation variable (*Y4-PP*) consists of 5 subvariables with 17 indicators; and the sustainable private forest management variable (*Y5-PHRB*) consists of 3 subvariables with 29 indicators.

The data from 60 respondents were then tabulated and reliability and validity tests were performed. The results of the validity and reliability tests that met the requirements were followed by a modelling process using WarpPLS 5.0.

4. Result and discussion

The relationship results between various variables (community institutions, stakeholder support, group dynamics, farmer participation) affecting sustainable community forest management can be seen in Figure 1.

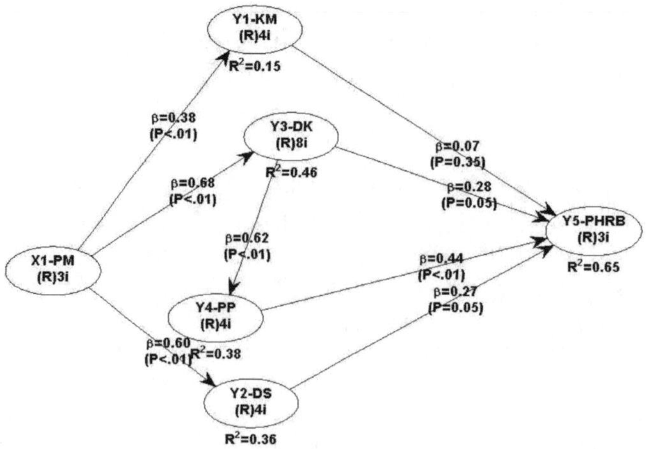

Figure 1. Various factors affecting sustainable private forest management.

Figure 1 shows the testing results on the direct effect model. The results can account for the direct relationship between variables, for instance the community empowerment direct effect on community institutions was significant with path coefficient of 0.38; the direct effect of *X1-PM* on *Y3-DK* was significant with path coefficient of 0.68, whereas the direct effect of *Y1-KM* on *Y5-PHRB* was not significant with p-value of 0.35 (>0.05). The suitability of the model can be seen with various indicators in Table 1.

Table 1. Model fit indices and p-value.

Indicators	Value	P value
Average path coefficient (APC)	0.417[a]	0.002
Average R-squared (ARS)	0.400[a]	0.003
Average adjusted R-squared (AARS)	0.370[a]	0.005
Average block VIF (AVIF)	1.638	acceptable if <= 5, ideally <= 3.3

[a]P < 0,01
[b]P < 0,05

The output in Table 1 shows that the model fit indicators have been met, i.e. APC with a value of 0.417, which means with significant with p-value 0.002 (***p<0.01) and ARS with a value of 0.400, which means significant with p-value if 0.000 (*** p<0.01). Likewise, AVIF indicator of 1.638 meets the ideal requirement (AVIF<=3.3).

Table 2. PLS results (path coefficient, *p-value*, R^2).

	Path to				
Variable	Y1-KM	Y2-DS	Y3-DK	Y4-PP	PHRB
X1-PM	0.38[a]				0.07[c]
X1-PM			0.68[a]		0.28[b]
X1-PM			0.68[a]	0.62[a]	0.44[a]
X1-PM		0.60[a]			0.27[b]
R^2	0.15	0.36	0.46	0.38	0.65

[a] P<0,01
[b] P<0,05
[c]nonsignificant

Table 2 shows that all the path coefficient of the predictor variable effect on the response variable have a positive and significant effect, except for *Y1-KM* variable on *PHRB* which was not significant with p-value of 0.35 (>0.05). WarpPLS output was the standardized data so that it can be interpreted that the greater the path coefficient the stronger the effect will be.Table 2 displays, on the whole, the model *PHRB* variable has a determination value (R^2) of 0.65. the relationship path that portrays the significant effect was demonstrated by the variable coefficient and its p-value, namely *Y2-DS, X1-PM, Y3-DK, Y4-PP* with a p-value of <0.001, whereas if *Y3-DK* goes directly to *PHRB* without going through *Y4-PP* it is significant at p=0.05.

Based on Table 3 it can be explained that the community empowerment variable (*X1-PM*) has a significant direct effect on the community institutions variable (*Y1-KM*) with coefficient (beta) value of 0.38 but *X1-PM* has a non-significant effect (p=0.35>0.10) on the sustainable private forest management (*Y5-PHRB*). The *X1-PM* variable has a significant effect on *Y5-PHRB* indirectly and through the institution dynamics (*Y3-DK*) of 0.422 and was significant with a p-value of <0.001, so that it can be said that *Y3-DK* is a mediation effect of *X1-PM* on *Y5-PHRB* (mediation of 2 segments). *X1-PM* has a significant effect on *Y5-PHRB* indirectly through *Y3-DK* and farmer participation (*Y4-PP*) of 0.186 and significant with a p-value of <0.001 (mediation of 3 segments). From all path coefficients, the predictor variable has a positive and significant effect. However, the path coefficient of community institution variable effects (*Y1-KM*) was not significant.

Table 3. The indirect, direct, and total effects.

Variable type				Type of influence		
Predictors	Response	Mediation 1	Mediation 2	Direct	Indirect	Total
X1-PM	Y1-KM			0.380		0.380[a]
X1-PM	Y2-DS			0.603		0.603[a]
X1-PM	Y3-DK			0.679		0.679[a]
X1-PM	Y4-PP	Y3-DK			(0.68x0.62)	0.422[a]
X1-PM	Y5-PHRB	Y3-DK	Y4-PP		(0.68x0.62x0.44)	0.186[a]
Y1-KM	Y5-PHRB			0.069		0.069[c]
Y2-DS	Y5-PHRB			0.271		0.271[b]
Y3-DK	Y5-PHRB			0.280		0.280[a]
Y3-DK	Y5-PHRB	Y4-PP			(0.62x0.44)	0.548[a]
Y4-PP	Y5-PHRB			0.443		0,443[a]

[a]P<0,01
[b]P<0,05
[c]nonsignificant

This was probably due to several things: 1) The community institutions were too large/extensive to be able to attend to the private forest activities, and 2) Not all community institutions are capable of able to support the private forest management activities effectively and efficiently.

The other results of research show that the farmer's orientations in dealing with private forest are as follows: (1) Economic reasons to meet their basic necessities of life; for their livelihood, they rely on working as laborers, private forest operation, and partly on financial assistance from successful relatives; (2) Ecological reasons to create more green and healthier environment for the community collectively; (3) Social reasons for social life by creating job opportunities for their welfare, and cultural reasons of the community who still uphold the principle that ownership of trees (especially teak) gave a higher social status.

The three reasons were a dimension of the sustainable private forest management system, and specifically, the third reason is an extremely supportive of the sustainable private forest management achievements.

5. Conclusion and recommendations

The impacts of climate change can be reduced by sustainable private forest management. To achieve the sustainability of private forests with attention to the community institutions factors, stakeholder support, community empowerment, group dynamics, and farmer participation.

This research has analyzed the various variables effects (community institutions, stakeholder support, community empowerment, group dynamics, and farmer participation) on sustainable private forest management. The PLS results displayed that the total predictor variable percentage as much as 65%. Based on Table 2, only the relationship of *Y1-KM* to *Y5-PHRB* which was not significant, but from data processing by removing/deleting the predictor variable, the R^2 of the response variable decreases to 62%. Various reasons why farmers postpone logging were as follows: harvesting which according to the trees age; environmental requirements; the need for certification; and culture. The most interesting reason was by owning large-sized and older age trees, the farmers have a high social status in society, and the cultural consideration was very influential in sustainable forest management.

Although with a sample of 60 the SEM-PLS software was already able to estimate highly enough, a further research needs to use samples in accordance with the recommendation of Maximum Likelihood Estimation (MLE). Bigger samples can provide a good basis for estimation and result in data stability. Big-sized samples (>250) can enhance the accuracy and consistency of the estimated results of SEM-PLS [17].

References

[1] Deutsches Institutfür Entwicklungspolitik (DIE)/ German Development Institute "The Sustainable Development Goals of the Post-2015 Agenda: Comments on the OWG and SDSN Proposals" *https://www.oecd.org* 2015 [Online] Available: https://www.google.co.id/search?q=German+Development+Institute+%2F+DeutschesInstitutf%C3%BCrEntwicklungspolitik+(GDI%2F+DIE).+2015.+The+Sustainable+Development+Goals+of+the+Post2015&oq=German+Development+Institute+%2F+DeutschesInstitutf%C3%BCrEntwicklungspolitik+(GDI%2F+DIE).+2015.+The+Sustainable+Development+Goals+of+the+Post-2015&aqs=chrome..69i57.1967j0j8&sourceid=chrome&ie=UTF-8 [Accessed: 13-09-2017]

[2] Soepijanto B 2014 *Forests of Indonesia are Independent, Sovereign and Personality* (Tangerang: Penerbit Wana Aksara) pp 70-5

[3] Awang S A, Wiyono E B and Sadiyo S 2007 *Private Forest Management Unit: Local Knowledge Construction Process* (Yogyakarta: Banyumili Art Network) pp 13-5

[4] Lembaga Ekolabel Indonesia (LEI) "PHBML Certification Guidelines" *https://www.lei.or.id* 2016 [Online] Available: https://www.google.co.id/search?q=lembaga+ekolabel+Indonesia.or.id&oq=lembaga+ekolabel+indonesia.or.id&aqs=chrome..69i57.24066j0j8&sourceid=chrome&ie=UTF-8 [Accessed: 13-09-2017]

[5] UMHR Wono Lestari Pajangan 2015 *Document of Submission of Certification of Sustainable Community Based Forest Management* (Yogyakarta: PHBML) pp 7-11

[6] Suprayitno A R 2008 Local Community Involvement: Efforts to Empower Communities Toward Sustainable Forest *Jurnal Penyuluhan* **4(2)** pp 135-8

[7] Karsidi R 1999 Studi on the success of occupation transformation from farmer to small-scale industry worker *Thesis Dissertation* (Bogor: Graduate School of Institut Pertanian Bogor) p 46

[8] Nasdian F T 2015 *Community Development* (Jakarta: Yayasan Pustaka Obor Indonesia) pp 96-7

[9] Ife J 2002 *Community Development: Creating Community Alternatives Vision Analysis and Practice* (Sydney: Addison Wesley Longman Australia Pty Ltd) p 53

[10] Rustiadi E, Sunsun S and Dyah R P 2011 *Planning and Regional Development* (Jakarta: Crestpent Press and Yayasan Pustaka Obor Indonesia) p 365

[11] Narayan D 2002 *Empowerment and Poverty Reduction: A Sourcebook* (Washington DC: The World Bank) p 14

[12] Cernea M M 1988 Put People first in Development: Sociological Variables in Rural Development *Publication of the World Bank* (Jakarta: UI-Press) pp 341-373

[13] Nitiwijaya Y S 2013 The factors affecting empowerment farmers surrounding forest *Thesis Dissertation* (Surakarta: Graduate School of Universitas Sebelas Maret) p 56

[14] Mardikanto T 2013 *Models of Community Empowerment* (Surakarta: UNS Press) pp 185-200

[15] Babbie E 1983 *The Practice of Social Research* 3th Edition (California: Wadsworth Publishing Company) p 209

[16] Widoyoko E P 2013 *Technique of Research Instrument Preparation* (Yogyakarta: Pustaka Pelajar) p 104

[17] Hair J, C Ringle and Sartstedt M 2013 *A Primer on Partial Least Squares Structural Equation Modelling (PLS-SEM)* (Los Angeles: Sage) p 16

IOP Conf. Series: Earth and Environmental Science **129** (2018) 012043 doi:10.1088/1755-1315/129/1/012043

The interrelationship of households economics activities of upland rice farmers in rain-fed farming in Ponjong Sub-district, Gunungkidul District, Indonesia

W D E Rini[1,2,5], M Harisudin[3], Supriyadi[4] and E S Rahayu[3]

[1] Department of Agribusiness, University of Pembangunan Nasional "Veteran" Yogyakarta, Jl. SWK 104, Condongcatur, Depok, Sleman, Yogyakarta 55283 Indonesia
[2] Doctoral Program in Agriculture Science, Graduate School, Sebelas Maret University, Jl. Ir. Sutami 36 A, Kentingan, Surakarta, Central Java 57126 Indonesia
[3]Department of Agribusiness, Sebelas Maret University, Jl. Ir. Sutami 36 A, Kentingan, Surakarta, Central Java 57126 Indonesia
[4]Department of Agrotechnology, Sebelas Maret University, Jl. Ir. Sutami 36 A, Kentingan, Surakarta, Central Java 57126 Indonesia
[5]Corresponding author: wulandaririni@ymail.com

Abstract. Gunungkidul is one of the regencies at Yogyakarta, Indonesia which is 90% occupied by dry land, and thus vulnerable to climate change impact. Since dryland relies on water only from rain to meet crop water requirement, part of land management is rainfed. This condition encourages farmers to make the right decision regarding their additional income to meet household needs. Under the limited land resources, farmers decided to plant upland rice once or twice a year. The aim of the study is to analyze the interrelationship of households economics activities of upland rice farmers in rain-fed farming based on production, labor allocation, and consumption. The research method is descriptive analysis, with research site Ponjong sub-district, determined by the purposive method. Sampling method using proportional random sampling. Economics model was determined by using simultaneous equation model, with 2 SLS estimation method. The results showed that the household economics model of upland rice farmers in the rainfed land can be explained by using farmers household model and there is a linkage between production, labor allocation, and consumption.

1. Introduction

Agriculture sector plays important roles since the development of agriculture sector aims to boost production result and quality, to strengthen farmers income and living standards, to multiple employment and business opportunities, and to support industrial sectors. One of the agriculture development purposes is to intensify farmers income. Their income development can be done through many factors, such as by increasing broad unity production or productivity. Besides extensification, the agricultural productivity increment through an intensification activity becomes a source of agriculture sector prosperity. Meanwhile, most of Indonesia agricultural activities are cultivated by them as family farming with land ownership which is narrow of less than 0.5 ha [1]. it is managed by them because of their low-income, especially in dryland farming. Its potential in this country is enormous, due to it reaches 9.7 million hectares [2].

Content from this work may be used under the terms of the Creative Commons Attribution 3.0 licence. Any further distribution of this work must maintain attribution to the author(s) and the title of the work, journal citation and DOI.
Published under licence by IOP Publishing Ltd

These last few years, food self-sufficiency and security began to be threatened. Simultaneously, there was a productivity level degradation especially towards rice and rice field conversion in Java which continues. Nowadays, food production circumstances are being aggravated by climate factor which is unfriendly. Farming development must notice the land character which is cultivated. According to the water state, types of land can be divided into irrigation, rain-fed, dryland, and swampland rice field. The farming achievement of those types is quite different, it depends on the plants technical and economic feasibility response towards the land characteristics which is harvested.

The problems which deal with the production factor employment is scarcity on fertile lands availability for crops production. Its crops land area development has already been difficult to be cultivated and there is a fact of rice field conversion to another use. Besides the problem of the land availability number, also faces problems with the land quality reduction problem due to erosion and pollution. The suboptimal land advancement will become an alternative for food provision in the future because many fertile irrigation rice fields switch functions to nonagricultural interests. Suboptimal lands include dryland, rain-fed land, tidal rice fields, and swampland.

The rain-fed land is the second granary after irrigation rice field. In Daerah Istimewa Yogyakarta, numerous rain-fed lands are found which have a high level of rainfall, however, it rarely performs such as in Gunungkidul district, two-seasons tropical climates area as the rainy and dry season. For farmers, the seasonal condition has a big influence towards farming activities. A 90% area of Gunungkidul district is dryland and less fertile also really depends on rainfalls, therefore it is vulnerable towards climate changes impacts. It is a district which has a site specification. Most of the farming areas belong to this area are rain-fed lands which depend on the climate cycles particularly rainfall.

The agricultural production land, dry wetlands are the farmers household income in Gunungkidul district. Although the water availability is limited, farmers attempt to manage the lands by cropping pattern which is adjusted to the land condition. That underline background is to conduct a research on farmers household's behaviors in the rain-fed land. It is where the highland rice farmers highland decides to manage it which is one-time of highland and two times of plateau cropping pattern highland. Moreover, the household's economics activities perspective is also being observed, the land use selection is influenced by rational farmers. The choice of farming land use is closely related to farmers decision such as being influenced by various aims namely: maintaining household food security; supporting cash income in fulfilling needs; minimizing risks, maximizing leisure which is related to time allocation, assuring family members good and prosperous condition, and also to achieve certain social class in his community [3].

The approach of farmers household's economics research must consider the farmers and their family behavior in taking production and consumption decisions which are related to one another that will influence the income levels, labor allocation, and consumption. The aim of the study is to analyze the interrelationship of households economics activities of upland rice farmers in rain-fed farming based on production, labor allocation, and consumption.

2. Research Methods

2.1. Research site and respondent's determination of upland rice farmers household

The applied research method was the descriptive analysis method. The research site determination was applied by employing a purposive method which resulting in Ponjong Sub-district with areas consideration are found in each zone in Gunungkidul District and there are farmers household conducting upland rice farming in a rain-fed land with one time and two times of cropping seasons in one year. The samples collecting method of upland was applied proportional random sampling method by taking 80 respondents of farmers household with 51 and 29 respondents of farmers household which planting upland two times upland and one time in one year respectively.

2.2. Models of household economics

upland rice farmers household economics qualification was researched the first step by applying the econometrics approach method. Econometrics model describes the relationship of each independent variables towards dependent variables especially about magnitude and sign of the prior functional parameter estimated coefficients which is according to economic theories [4]. In order to identify the interconnectedness among production, labor allocation, and consumption also other aspects related to households decision, and agricultural household economics model was formulated in simultaneous equations using 16 formulas which consist of 10 structural and 6 identity equation modal.

The equation model was spesified into: production, labor allocation, income, and outcome. The model built specification in this research can be explained as the following:

2.2.1. Production:

$$QPG = a_0+a_1LL+a_2JB+a_3JPU+a_4JPN+a_5CKDKP+a_6CKLKP+a_7DPG+E_1 \quad (1)$$

Notes:

QPG =upland rice production(kg/year)
LL =land area(ha)
JB =the number of seeds(kg/year)
JPU =the number of urea fertilizers(kg/year)
JPN =the number of NPK fertilizers(kg/year)
CKDKP =labor allocation in upland rice family (Labor Days Requiraments=LDR/year)
CKLKP =labor allocation outside upland rice family (LDR/year)
DPG =upland ricedummy (0= pattern 1x of Uplandrice in one year, 1= pattern 2x of upland rice in one year)

The expected estimate parameter a1,a2,a3,a4,a5,a6,a7>0

2.2.2. Labor allocation:

$$CKDKP = b0+b1LL+b2CKDKNP+b3CKDOF+b4PDRT+b5DPG+E1 \quad (2)$$
$$CKDKUT = CKDKP+CKDKNP \quad (3)$$
$$CKLKP = c0+c1LL+c2CKDKP+c3PDRT+c4DP+E3 \quad (4)$$
$$CKLK = CKLKP+CKLKNP \quad (5)$$
$$CKDKOF = d0+d1LL+d2CKDKUT+d3CKDKNPt+d4DPG+E4 \quad (6)$$
$$CKDKNPt = e0+e1LL+e2CKDKUT+e3CKDKOF+e4JTKDK+e5DPG+E5 \quad (7)$$

Notes:

CKDKNP =labor allocation in non-rice family (LDR/year)
CKDKOF =labor allocation in off farm family (LDR/year)
PDRT =household income (IDR/year)
CKDKUT =labor allocation in farming activites family (LDR/year)
CKDKNPt =labor allocation in non farming family (LDR/year)
JTKDK =the number of labors in family(people)

The expected estimate parameterb1,b5,c1,e4,c4,d4,e5>0 ;b2,b3,b4,c2,c3,d1,d2,d3,e1,e2,e3 <0

2.2.3. Income :

$$BPG = BSP+BTK+BLP \quad (8)$$
$$PDPG = PNPG-BPG \quad (9)$$
$$PDOF = f0+f1PDPG+f2PDNP+f3CKDKOF+f4DPG+E6 \quad (10)$$
$$PDNP = g0+g1PDPG+g2CKDKNP+g3PDIK+g4DPG+E7 \quad (11)$$
$$PDRT = PDPG+PDOF+PDNP \quad (12)$$

Notes:

BPG =the costs on upland rice farming(IDR/year)
BSP =the costs on upland rice means of production(IDR/year)
BTK =the costs on labors of upland rice farming(IDR/year)

BLP =other costs on upland rice farming(IDR/year)
PDPG =income from upland rice farming(IDR/year)
PNPG =acceptance from upland rice farming(IDR/year)
PDOF =income of off farm(IDR/year)
PDNP =income from non rice(IDR/year)
PDIK =spending for education(IDR/year)
PDRT =household income(IDR/year)
The expected estimate parameter f1,f2,g1 <0 ; f3,f4,g2,g2,g4 > 0

2.2.4. Consumption:

PPPG =h0+h1PNPG+h2PDRT+h3JART+h4DPG+E8 (13)
PKP =i0+i1PDRT+i2JART+i3PKNP+i4DPG+E9 (14)
PKNP =j0+j1PDRT+j2JAS+j3DPG+E10 (15)
PKT =PPPG+PKP+PKNP (16)

Notes:
PPPG =food expenditures from upland rice (IDR/year)
JART =the number of household members (IDR/year)
PKP =consumption expenditures (IDR/year)
PKNP =non food consumption expenditures (IDR/tahun)
JAS =the number of students(people)
PKT =total consumption expenditures(IDR/year)
The expected estimate parameter h1,h2,h3,h4,i1,i2,i4,j1,j2,j3 > 0 ; i3 < 0

2.3. Identification and validation of household's economics model

In order to estimate its parameter, a simultaneous equation model must be identified. The equation model identification result showed an overidentified equation therefore the applied method to estimate was a method of two-stage least squares (2SLS). In order to recognizer the model prediction power, Root Means Squares Error (RMSE), Root Means Squares Percent Error (RMSPE), and Theil's Inequality Coefficient (U) were applied [5].

3. Result and discussion

3.1. Model of rain-fed land upland rice farmers household's economics

The economics model of upland rain-fed land rice farmers household economics was constructedin this research which consists of 10 structural and 6 identity equations. The model estimated result was the following :

3.1.1. Production :

QPG = -627.4161+0.525116LL+15.40981JB-0.370556JPU+0.482637JPN+7.412084CKDKP +2.133793CKLKP+559.7129DPG

3.1.2. Labor Allocation :

CKDKP = 1.770143+4.08x10^{-05}LL+0.039612CKDKNP+0.002714CKDOF+ 1.25x10^{-10}PDRT+0.928857DPG

CKLKP = -0.922965+0.012783LL- 0.227392CKDKP -1.86x10-7PDRT+21.00265DPG

CKDKOF = 7.729850+0.000168LL-1.456487CKDKUT+0.098997CKDKNPt+ 0.669393DPG

CKDKNPt= 14.99920+0.000661LL+0.364512CKDUT-0.113311CKDOF-0.900158JTKDK - 8.510236DPG

IOP Conf. Series: Earth and Environmental Science **129** (2018) 012043 doi:10.1088/1755-1315/129/1/012043

3.1.3. Income:

PDOF =18.56963-0.060893PDPG-0.142680PDNP+0.222362CKDOF-1359695DPG

PDNP =8822065-0.075794PDPG+36020.77CKDKNP+0.336804PDIK+2925290DPG

3.1.4. Consumption :

PPPG = -181910.6+0.560528PNPG+0.007387PDRT+51233.05JART-224091.0DPG

PKP =8626395+0.027729PDRT+927653.0JART- 0.128651PKNP-523217.2DPG

PKNP =14.94591+3.86x10^{-09}PDRT+0.511103JAS-0.284903DPG

The estimated result towards 10 structural model equations which were proposed to demonstrate that the determination coefficient value (R^2) that resulted from its varied parameter from0.284370 to 0.937620. It can be presented in table1.

Table 1. The result of rain-fed land upland rice farmers households economics model testing in Ponjong Sub-district, Gunungkidul District.

No	Activities	Equation	R^2	F	Prob-F
1.	Production	QPG	0.937620	167.9390	0.000
2.	Labor allocation	CKDKP	0.904438	150.5371	0.000
3.		CKLKP	0.738829	56.1638	0.000
4.		CKDKOF	0.823234	4.6420	0.000
5.		CKDKPt	0.618998	31.5845	0.000
6.	Income	PDOF	0.609177	23.0237	0.000
7.		PDNP	0.280585	8.1511	0.000
8.	Consumption	PPPG	0.907485	194.7299	0.000
9.		PKP	0.284370	7.7190	0.000
10.		PKNP	0.771133	89.7261	0.000

It can be found in Table 1 from 8 equations, the R^2 value was > 0,5 and 2 equations, it was < 0.5. It exhibits the average endogenous variables variation could be explained more than 50 % by exogenous included variables in the rain-fed highland rice household economics model. The F testing overall result showed that sticking all models arranging variables provided impacts towards endogenous variables in 1% of significance level which means that exogenous variables included in each equation gave impacts towards endogenous variables. Statistical test results of rain-fed upland rice farmers household economics model prediction power could be presented in table 2.

Table 2. Statistical test results of RMSE, RMSPE, and U-Theil values in the rain-fed land upland rice farmers household economics model, in Ponjong Sub-district, Gunungkidul District.

Activities	Equation	RMSE	RMSPE	UM	US	UC	Notes
Production	QPG	187.1953	18.3532	0.0000	0.0151	0.9848	Valid
Labor allocation	CKDKP	0.1589	4.0527	0.0000	0.0234	0.9765	Valid
	CKLKP	7.1660	26.3601	0.0000	0.0682	0.9317	Valid
	CKDKOF	0.2752	4.7795	0.0000	0.0098	0.9901	Valid
	CKDKNPt	2.6518	18.3742	0.0000	0.0063	0.9836	Valid
Income	PDOF	0.5462	2.5471	0.0000	0.2193	0.7806	Valid
	PDNP	4599366	57.8095	0.0000	0.5638	0.4361	Valid
Consumption	PPPG	464032.7	19.8097	0.0000	0.0229	0.9770	Valid
	PKP	1823257	13.0548	0.0000	0.3192	0.5807	Valid
	PKNP	0.2355	1.2158	0.0000	0.0620	0.9379	Valid

It can be seen in Table 2 that if the RMSPE and U-Theil values as UM and US approach, 0 and UC approach 1, therefore the equation which is obtained is quite valid to be used in the projection and policy simulation. Meanwhile, in equation consistency testing, it is given value based on consistency reflected test in RMSE. When the number which is resulted in more than one, therefore the equation produces an untrusted result. When the number which contributes minimal result, approaches zero, and it is quite consistent and feasible to become projection and simulation basics. It displayed that for 6 equations, the RMSE value was > 1 which meaning that it could not be trusted and 4 equations, the RMSE value was < 1 which it could be trusted. It can be seen from the UM and US values were approaching zero and the UC value was closed to 1 which was valid and feasible as the simulation policy basis.

3.2. The interrelationship of production, the labor allocation, and household consumption of rain-fed rice farmers to the climate change

The production linkage, labor allocation, and household consumption of rainfed farmers to the climate change was in the cropping pattern. It is where the farmers of upland rice farmers decided to conduct different cropping pattern which is the one-time crop and two times crops of upland rice in one year. Therefore, there were dummy variables that show different cropping pattern due to the climate change.

The research production result, in this case of its upland rice equation model, showed that there was a different sign from the expected estimate parameter which was negative in the urea fertilizers number, which indicates that the urea fertilizers precisely addition would decrease rice production. This condition can be explained that urea fertilizers are sources of nitrogen (N) in ammonium. Urea is easily decomposed by urease enzyme which is produced by a microorganism in wet condition or humid so, the urea application in rain-fed land becomes ineffective unless the water condition is sufficient, therefore, a microorganism which producing urease enzyme can live. Other production equation variables were in accordance with the expected parameter. The rice cropping pattern dummy variables were indicated a positive value which implying different rice production cropping pattern. It can be explained that upland rice farmers household has one-time cropping pattern of upland rice in one year has become farmers routines. This is because by having experiences of previous years related to the rainy season condition in the second planting season, rainfalls rarely occur so they are no longer plant upland rice, but *palawija* which does not need a lot of water. For upland rice farmers which apply two times cropping pattern in one year, they would get benefits from this. This is because 2016 climate changes in the second planting season still often appeared. Therefore, in the second season, they could still get production like previous planting season.

The estimated results of labor allocation indicated that the labor allocation equations in the highland rice farming family. Its nonrice family labor allocation, of off-farm family and household incomes did not match the expected estimation, the result displayed a positive sign. It happened due to the farmers household family members did not only allocate their working time in rice and non-rice farming but also assign their times to find other incomes sources by wandering seasonally, trading, working as bricklayers, and other professions. Therefore, the relation between labor allocation in rice family, non-rice family, and off-farm family could complete each other, and also did household income members. Dummy cropping pattern showed a positive value which means that there was a difference in the family of upland rice cropping pattern family. The labor allocation equation of upland rice outside the family of, it was obtained that all signs were in accordance with the estimated parameter. The dummy cropping pattern, it was indicated positive sign that there was a labor use difference of outside family. From the off-farm equation, it was obtained two signs of variables which were not in accordance with the estimated parameter such as, land area and labor allocation variable in nonfarming activities family which were supposed to be negative, but its result indicated positive sign. It occurred because off-farm activities were only conducted by the head of the family so the land area and labor allocation in nonfarming activities, the result indicated positive sign. Dummy cropping patterns variables indicated a positive sign which means that there was a difference of off-farm labor use. The labor allocation equation in nonfarming family activities, was obtained three signs of variables which were not in accordance with the estimated parameter such as, land area and labor allocation variables in farming which were supposed to indicate negative sign however, it contributed a positive result and the family labors number variables which were supposed to be positive, but the result was negative, it was gained negative result. It has related one another that the nonfarming activities condition was conducted only by the head of the family. It indicated negative sign which means that there was no difference labor allocation in agriculture for nonfarming activities.

The income estimated result, its equation of off-farm, indicated that all signs of estimated parameter were in accordance with the expectation. Dummy cropping pattern variables indicated positive sign, which means that there was a different off-farm income. In the non-rice income equation, was obtained that all estimated parameter signs were in accordance with the expectation.It indicated positive sign, it means that there was a different non-rice income.

The consumption estimated result was upland rice food expenditures equation indicated that all estimate parameter signs were in accordance with the expectation which was highland rice acceptance variables having a positive value. Dummy cropping pattern variables indicated negative sign, it means that there was no difference in upland rice food expenditures. The food expenditures equation was obtained that all estimate parameter signs were in accordance with the expectation. It indicated negative signs which mean that there was no difference in upland rice food consumption expenditures of. By non-rice consumption expenditures equation was obtained a positive value of household income variables parameter which was in accordance with the expectation. It pointed out that there was no difference in non-rice consumption expenditures.

4. Conclusion

Based on the research result, a conclusion can be drawn that the model of rain-fed land upland rice farmers household economics could explain the interconnectedness among production, labor allocation, and consumption with the climate change. Its implication is that every activity in an economic activity will influence other economics activities

5. Suggestion

Associated with the limited condition of land especially the water availability which depends on rainfalls, farmers need to pay attention to climate condition which cannot be predicted therefore, they can determine the decision of choosing cropping pattern which will be conducted because it will influence production, labor allocation, income, and farmers household expenditures. The government needs to intervene in providing information and counseling towards upland rice farmers about the climate changes and their condition in order to optimize the available resources potential.

Acknowledgement

Thanks to the managers of Statistics and Data Processing Laboratory in Faculty of Agriculture UPN "Veteran" Yogyakarta

References

[1] Suratiyah K 2011 *Farming Science* (Jakarta: Penebar Swadaya) pp 1-4

[2] Ruslan K "Sensus Pertanian" *https://www.kompasiana.com* 2013 [Online] Available:https://www.kompasiana.com/kadirsaja/sensuspertanian [Accessed:19-6-2013]

[3] Kokoye S E H, Tovignan S D, Yabi J A and Yegbemey R N 2013 Econometric modeling of farm household land allocation in the municipality of banikoara in northern Benin *Journal Land Use Policy* **34** pp 72-9

[4] Koutsoyiannis A 1977 *Theory of Econometrics* (New York: Harper and Row Publisher Inc.) pp 698

[5] Pindyck R S and D L Rubinfield 1991 *Econometric Models and Economic Forecast.* Third Edition (New York: McGraw-Hill Inc.) pp 340-5